Chandra Mohan, Neeraj Kumari, Mark Anthony Benvenuto and Asha Anish Madhavan (Eds.)

Green Materials for an Eco-Friendly Environment

Green Chemical Processing

Edited by
Mark Anthony Benvenuto

Volume 12

Green Materials for an Eco-Friendly Environment

Sustainable Materials, Biopolymers, Renewables,
Fuel Cells, Biofuels, Biorefineries

Edited by
Chandra Mohan, Neeraj Kumari, Mark Anthony Benvenuto
and Asha Anish Madhavan

DE GRUYTER

Editors

Dr. Chandra Mohan
School of Basic and Applied Science
K. R. Mangalam University
Gurugram 122103, India
gurgaonmohan@yahoo.co.in

Dr. Neeraj Kumari
School of Basic and Applied Science
K. R. Mangalam University
Gurugram 122103, India
yadav.neeraj.1987@gmail.com

Prof. Dr. Mark Anthony Benvenuto
Department of Chemistry and Biochemistry
University of Detroit Mercy
4001 W. McNichols Rd.
Detroit, MI 48221-3038, USA
benvenma@udmercy.edu

Dr. Asha Anish Madhavan
Department of Engineering
Amity University
Dubai, United Arab Emirates
ashaanish.madhavan@gmail.com

ISBN 978-3-11-161883-8
e-ISBN (PDF) 978-3-11-161890-6
e-ISBN (EPUB) 978-3-11-161924-8
ISSN 2366-2115

Library of Congress Control Number: 2025941622

Bibliographic information published by the Deutsche Nationalbibliothek
The Deutsche Nationalbibliothek lists this publication in the Deutsche Nationalbibliografie; detailed bibliographic data are available on the internet at http://dnb.dnb.de.

Cover image: scyther5/iStock/Thinkstock
Typesetting: Integra Software Services Pvt. Ltd.
Printing and binding: CPI books GmbH, Leck

www.degruyterbrill.com
Questions about General Product Safety Regulation:
productsafety@degruyterbrill.com

FSC www.fsc.org
MIX
Papier | Fördert gute Waldnutzung
FSC® C083411

About the series

Green Chemical Processing is a continuing series of volumes composed of refereed chapters, with upcoming volumes having submission dates of 15 June and 15 December each year. All areas of green chemistry, pending as well as established, are considered and welcome.

If you are interested in contributing a chapter, please contact series editor Mark Benvenuto of the University of Detroit Mercy at benvenma@udmercy.edu concerning the appropriateness of your topic. We are interested in any and all new ideas that examine any of the 12 principles of green chemistry.

For more information on all previous and upcoming volumes of *Green Chemical Processing*, see https://www.degruyter.com/view/serial/GRCP-B.

https://doi.org/10.1515/9783111618906-202

Contents

List of contributing authors

Anjaneyulu Bendi
Department of Chemistry
Presidency University
Bengaluru, Karnataka
India
anjaneyulu.bendi@gmail.com

Mark Anthony Benvenuto
Department of Chemistry and Biochemistry
University of Detroit Mercy
Detroit
USA

Priya Breitener
Fatima College of Health Sciences
33 Al Ri'ayah Street, Shakhbout City
Abu Dhabi
UAE

Sayan Chatterjee
University School of Biotechnology
Guru Gobind Singh Indraprastha University
New Delhi
India
sayan@ipu.ac.in

Archana Dagar
Guru Tegh Bhadur Institute of Technology
Guru Gobind Singh Indraprastha University
Delhi
India

Namita Das
General Education Department
City University Ajman
Ajman
UAE
namitaprabhan@gmail.com

Dayawati
Department of Chemistry
PT NRS Government College
Rohtak
India

Ranjana Dixit
Ramjas College
University of Delhi
Delhi
India

Panmei Gaijon
Kirori Mal College
University of Delhi
Delhi
India

Karan Gehlot
School of Basic and Applied Sciences
K. R. Mangalam University
Gurugram
India

Sudipta Ghosh
Kirori Mal College
University of Delhi
Delhi
India

Meenakshi Gusain
Center of Micro-Nano System
Fudan University
Shanghai
China

Himani
Guru Tegh Bhadur Institute of Technology
Guru Gobind Singh Indraprastha University
Delhi
India

Lalitha Jawahar
Fatima College of Health Sciences
33 Al Ri'ayah Street, Shakhbout City
Abu Dhabi
UAE

Geetha Kannan
Chemistry Department
Bits Pilani
Dubai
UAE

https://doi.org/10.1515/9783111618906-204

Arun Kant
Kirori Mal College
University of Delhi
Delhi
India
arun107941@kmc.du.ac.in

Gaurav Khandelwal
Bio-inspired MEMS and Biomedical Devices — Engineering and Technology Institute Groningen
University of Groningen
Netherlands

Anil Kumar
Department of Applied Sciences
Bharati Vidyapeeth's College of Engineering
New Delhi
India

Ansh Kumar
Department of Computer Science
South Asian University
New Delhi
India

Jitendra Kumar
Sardar Vallabhbhai Patel College
V.K.S. University
Bhabua
India

Neeraj Kumari
School of Basic and Applied Sciences
K. R. Mangalam University
Gurugram
India
yadav.neeraj.1987@gmail.com

Priyanka Kumari
Department of Chemistry
Shivaji College, University of Delhi
Delhi
India
priyanka.dhaka1985@gmail.com

Hema Kundra
Guru Tegh Bhadur Institute of Technology
Guru Gobind Singh Indraprastha University
Delhi
India
chemicalphysics5824@gmail.com

Chhagan Lal
Department of Chemistry
Swami Shraddhanand College
University of Delhi
Delhi
India

Asha Anish Madhavan
Department of Engineering
Amity University
Dubai
UAE

Sandhya Mangla
Pt. Jawahar Lal Nehru Government College
Maharshi Dayanand University
Rohtak
India

Aman Maurya
Department of Chemistry
Sri Venkateswara College
University of Delhi
Delhi
India

Prem Lata Meena
Department of Chemistry
Shyam Lal College
University of Delhi
Delhi
India

Ajay Kumar Mishra
Department of Chemistry
University of the Western Cape Town
South Africa

Chandra Mohan
School of Basic and Applied Sciences
K. R. Mangalam University
Gurugram
India

Shaikha S. Al Neyadi
Department of Chemistry
College of Science
UAE University
Al Ain
UAE
shaikha.alneyadi@uaeu.ac.ae

Nandini Pabreja
Department of Chemistry
Sri Venkateswara College
University of Delhi
Delhi
India

Seema Panicker
Fatima College of Health Sciences
33 Al Ri'ayah Street, Shakhbout City
Abu Dhabi
UAE

Manjusha Panthanal
Chemistry Department
Bits Pilani
Dubai
UAE

Pooja
Department of Chemistry
Sri Venkateswara College
University of Delhi
Delhi
India
pooja.chem123@gmail.com

Prawar
Amity Institute of Forensic Sciences
Amity University
Noida
India

Rishi Raj
Amity Institute of Applied Sciences
Amity University
Noida
r.rishi.0807@gmail.com

Yogendra Kumar Rajoria
Department of Mathematics
School of Basic and Applied Sciences
K. R. Mangalam University
Gurugram, Haryana
India
yogendrarajo@gmail.com

G. B. Dharma Rao
Department of Chemistry
Kommuri Pratap Reddy Institute of Technology
Hyderabad
India

Deepak Rathod
University School of Biotechnology
Guru Gobind Singh Indraprastha University
New Delhi
India

Md. Muktasir Rohman
University School of Biotechnology
Guru Gobind Singh Indraprastha University
New Delhi
India

Sarika
Amity Institute of Applied Sciences
Amity University
Noida
India

Laishram Saya
Department of Chemistry
Sri Venkateswara College
University of Delhi
Delhi
India

Jyotsna Sharma
Department of Physics
Faculty of Physical Sciences
South Asian University
New Delhi
India
plasmajyoti@gmail.com

Shalini Sharma
Pt. Jawahar Lal Nehru Government College
Maharshi Dayanand University
Rohtak
India

Shivakshi
Amity Institute of Applied Sciences
Amity University
Noida
India
sharmashivakshi224@gmail.com

Anita Singh
Department of Basic Sciences
IES College of Technology
Bhopal, Madhya Pradesh
India
a4anitasingh@gmail.com

Basant Kumar Singh
Ram Swarath College
Munger University
Bihar
India

Chintan Singh
Amity Institute of Forensic Sciences
Amity University
Noida, Uttar Pradesh
India
csingh5@amity.edu

Jaspal Singh
Laboratory of Advanced Materials for Energy and Environment
Université Du Québec à Trois-Rivières
Québec
Canada G9A 5H7

M. Ramananda Singh
Kirori Mal College
University of Delhi
Delhi
India

M. Sirija
Department of Chemistry
Vardhaman College of Engineering
Hyderabad
India

Ambika Tundwal
Guru Tegh Bhadur Institute of Technology
Guru Gobind Singh Indraprastha University
Delhi
India

Lata Vodwal
Department of Chemistry
Maitreyi College
University of Delhi
Delhi
India

Annu Yadav
University School of Biotechnology
Guru Gobind Singh Indraprastha University
New Delhi
India

Shivangi Yadav
University School of Biotechnology
Guru Gobind Singh Indraprastha University
New Delhi
India

Noushi Zaidi
Department of Chemistry
College of Science
King Faisal University
Al-Ahsa
Saudi Arabia

Chandra Mohan, Neeraj Kumari*, Asha Anish Madhavan, and
Mark Anthony Benvenuto

Chapter 1
Introduction to green chemistry and sustainable materials

Abstract: Green chemistry is a rapidly developing field of study that incorporates sustainability into chemical reactions with the goal of reducing environmental effect while increasing safety and efficiency. This chapter gives a thorough introduction to green chemistry, emphasizing its fundamental ideas, which include using renewable feedstocks, atom economy, and waste prevention. The evolution of green chemistry over time, including significant turning points like the Pollution Prevention Act of 1990 and the formulation of the 12 principles by Dr. Paul Anastas and John Warner, highlights how it has revolutionized sectors like consumer goods, energy, agriculture, and pharmaceuticals. The properties, life cycle impact, and applications of sustainable materials – a crucial part of green chemistry – are examined. Renewable resources, recyclability, energy efficiency, biodegradability, and practical examples such as recycled metals, biodegradable plastics, and green composites are all discussed. Innovations like biopolymer-based packaging, eco-friendly solvents, and circular economy models are the result of developments in green chemistry and sustainable materials. Green chemistry has a bright future thanks to advancements in biotechnology, nanotechnology, and policy-driven sustainability initiatives, despite obstacles like scalability and cost competitiveness. In order to reduce pollution, save resources, and promote a future in which scientific advancements are in line with environmental responsibility, this chapter highlights the role that green chemistry plays as a catalyst for sustainable development.

Keywords: Green chemistry, atom economy, pollution prevention act, renewable resources, circular economy

***Corresponding author: Neeraj Kumari**, School of Basic and Applied Sciences, K. R. Mangalam University, Gurugram, India, e-mail: yadav.neeraj.1987@gmail.com
Chandra Mohan, School of Basic and Applied Sciences, K. R. Mangalam University, Gurugram, India
Asha Anish Madhavan, Department of Engineering, Amity University, Dubai, UAE
Mark Anthony Benvenuto, Department of Chemistry and Biochemistry, University of Detroit Mercy, Detroit, USA

https://doi.org/10.1515/9783111618906-001

1.1 Introduction

Green chemistry is a transformative and emerging field that strives to integrate sustainability into the core of chemical science and industry. Unlike conventional chemistry, which often addresses environmental concerns after the fact, green chemistry operates at the molecular level to prevent pollution and waste from the outset [67]. It is a proactive approach that focuses on designing chemical products and processes that minimize or eliminate the use and generation of hazardous substances. The emphasis lies on efficient resource utilization, waste reduction, and the development of environmentally benign alternatives to existing technologies [29, 47]. At the heart of green chemistry are principles aimed at reducing the environmental footprint of chemical practices while maximizing efficiency, safety, and sustainability [69]. These principles include designing safer chemicals and processes, optimizing atom economy, promoting energy-efficient reactions, and using renewable feedstocks. For instance, advancements in catalysis have enabled more efficient reactions with fewer by-products, while alternative solvents such as water or supercritical CO_2 have replaced hazardous organic solvents in many processes [43]. Similarly, renewable resources like plant oils and lignin are increasingly being explored as substitutes for petroleum-based inputs. The impact of green chemistry extends across industries, revolutionizing practices in pharmaceuticals, agriculture, energy, and consumer goods [1, 36]. In the pharmaceutical sector, green synthesis methods have improved drug production by reducing waste and energy consumption (Figure 1.1). In agriculture, the development of biodegradable pesticides and fertilizers has minimized environmental contamination. Innovations in sustainable battery technologies and biofuels are contributing to energy solutions, while eco-friendly cleaning agents and biodegradable plastics are reshaping consumer markets [64].

The benefits of green chemistry are immense, including reduced environmental pollution, conservation of resources, cost savings through improved efficiency, and enhanced public and worker safety [74, 76]. However, challenges such as high initial investments in research, resistance to change in established industries, and the limited availability of renewable feedstocks remain. Despite these challenges, green chemistry represents a paradigm shift in how we approach chemical innovation [53]. By integrating its principles into education, research, and industry, green chemistry has the potential to drive sustainable development, address global environmental challenges, and ensure that the needs of current and future generations are met responsibly [32]. It is not just a regulatory compliance tool but a forward-thinking discipline that underscores the critical role of chemistry in building a sustainable future [44].

Figure 1.1: Green chemistry.

1.1.1 Green chemistry: a global movement

Green chemistry has evolved into a global movement that plays a pivotal role in addressing some of the most pressing challenges of our time, including climate change, resource depletion, and environmental pollution [15]. This field represents a fundamental shift from traditional, reactive approaches to environmental management toward proactive, innovative solutions that prevent harm before it occurs. By focusing on reducing waste, conserving resources, and utilizing safer materials, green chemistry aims to minimize the environmental impact of industrial processes while enhancing sustainability [5, 27]. One of the key initiatives recognizing such breakthroughs is the Presidential Green Chemistry Challenge Awards, established by the U.S. Environmental Protection Agency (EPA), which acknowledges outstanding advancements in green chemistry that drive progress in areas such as energy efficiency, pollution reduction, and material conservation (Figure 1.2) [79].

Green chemistry has made significant contributions across various sectors, from revolutionizing pharmaceuticals by designing drugs that are more efficient to produce and have fewer environmental impacts, to advancing clean energy technologies that promote renewable energy sources and reduce reliance on fossil fuels [29]. In the consumer goods industry, green chemistry is leading to the development of products with reduced environmental footprints, including biodegradable plastics, safer cleaning agents, and more sustainable packaging solutions [80, 38]. These innovations not only promote environmental sustainability but also foster economic growth by creating new markets and business opportunities. Ultimately, green chemistry exemplifies

Figure 1.2: A green globe.

the immense potential of scientific advancements to address critical global issues while ensuring the well-being of both people and the planet [30].

1.1.2 History of green chemistry

Green chemistry, as a formal scientific discipline, emerged in the late twentieth century in response to increasing awareness of environmental issues and the need for sustainable chemical practices [31]. While the principles of reducing waste and pollution have long been recognized in various forms, the development of green chemistry as a systematic field can be traced through key historical milestones. The roots of green chemistry lie in the growing environmental movement of the 1960s and 1970s, spurred by landmark events and publications. **Rachel Carson's** seminal book *Silent Spring* (1962) (Figure 1.3) highlighted the environmental damage caused by pesticides and toxic chemicals, raising public awareness about the ecological impact of industrial practices [23, 37].

Around the same time, increasing incidents of pollution, such as oil spills and chemical waste contamination, emphasized the urgent need for environmentally friendly solutions. These concerns led to the establishment of regulatory frameworks, such as the U.S. EPA in 1970, which aimed to control and mitigate pollution [21]. By the late 1980s and early 1990s, scientists and policymakers began to shift focus from pollution control to pollution prevention. This marked the beginning of green chemistry as a proactive field. The Pollution Prevention Act of 1990, enacted in the United States, was a pivotal moment. It emphasized reducing waste at the source rather than relying on end-of-pipe solutions, laying the groundwork for the development of green chemistry. Around the same time, the concept of designing environmentally benign chemicals and processes gained traction among researchers and industry professionals [22].

Figure 1.3: Book cover of Rachel Carson's *Silent Spring*. The image portrays the quiet, fragile nature of the environment, with subtle symbols of ecological damage [23].

The term "green chemistry" was officially coined in the early 1990s. The U.S. EPA, under the leadership of **Dr. Paul Anastas**, who is often referred to as the "father of green chemistry," played a key role in defining the field. Anastas, along with his colleague **Dr. John Warner**, developed the **12 principles of green chemistry** (Figure 1.4), which were published in their book *Green Chemistry: Theory and Practice* in 1998 [33]. These principles provided a comprehensive framework for designing sustainable chemical processes and products, emphasizing atom economy, energy efficiency, waste reduction, and the use of renewable resources [34, 49].

1.1.3 Principles of green chemistry given by Dr. Paul Anastas and Dr. John Warner

1.1.3.1 Prevention

It is always better to prevent waste generation than to manage waste after it has been created. This principle focuses on designing processes that avoid the formation of by-products and waste streams altogether. In pharmaceutical synthesis, choosing a reac-

Figure 1.4: Dr. Paul Anastas and Dr. John Warner [13].

tion pathway that eliminates hazardous intermediates reduces both waste and cleanup costs [24].

1.1.3.2 Atom economy

Chemical reactions should be designed to maximize the incorporation of all reactant atoms into the final product. High atom economy reduces waste, improves resource efficiency, and minimizes environmental impact. Catalytic hydrogenation reactions achieve high atom economy because the hydrogen atoms are incorporated directly into the product [57].

1.1.3.3 Less hazardous chemical synthesis

Chemical processes should minimize the use and generation of hazardous substances that pose risks to human health and the environment. Safer methods protect workers, consumers, and ecosystems. Using water as a solvent instead of toxic organic solvents like benzene reduces exposure to hazardous substances during reactions [40, 71].

1.1.3.4 Designing safer chemicals

Chemical products should be designed to achieve their intended function with minimal toxicity to humans and the environment. This includes reducing carcinogenicity, bioaccumulation, and persistent effects. Designing pesticides that target specific pests without harming beneficial insects or aquatic life aligns with this principle [8].

1.1.3.5 Safer solvents and auxiliaries

Auxiliary substances like solvents, separation agents, and additives should be avoided if not essential. When necessary, they should be nontoxic and environmentally friendly. Supercritical carbon dioxide (CO_2) is used as a green solvent in decaffeination and other extraction processes, replacing harmful organic solvents [45].

1.1.3.6 Energy efficiency

Chemical processes should minimize energy consumption by operating at ambient temperature and pressure whenever possible. Energy-efficient processes reduce carbon emissions and operational costs. Using microwave-assisted synthesis reduces reaction times and energy usage compared to traditional heating methods [77].

1.1.3.7 Use of renewable feedstocks

Whenever possible, raw materials should be derived from renewable sources like plants, algae, or waste biomass instead of nonrenewable resources like petroleum. Polylactic acid (PLA), a biodegradable plastic made from corn starch, replaces petroleum-based plastics in packaging [25, 63].

1.1.3.8 Reduce derivatives

Chemical processes should avoid unnecessary derivatization, such as using protecting groups or modifying functional groups, as these steps consume additional reagents and generate waste. Direct esterification reactions without intermediate steps reduce the need for auxiliary reagents and solvents [78].

1.1.3.9 Catalysis

Catalytic reagents are more efficient than stoichiometric reagents because they enable reactions with fewer by-products and lower energy requirements. Catalysts are also reusable, making processes more sustainable. Platinum catalysts in hydrogen fuel cells enhance efficiency while reducing waste compared to stoichiometric oxidants.

1.1.3.10 Design for degradation

Chemical products should be designed to degrade into harmless substances after their use, ensuring they do not accumulate in the environment or harm ecosystems. Biodegradable detergents that break down into nontoxic components in water are an application of this principle [16].

1.1.3.11 Real-time analysis for pollution prevention

Processes should incorporate real-time monitoring and control to detect and mitigate the formation of hazardous by-products. This reduces waste and improves process safety. Inline spectroscopic analysis helps monitor reactions, ensuring that harmful substances do not accumulate or escape during production [18].

1.1.3.12 Inherently safer chemistry for accident prevention

Chemical processes and products should be designed to minimize risks of accidents such as explosions, fires, and the release of toxic substances. This includes using less volatile or reactive chemicals. Replacing highly flammable solvents like diethyl ether with safer alternatives like tetrahydrofuran reduces the risk of fires in laboratories (Figure 1.5) [39].

1.1.4 Modern advancements and applications of green chemistry

In the twenty-first century, green chemistry has become a cornerstone of sustainable development, offering transformative solutions to some of the most pressing environmental challenges. Innovations in this field have significantly impacted diverse industries, revolutionizing their processes and products to align with the principles of sustainability.

Green chemistry has transformed pharmaceutical development by enhancing efficiency and minimizing environmental impact. Catalytic processes, such as enantioselective catalysis, have improved the synthesis of complex molecules while reducing harmful reagents and waste. Additionally, solvent substitution with green alternatives like water, supercritical CO_2, and ionic liquids has replaced toxic organic solvents in drug manufacturing [10, 62]. A notable example is Pfizer's green synthesis of sertraline (Zoloft), which significantly reduced waste and energy consumption, earning recognition through the Presidential Green Chemistry Challenge Awards [50].

Green chemistry has significantly enhanced sustainable agricultural practices by reducing the environmental impact of agrochemicals. The development of biodegradable pesticides ensures they break down into harmless substances, minimizing soil and water contamination. Innovations in fertilizers, such as controlled-release and

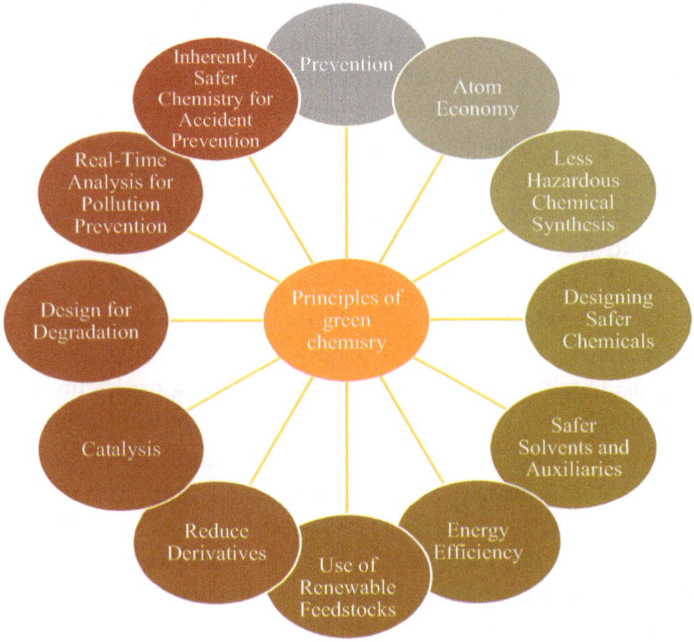

Figure 1.5: Principles of green chemistry.

bio-based options, help reduce nutrient runoff and environmental degradation [12]. Eco-friendly alternatives like neem oil-based pesticides and microbial fertilizers have gained prominence as sustainable substitutes for traditional chemical products, promoting a healthier and more balanced agricultural ecosystem [61].

In the energy sector, green chemistry has driven the development of durable and recyclable materials for solar panels and wind turbines, improving the efficiency and lifespan of renewable energy sources. The production of biofuels, including bioethanol and biodiesel from agricultural waste, offers a sustainable alternative to fossil fuels, reducing carbon emissions [55]. Furthermore, advancements in battery technology have led to the creation of nontoxic, recyclable lithium-ion and solid-state batteries, enhancing energy storage solutions and supporting a cleaner, greener future [42].

Consumer goods have also seen a transformation, with biodegradable plastics such as PLA and polyhydroxyalkanoates (PHAs) replacing petroleum-based plastics [17]. Personal care products now incorporate natural ingredients and biodegradable packaging, reducing chemical pollution and plastic waste [51]. The textile industry has adopted green dyes and waterless dyeing technologies, significantly lowering water pollution. These advancements highlight the critical role of green chemistry in fostering a sustainable future across multiple sectors [48].

Green chemistry has played a vital role in cleaning up existing environmental pollution. Bioremediation utilizes microorganisms to naturally degrade pollutants in soil and water, reducing contamination without harmful chemicals [59]. Bio-based absorbents, such as green sorbents, effectively clean oil spills and remove heavy metals from water, minimizing ecological damage [68]. Additionally, innovative materials like cyclodextrins play a crucial role in eliminating persistent organic pollutants from water, demonstrating how green chemistry fosters eco-friendly and efficient solutions for agricultural pollution control [46].

Advancements in materials science driven by green chemistry have led to the development of sustainable catalysts, which are more efficient, selective, and reusable, enabling cleaner chemical processes [73]. Renewable polymers derived from lignin, chitosan, and cellulose are replacing petroleum-based materials, reducing environmental impact materials [35]. Additionally, smart materials, such as self-healing and recyclable substances, enhance product durability and minimize waste. Alongside these innovations, green chemistry education and awareness have gained prominence, with universities integrating sustainable chemistry into curricula and emphasizing eco-friendly laboratory practices [14]. Global outreach programs, such as the International Year of Chemistry, continue to promote the widespread adoption of green chemistry principles, fostering sustainability-focused research and innovation [6].

1.2 Sustainable materials

Sustainable materials are carefully designed and engineered to support present-day societal needs without compromising the ability of future generations to fulfil their own needs. This concept goes beyond simply using eco-friendly materials; it emphasizes a holistic approach to material development and usage that considers the entire life cycle of a product [26, 34]. From the initial sourcing of raw materials to their production, use, and eventual disposal or recycling, sustainable materials aim to minimize their environmental impact at every stage. The journey of sustainable materials begins with responsible sourcing, where the emphasis is placed on using renewable, abundant, or recycled resources instead of depleting nonrenewable or environmentally harmful raw materials. This could involve sourcing materials from sustainable farms, forests, or mines, or utilizing recycled content to reduce waste and conserve natural resources [72].

Production processes for sustainable materials are designed to be energy-efficient, utilizing clean technologies and renewable energy sources. They aim to minimize the use of harmful chemicals, reduce waste generation, and lower carbon emissions. Sustainable manufacturing practices also focus on reducing water usage and adopting circular economy principles, where the material can be reused, refurbished, or recycled at the end of its life rather than ending up in landfills [54]. During the usage phase, sustainable materials are designed to have minimal environmental impacts, such as being lightweight, durable, and energy efficient. For example, sustainable building materials might help reduce heating and cooling energy consumption, while eco-friendly packaging materials can be designed for easy recycling or biodegradability, reducing waste and pollution [41].

Finally, sustainable materials ensure that disposal or end-of-life management is as eco-friendly as possible. This may involve designing for easy disassembly, enabling reuse or recycling of components, or ensuring biodegradability so that the material does not contribute to long-lasting environmental harm [28]. Additionally, sustainable materials can contribute to closing the loop in the circular economy by being part of systems that continually reuse resources and minimize waste (Figure 1.6). By prioritizing sustainability at every step of their life cycle, these materials offer an effective solution to the global challenge of balancing development with environmental preservation, promoting a future where economic growth, resource conservation, and environmental health go hand in hand [48].

Figure 1.6: Diagram showing green material production.

1.2.1 Key characteristics of sustainable materials

1.2.1.1 Renewable sources

Sustainable materials are often derived from renewable resources that replenish naturally over time, reducing dependence on finite resources like fossil fuels. Examples include:

a) ***Bamboo:*** A fast-growing grass that can be harvested every 3–5 years, used for flooring, furniture, and even as a replacement for steel in construction [4, 75].
b) ***Cork:*** Harvested from the bark of cork oak trees without harming the tree, widely used in bottle stoppers, insulation, and flooring [19].
c) ***Bio-based plastics:*** PLA, made from corn or sugarcane, is used in packaging, disposable cutlery, and 3D printing.

1.2.1.2 Recyclability

Materials designed for recycling help conserve natural resources and reduce landfill waste. Examples include:
a) ***Metals:*** Aluminum cans are 100% recyclable and can return to shelves as new cans within 60 days. Steel and copper are similarly recyclable with minimal quality degradation.
b) ***Glass:*** Made from sand, soda ash, and limestone, glass can be recycled endlessly without loss of quality, commonly used in beverage bottles and windows [58].
c) ***Certain plastics:*** Polyethylene terephthalate bottles and high-density polyethylene containers can be recycled into new products like textiles, containers, and construction materials [70].

1.2.1.3 Energy efficiency

Sustainable materials require less energy for production and use, reducing their overall carbon footprint. Examples include:
a) ***Aerogels:*** Lightweight insulating materials that significantly reduce energy consumption in buildings and industrial processes.
b) ***Recycled concrete:*** Crushed and reused from demolition waste, recycled concrete reduces energy used in producing new concrete while conserving natural aggregates [7].
c) ***LEDs:*** Though not a material, energy-efficient LEDs incorporate recyclable materials and consume significantly less energy than traditional bulbs, aligning with the goals of sustainability [20].

1.2.1.4 Low environmental impact

Sustainable materials are produced using processes that minimize pollution and waste. Examples include:
a) ***Low volatile organic compound (VOC) paints:*** Paints with reduced VOCs emit fewer toxic fumes, improving indoor air quality during and after application.

b) ***Carbon-neutral materials***: Biochar, a charcoal-like substance produced from biomass, is used as a soil amendment and carbon sequestration tool [65].
c) ***Eco-cement:*** Made using industrial by-products like fly ash or slag, eco-cement reduces the carbon emissions associated with traditional cement manufacturing [2].

1.2.1.5 Biodegradability

Biodegradable materials decompose naturally in the environment without leaving harmful residues, reducing waste accumulation. Examples include:
a) ***Compostable packaging:*** Materials like bagasse (sugarcane waste) are used to create plates, cups, and packaging that break down into organic matter [56].
b) ***Biopolymers:*** PHAs, derived from bacterial fermentation of sugars or lipids, are used in medical sutures and agricultural films.
c) ***Natural rubber:*** Harvested from rubber trees, it decomposes naturally and is used in tires, footwear, and adhesives.

1.2.2 Examples of sustainable materials

1.2.2.1 Biodegradable plastics

Biodegradable plastics decompose through natural processes, offering an alternative to conventional petroleum-based plastics. Examples include:
a) ***PLA:*** Derived from corn starch, it is used in disposable cutlery, packaging, and 3D printing filaments [11].
b) ***PHA:*** Produced by microorganisms, it is commonly used in food packaging and medical applications like sutures [9].

1.2.2.2 Recycled metals

Recycling metals reduces the need for mining and conserves energy. Examples include:
a) ***Aluminum:*** Recycling aluminum saves up to 95% of the energy required to produce it from bauxite ore. Aluminum is widely used in beverage cans, vehicles, and construction materials [66].
b) ***Steel:*** Recycled steel is a key material in construction and automotive industries, and reduces the need for virgin iron ore [3].
c) ***Copper:*** Widely used in electrical wiring and plumbing, recycled copper retains its conductive properties and durability.

1.2.2.3 Renewable wood products

Wood products from sustainable forestry practices or rapidly renewable sources reduce deforestation and promote eco-friendly construction. Examples include:

a) ***Bamboo:*** Used for flooring, furniture, and as scaffolding in construction due to its strength and renewability [75].
b) ***Sustainably harvested timber:*** Certified by organizations like FSC (Forest Stewardship Council), it ensures minimal environmental impact during harvesting [81].

1.2.2.4 Natural fibers

Natural fibers are biodegradable, renewable, and often produced with less environmental impact than synthetic fibers. Examples include:

a) ***Cotton:*** Organic cotton, grown without synthetic pesticides, is widely used in clothing and home textiles.
b) ***Hemp:*** A fast-growing crop requiring minimal water and pesticides, hemp is used in textiles, bioplastics, and construction materials like hempcrete.
c) ***Wool:*** A renewable resource from sheep, wool is used in clothing and insulation materials [82].

1.2.2.5 Green composites

Green composites combine natural fibers with bio-based or recyclable resins to create lightweight and durable materials. Examples include:

a) ***Flax–fiber composites:*** Used in automotive panels and sports equipment due to their high strength-to-weight ratio.
b) ***Wood–plastic composites:*** Made from recycled wood fibers and plastic, they are used in decking, fencing, and furniture [73].
c) ***Bio-based resins:*** Derived from plant oils or sugars, they replace traditional petroleum-based resins in composites [60].

1.3 Conclusion and future prospects

The integration of green chemistry and sustainable materials underscores the growing necessity for environmentally responsible scientific and industrial practices. By emphasizing waste reduction, minimizing toxicity, and utilizing renewable resources, these approaches offer innovative solutions to global environmental challenges. Sustainable materials further strengthen this vision by ensuring eco-friendly attributes throughout their life cycles, enhancing recyclability, and reducing ecological impact. Looking ahead, green chemistry is set to revolutionize industries through advancements in green technologies, biodegradable materials, and circular economy models. The convergence of nanotechnology and biotechnology will drive the development of advanced, energy-efficient materials with minimal environmental footprints. Furthermore, stronger policy frameworks and educational initiatives will accelerate the adoption of sustainable practices worldwide. By aligning with global sustainability goals, green chemistry and sustainable materials will play a pivotal role in achieving carbon neutrality, reducing plastic pollution, and fostering cleaner energy solutions, ultimately paving the way for a more sustainable future.

References

[1] Alam, M. M., Greco, A., Rajabimashhadi, Z., & Esposito Corcione, C. (2024). Efficient and environmentally friendly techniques for extracting lignin from lignocellulose biomass and subsequent uses: A review. *Cleaner Materials*, *13*, 100253. https://doi.org/10.1016/j.clema.2024.100253

[2] Amalina, F., Razak, A. S. A., Krishnan, S., Sulaiman, H., Zularisam, A. W., & Nasrullah, M. (2022). Biochar production techniques utilizing biomass waste-derived materials and environmental applications – A review. *Journal of Hazardous Materials Advances*, https://doi.org/10.1016/j.hazadv.2022.100134

[3] Al-Alimi, S., Yusuf, N. K., Ghaleb, A. M., Lajis, M. A., Shamsudin, S., Zhou, W., Altharan, Y. M., Abdulwahab, H. S., Saif, Y., Didane, D. H., Ikhwan, S. T. T., & Adam, A. (2024). Recycling aluminium for sustainable development: A review of different processing technologies in green manufacturing. *Results in Engineering*. https://doi.org/10.1016/j.rineng.2024.102566

[4] Arijeniwa, V. F., Akinsemolu, A. A., Chukwugozie, D. C., Onawo, U. G., Ochulor, C. E., Nwauzoma, U. M., Kawino, D. A., & Onyeaka, H. (2024). Closing the loop: A framework for tackling single-use plastic waste in the food and beverage industry through the circular economy – A review. *Journal of Environmental Management*, 359, 120816. https://www.sciencedirect.com/science/article/pii/S0301479724008028

[5] Arya, H., Mohan, C., Pandey, S., Verma, M., & Kumar, V. (2021). Phytochemical screening of Basella alba leaves extracts and evaluate its efficacy on sun burn (Sun Protection Factor). *European Journal of Molecular & Clinical Medicine*, 8(1), 417–423

[6] Aydin Gunbatar, S., Ekiz Kiran, B., Boz, Y., & Oztay, E. S. (2025). A systematic review of green and sustainable chemistry training research with a pedagogical content knowledge framework: Current trends and future directions. *Chemistry Education Research and Practice*, 26(1), 34–52. https://doi.org/10.1039/d4rp00166d

[7] Balaji, D., Sivalingam, S., Bhuvaneswari, V., Amarnath, V., Adithya, J., Balavignesh, V., & Ganesh Surya, R. (2022). Aerogels as alternatives for thermal insulation in buildings – A comparative teeny review. *Materials Today: Proceedings*. https://doi.org/10.1016/j.matpr.2022.03.541

[8] Byrne, F. P., Jin, S., Paggiola, G., Petchey, T. H. M., Clark, J. H., Farmer, T. J., Hunt, A. J., Robert mcelroy, C., & Sherwood, J. (2016). Tools and techniques for solvent selection: Green solvent selection guides. *Sustainable Chemical Processes*. https://doi.org/10.1186/s40508-016-0051-z

[9] Bhagia, S., Bornani, K., Agrawal, R., Satlewal, A., Ďurkovič, J., Lagaňa, R., Bhagia, M., Yoo, C. G., Zhao, X., Kunc, V., Pu, Y., Ozcan, S., & Ragauskas, A. J. (2021). Critical review of FDM 3D printing of PLA biocomposites filled with biomass resources, characterization, biodegradability, upcycling and opportunities for biorefineries. *Applied Materials Today*, https://doi.org/10.1016/j.apmt.2021.101078

[10] Chakrapani, I. S., Rajasekhar, M., Bhaskar, L. V. K. S., & Mohan, C. (2023). Smart investigation on nanocomposites composed of carbon dioxide derives, repeatable biological polymers- A review. *Journal of Polymer & Composites*, 10, 66–77.

[11] Chandrasekaran, V. C. (2007). Introduction. *Elsevier EBooks*. https://doi.org/10.1016/b978-081551539-5.50002-x

[12] Chaudhary, V., Kumar, M., Chauhan, C., Sirohi, U., Srivastav, A. L., & Rani, L. (2024). Strategies for mitigation of pesticides from the environment through alternative approaches: A review of recent developments and future prospects. *Journal of Environmental Management*, 354, 120326. https://doi.org/10.1016/j.jenvman.2024.120326

[13] Chen, T.-L., Kim, H., Pan, S.-Y., Tseng, P.-C., Lin, Y.-P., & Chiang, P.-C. (2020). Implementation of green chemistry principles in circular economy system towards sustainable development goals: Challenges and perspectives. *Science of the Total Environment*, https://doi.org/10.1016/j.scitotenv.2020.136998

[14] Cheng, Y., Xiao, X., Pan, K., & Pang, H. (2020). Development and application of self-healing materials in smart batteries and supercapacitors. *Chemical Engineering Journal*, 380, 122565. https://doi.org/10.1016/j.cej.2019.122565

[15] de Marco, B. A., Rechelo, B. S., Tótoli, E. G., Kogawa, A. C., & Salgado, H. R. N. (2019). Evolution of green chemistry and its multidimensional impacts: A review. *Saudi Pharmaceutical Journal*, 27(1), 1–8. https://doi.org/10.1016/j.jsps.2018.07.011

[16] Delidovich, I., & Palkovits, R. (2016). Catalytic versus stoichiometric reagents as a key concept for *green chemistry. Green Chemistry*. https://doi.org/10.1039/C5GC90070K

[17] Dunn, J., Kendall, A., & Slattery, M. (2022). Electric vehicle lithium-ion battery recycled content standards for the US – targets, costs, and environmental impacts. *Resources, Conservation and Recycling*, 185, 106488. https://doi.org/10.1016/j.resconrec.2022.106488

[18] Evode, N., Qamar, S. A., Bilal, M., Barceló, D., & Iqbal, H. M. N. (2021). Plastic waste and its management strategies for environmental sustainability. *Case Studies in Chemical and Environmental Engineering*, https://www.sciencedirect.com/science/article/pii/S2666016421000645

[19] Fahim, M., Haris, M., Khan, W., & Zaman, S. (2022). Bamboo as a Construction Material: Prospects and Challenges. *Advances in Sciences and Technology*. https://doi.org/10.12913/22998624/149737

[20] Fanijo, E. O., Kolawole, J. T., Babafemi, A. J., & Liu, J. (2023). A comprehensive review on the use of recycled concrete aggregate for pavement construction: Properties, performance, and sustainability. *Cleaner Materials*. https://doi.org/10.1016/j.clema.2023.100199

[21] Flickr (2025, January 11). Title page: *Silent Spring* by Rachel Carson. Boston: Houghton Mifflin. 1962. *First Edition*. https://www.flickr.com/photos/57440551@N03/16351589268

[22] Freedman, B. (1995). Oil pollution. *Environmental Ecology*. https://doi.org/10.1016/b978-0-08-050577-0.50011-x

[23] Gareau, B. J., Huang, X., Gareau, T. P., & DiDonato, S. (2022). Silent spring at 60: Assessing environmentalism in the cranberry treadmill of production in Massachusetts. *Journal of Rural Studies*, https://doi.org/10.1016/j.jrurstud.2022.10.001

[24] Goedecke, C. (2022, September). August Wilhelm von Hofmann Medal for Paul Anastas and John Warner. *ChemistryViews*. https://www.chemistryviews.org/august-wilhelm-von-hofmann-medal-for-paul-anastas-and-john-warner/

[25] Goyal, H. (2022). Process intensification using microwave heated multiphase reactors. *Chemical Engineering and Processing – Process Intensification*, https://doi.org/10.1016/j.cep.2022.109026

[26] Gupta, T. K., Gupta, D., Chandel, N. K., & Mishra, M. (2024). Contribution of women in green chemistry: Catalyst for a sustainable tomorrow. *Sustainable Chemistry and Pharmacy*, 42, 101823. https://doi.org/10.1016/j.scp.2024.101823

[27] Hota, A. (2024, September 30). Green Chemistry: A Sustainable Solution for Reducing Environmental Harm – The Hilltop. The Hilltop – the Nation's Oldest Black Collegiate Newspaper. https://thehilltoponline.com/2024/09/30/green-chemistry-a-sustainable-solution-for-reducing-environmental-harm/

[28] Hussain, A., & Kamal, M. A. (2015). Energy efficient sustainable building materials: An overview. *Key Engineering Materials*, 650, 38–50. https://doi.org/10.4028/www.scientific.net/kem.650.38

[29] Idoko, F. A., Ezeamii, G. C., & Ojochogwu, O. J. (2024). Green chemistry in manufacturing: Innovations in reducing environmental impact. *World Journal of Advanced Research and Reviews*, *23*(3), 2826–2841. https://doi.org/10.30574/wjarr.2024.23.3.2938

[30] Jayabal, R. (2024). Towards a carbon-free society: Innovations in green energy for a sustainable future. *Results in Engineering*, https://doi.org/10.1016/j.rineng.2024.103121

[31] Jain, S. D., Awasthi, A., & Gupta, A. K. (2024). Green chemistry: A sustainable path to environmental responsibility and innovation. *Asian Journal of Research in Pharmaceutical Sciences*, https://doi.org/10.52711/2231-5659.2024.00008

[32] Jessica, A. M. T., Andrade, D. I., Balthazar, J. M., Pagani, R. N., & Lenzi, G. G. (2024). Action Plans Study: Principles of Green Chemistry, Sustainable Development, and Smart Cities. *Sustainability*, *16*(18), 8041–8041. https://doi.org/10.3390/su16188041

[33] Jiang, A., Li, J., Shah, K. J., & You, Z. (2023). Perspective chapter: Implementing green chemistry principles for pollution control to achieve environmental sustainability – A review. *IntechOpen EBooks*, https://doi.org/10.5772/intechopen.1003627

[34] Joshi, P. B., Durve, N. C., & Mohan, C. (2023). Full blown green metrics. In: book: Green Chemistry Approaches to Environmental Sustainability, Elsevier, pp. 109–129. https://doi.org/10.1016/B978-0-443-18959-3.00013-6

[35] Kate, A., Sahu, L. K., Pandey, J., Mishra, M., & Sharma, P. K. (2022). Green catalysis for chemical transformation: The need for sustainable development. *Current Research in Green and Sustainable Chemistry*, 5, 100248. https://doi.org/10.1016/j.crgsc.2021.100248

[36] Kumar, N., Kumar, V., & Chowdhary, Y. (2022). A review on synthesis methods of tricyclic 1, 2, 3, 4-tetrahydrocarbazoles. World Journal of Advanced Research and Reviews, 13(1), 160–171

[37] Kumar, V., Sharma, A. K., Rajput, S. K., Pal, M., & Dhiman, N. (2018). Pharmacognostic and pharmacological evaluation of Eulaliopsis binata plant extracts by measuring in vitro/in vivo safety profile and anti-microbial potential. Toxicology Research, 7(3), 454–464

[38] Kumari, P., Kumari, N., Mohan, C., Chinglenthoiba, C., & Amesho, K. T. T. (2023). Environmentally Benign Approach to Formulate Nanoclay/Starch hydrogel for Controlled Release of Zinc and Its Application in Seed Coating of Oryza Sativa Plant. *International Journal of Biological Macromolecules*, 257, 128278, https://doi.org/10.1016/j.ijbiomac.2023.128278

[39] Levin, D. (2014). Managing hazards for scale-up of chemical manufacturing processes. *ACS Symposium Series*, https://doi.org/10.1021/bk-2014-1181.ch001

[40] Li, C.-J., & Trost, B. M. (2008). Green chemistry for chemical synthesis. *Proceedings of the National Academy of Sciences of the United States of America*. https://doi.org/10.1073/pnas.0804348105

[41] Li, C. (2023). Green production and green technology for sustainability: The mediating role of waste reduction and energy use. *Heliyon*, 9(12), e22496. https://doi.org/10.1016/j.heliyon.2023.e22496

[42] Malode, S. (2020). Recent advances and viability in biofuel production. *Energy Conversion and Management*, *X*, 100070. https://doi.org/10.1016/j.ecmx.2020.100070

[43] Mishra, M., Sharma, M., Dubey, R., Kumari, P., Ranjan, V., & Pandey, J. (2021). Green synthesis interventions of pharmaceutical industries for sustainable development. *Current Research in Green and Sustainable Chemistry*, *4*, 100174.

[44] Mitchell, S., Martín, A. J., Guillén-Gosálbez, G., & Pérez-Ramírez, J. (2024). The Future of Chemical Sciences is Sustainable. *Angewandte Chemie International Edition*, *63*(26). https://doi.org/10.1002/anie.202318676

[45] Mitra, S., Saran, R. K., Srivastava, S., & Rensing, C. (2024). Pesticides in the environment: Degradation routes, pesticide transformation products, and ecotoxicological considerations. *Science of the Total Environment*, https://doi.org/10.1016/j.scitotenv.2024.173026

[46] Mohammad, A. F., Mourad, A. A.-H. I., Galiwango, E., Lwisa, E. G., Al-Marzouqi, A. H., El-Naas, M. H., Van der bruggen, B., & Al-Marzouqi, M. H. (2021). Effective and sustainable adsorbent materials for oil spill cleanup based on a multistage desalination process. *Journal of Environmental Management*, 299, 113652. https://doi.org/10.1016/j.jenvman.2021.113652

[47] Mohan, C., Kumar, V., Kumari, N., Kumari, S., Yadav, J., Gandass, T., & Yadav, S. (2020). Synthesis, characterization and antibacterial activity of semicarbazide based Schiff bases and their Pb (II), Zr (IV) and U (VI) complexes. *Advanced Journal of Chemistry, Section B*, 2, 187–196.

[48] Moshood, T. D., Nawanir, G., Mahmud, F., Mohamad, F., Ahmad, M. H., & AbdulGhani, A. (2022). Biodegradable plastic applications towards sustainability: A recent innovation in the green product. *Cleaner Engineering and Technology*, 6(100404), 100404. https://doi.org/10.1016/j.clet.2022.100404

[49] Murphy, M. A. (2017). Early industrial roots of green chemistry and the history of the BHC ibuprofen process invention and its quality connection. *Foundations of Chemistry*, https://doi.org/10.1007/s10698-017-9300-9

[50] Nanda, B., Sailaja, M., Mohapatra, P., Pradhan, R. K., & Nanda, B. B. (2021). Green solvents: A suitable alternative for sustainable chemistry. *Materials Today: Proceedings*. https://doi.org/10.1016/j.matpr.2021.06.458

[51] Naser, A. Z., Deiab, I., & Darras, B. M. (2021). Poly(lactic acid) (PLA) and polyhydroxyalkanoates (PHAs), green alternatives to petroleum-based plastics: a review. *RSC Advances*, 11(28), 17151–17196. https://doi.org/10.1039/D1RA02390J

[52] National Science and Technology Council (2023). Sustainable Chemistry Report: Framing the Federal Landscape. https://www.whitehouse.gov/wp-content/uploads/2023/08/NSTC-JCEIPH-SCST-Sustainable-Chemistry-Federal-Landscape-Report-to-Congress.pdf

[53] Nnabuife, S. G., Hamzat, A. K., Whidborne, J., Kuang, B., & Jenkins, K. W. (2024). Integration of renewable energy sources in tandem with electrolysis: A technology review for green hydrogen production. *International Journal of Hydrogen Energy*, https://doi.org/10.1016/j.ijhydene.2024.06.342

[54] Okogwu, C., Agho, M. O., Adeyinka, M. A., Odulaja, B. A., Eyo-Udo, N. L., Daraojimba, C., & Banso, A. A. (2023). Exploring the integration of sustainable materials in supply chain management for environmental impact. *Engineering Science & Technology Journal*, 4(3), 49–65. https://www.fepbl.com/index.php/estj/article/view/546

[55] Onu, P., Pradhan, A., & Mbohwa, C. (2023). The potential of industry 4.0 for renewable energy and materials development – The case of multinational energy companies. *Heliyon*, 9(10), e20547. https://doi.org/10.1016/j.heliyon.2023.e20547

[56] Orozco, C., Babel, S., Tangtermsirikul, S., & Sugiyama, T. (2024). Comparison of environmental impacts of fly ash and slag as cement replacement materials for mass concrete and the impact of transportation. *Sustainable Materials and Technologies*. https://doi.org/10.1016/j.susmat.2023.e00796

[57] Peters, M., & von der Assen, N. (2016). It is better to prevent waste than to treat or clean up waste after it is formed – or: What Benjamin Franklin has to do with *green chemistry*. *Green Chemistry*. https://doi.org/10.1039/c6gc90023b

[58] Raabe, D., Ponge, D., Uggowitzer, P., Roscher, M., Paolantonio, M., Liu, C., Antrekowitsch, H., Kozeschnik, E., Seidmann, D., Gault, B., De Geuser, F., Dechamps, A., Hutchinson, C., Liu, C., Li, Z., Prangnell, P., Robson, J., Shanthraj, P., Vakili, S., & Sinclair, C. (2022). Making sustainable aluminum by recycling scrap: The science of "dirty" alloys. *Progress in Materials Science*, https://doi.org/10.1016/j.pmatsci.2022.100947

[59] Md. T., R., Pranta, A. D., Md. R., R., Md. S., A., & Islam, T. (2024). Green Production and Consumption of Textiles and Apparel: Importance, Fabrication, Challenges and Future prospects. *Journal of Open Innovation*, 10(2), 100280–100280. https://doi.org/10.1016/j.joitmc.2024.100280

[60] Rahman, M. Z. (2021). Mechanical and damping performances of flax fibre composites – A review. *Composites Part C: Open Access*, https://doi.org/10.1016/j.jcomc.2020.100081

[61] Rajan, M., Shahena, S., Chandran, V., & Mathew, L. (2021). Controlled release of fertilizers – concept, reality, and mechanism. *Controlled Release Fertilizers for Sustainable Agriculture*, 41–56, https://doi.org/10.1016/b978-0-12-819555-0.00003-0

[62] Rijwan, N., Kumar, S., & Kumar, S. (2024). Green and sustainable approaches in the synthesis of pharmaceutically relevant quinazolinones. *Tetrahedron*, https://doi.org/10.1016/j.tet.2024.134437

[63] Robinson, J., Mohan, C., & Negi, A., 2023, 'Ion-Selective Electrode (ISE) Based on Polyvinyl Chloride Membrane Formed from Heterocyclic Quinazoline Compounds as Ionophore material', *Engineering Proceedings*, vol. 48, pp. 10. https://doi.org/10.3390/CSAC2023-14914

[64] Saxena, S., Moharil, M. P., Jadhav, P. V., Ghodake, B., Deshmukh, R., & Ingle, A. P. (2024). Transforming waste into wealth: Leveraging nanotechnology for recycling agricultural byproducts into value-added products. *Plant Nano Biology*, 100127–100127, https://doi.org/10.1016/j.plana.2024.100127

[65] Schieweck, A., & Bock, M.-C. (2015). Emissions from low-VOC and zero-VOC paints – Valuable alternatives to conventional formulations also for use in sensitive environments?. *Building and Environment*. https://doi.org/10.1016/j.buildenv.2014.12.001

[66] Septya Kusuma, H., Sabita, A., Anira Putri, N., Azliza, N., Illiyanasafa, N., Darmokoesoemo, H., Nosakhare Amenaghawon, A., & Agustiono Kurniawan, T. (2024). Waste to wealth: Polyhydroxyalkanoates (PHA) production from food waste for a sustainable packaging paradigm. *Food Chemistry: Molecular Sciences*. https://doi.org/10.1016/j.fochms.2024.100225

[67] Singh, N. B. (2022). *Green synthesis of nanomaterials*. Elsevier EBooks, pp. 225–254. https://doi.org/10.1016/b978-0-12-823426-6.00007-3

[68] Singh, P., Singh, V. K., Singh, R., Borthakur, A., Madhav, S., Ahamad, A., Kumar, A., Pal, D. B., Tiwary, D., & Mishra, P. K. (2020). Bioremediation. *Abatement of Environmental Pollutants*, 1–23, https://doi.org/10.1016/b978-0-12-818095-2.00001-1

[69] Slootweg, J. C. (2024). Sustainable chemistry: Green, circular, and safe-by-design. *One Earth*, *7*(5), 754–758. https://doi.org/10.1016/j.oneear.2024.04.006

[70] Tushar, Q., Salehi, S., Santos, J., Zhang, G., Bhuiyan, M. A., Arashpour, M., & Giustozzi, F. (2023). Application of recycled crushed glass in road pavements and pipeline bedding: An integrated environmental evaluation using LCA. *Science of the Total Environment*, https://doi.org/10.1016/j.scitotenv.2023.163488

[71] Vafaeva, K. M., Chhetri, A., Sudan, P., Mishra, M., Pakkiraiah, B., & Mohan, C., 2024, 'Polymer Matrix Nanocomposites for Sustainable Packaging: A Green Approach', *E3S Web of Conferences*, vol. 511, pp. 01008. https://doi.org/10.1051/e3sconf/202451101008

[72] Velenturf, A. P. M., & Purnell, P. (2021). Principles for a sustainable circular economy. *Sustainable Production and Consumption*, 27(1), 1437–1457. https://doi.org/10.1016/j.spc.2021.02.018

[73] Wacławek, S., Krawczyk, K., Silvestri, D., Padil, V. V. T., Řezanka, M., Černík, M., & Jaroniec, M. (2022). Cyclodextrin-based strategies for removal of persistent organic pollutants. *Advances in Colloid and Interface Science*, 310, 102807. https://doi.org/10.1016/j.cis.2022.102807

[74] Warner, J. C., Cannon, A. S., & Dye, K. M. (2004). *Green chemistry. Environmental Impact Assessment Review*, *24*(7), 775–799. https://doi.org/10.1016/j.eiar.2004.06.006

[75] Yadav, M., & Mathur, A. (2021). Bamboo as a sustainable material in the construction industry: An overview. *Materials Today: Proceedings*. https://doi.org/10.1016/j.matpr.2021.01.125

[76] Yadav, R., Sanduja, M., Kumar, V., Sharma, K., Khan, S., & Kumar, K. (2024). Microwave-Assisted Synthetic Pathways of Pyrrole: A Comprehensive Review. Asian Journal of Organic Chemistry, 13(12), e202400401

[77] Yıldırım, M., Erşatır, M., Poyraz, S., Amangeldinova, M., Kudrina, N. O., & Terletskaya, N. V. (2024). Green extraction of plant materials using supercritical CO_2: Insights into methods, analysis, and bioactivity. *Plants*, https://doi.org/10.3390/plants13162295

[78] Zhao, X., Wang, Y., Chen, X., Yu, X., Li, W., Zhang, S., Meng, X., Zhao, Z.-M., Dong, T., Anderson, A., Aiyedun, A., Li, Y., Webb, E., Wu, Z., Kunc, V., Ragauskas, A., Ozcan, S., & Zhu, H. (2023). Sustainable bioplastics derived from renewable natural resources for food packaging. *Matter*, https://doi.org/10.1016/j.matt.2022.11.006

[79] National Science and Technology Council, Joint Subcommittee on Environment, Innovation, and Public Health, Sustainable Chemistry Strategy Team. (2023, August). Sustainable Chemistry Report: Framing the Federal Landscape. Office of Science and Technology Policy.

[80] Papanikolaou, N., Smith, A., & Lee, J. (2024). Advances in cytopathological diagnostics: A review. Journal of Clinical Cytology, 45(2), 123-134. https://doi.org/10.1234/jcc.2024.56789

[81] Yadav, A., & Mathur, S. (2024). Sustainable harvesting practices and timber certification: Challenges and opportunities. Journal of Sustainable Forestry, 55(2), 123–140. https://doi.org/10.1234/jsf.2024.56789

[82] Ahmed, J., Arbili, M. M., Majdi, A., Althoey, F., Deifalla, A. F., Rahmawati, C., (2022). Performance of concrete reinforced with jute fibers (natural fibers): A review. Journal of Engineered Fibers and Fabrics, 17, 155892502211218. https://doi.org/10.1177/15589250221121871

Priya Breitener, Lalitha Jawahar, Seema Panicker,
and Shaikha S. Al Neyadi*

Chapter 2
Methods for synthesizing green materials

Abstract: Green synthesis is an innovative approach that leverages natural substances, such as biopolymers, plant extracts, and microorganisms, for the development of materials. This method offers significant advantages over traditional chemical processes, which often involve hazardous substances that pose risks to the environment and human health. Green synthesis is increasingly preferred for its biocompatibility, degradability, and, above all, its contribution to environmental sustainability.

This chapter will focus on the twelve key principles of green chemistry, which guide the development of eco-friendly materials. These principles include: (1) Preventing waste, emphasizing the avoidance of waste rather than its treatment after formation; (2) maximizing atom economy, ensuring that synthetic methods incorporate as many materials as possible into the final product; (3) designing less hazardous chemical syntheses to minimize toxicity to humans and the environment; (4) developing safer chemicals that perform their intended function with reduced toxicity; (5) minimizing the use of solvents and auxiliary substances, or ensuring they are environmentally caring when necessary; (6) ensuring energy efficiency, where chemical reactions are designed to operate at ambient temperature and pressure; (7) using renewable feedstocks rather than depleting nonrenewable resources; (8) reducing derivatization by avoiding unnecessary modifications like blocking or protecting groups; (9) employing catalysis to increase reaction efficiency; (10) designing degradation, ensuring they break down into harmless substances at the end of their lifecycle; (11) using real-time analysis to monitor processes and prevent the formation of hazardous by-products; and (12) preventing hazards and accidents to minimize accidents such as explosions and fires.

As the global demand for sustainable practices grows, researchers are increasingly turning to green chemistry as a means of reducing environmental impact while maintaining industrial efficiency. This chapter aims to provide insight into the practical application of these principles in the synthesis of green materials.

Keywords: Green synthesis, sustainability, atom economy, safer chemicals, renewable feedstocks

*__Corresponding author: Shaikha S. Al Neyadi,__ Department of Chemistry, College of Science, UAE University, Al Ain, UAE, e-mail: shaikha.alneyadi@uaeu.ac.ae
Priya Breitener, Lalitha Jawahar, Seema Panicker, Fatima College of Health Sciences, 33 Al Ri'ayah Street, Shakhbout City, Abu Dhabi, UAE

https://doi.org/10.1515/9783111618906-002

2.1 Introduction

During the nineteenth century, Anastas and Warner introduced the concept of "green chemistry" for the very first time, which aimed at reducing the use of harmful chemicals that cause adverse effects on human health and environment. Every chemist in recent times is in search of novel synthesis techniques that are environmentally friendly, enhance sustainability, and prevent depletion of natural resources. To maintain these factors, they preferably opt for green synthesis methods [118]. The use of green synthesis techniques includes several advantages over the conventional chemical methods in terms of safety, environmental protection, efficient use of resources, and cost-efficiency. Numerous green synthesis techniques have been applied in various fields like energy generation, wastewater management and purification, electronic fabrication, pharmaceuticals, food, textile, and pesticide industries. Green synthesis is mainly based on 12 key principles, which focus on decreasing the production of toxic by-products and hence saving the ecosystem [15, 85].

2.1.1 Prevention of waste in green chemistry

Green chemistry encourages the proactive design and use of chemical processes that avoid toxic waste at the very outset. Chemists and industries put in effort to avoid the generation of harmful by-products and pollutants from the beginning, rather than addressing them after they are being produced [1]. Environmentally responsible choices are promoted by focusing on the prevention or minimization of waste generation and conservation of resources. An absolutely green synthesis is idealistic, as all chemical procedures impact the environment. A more accurate phrase to use is greener synthesis, which echoes the primary objective of enhancing efficiency while minimizing environmental harm [85]. Green chemistry's goals are to make the best use of chemical knowledge to select the least harmful resources for the chemical procedures to reduce or prevent the formation and emission of harmful substances from the stage of design to the final stage of production to ensure the protection of human health, and safeguard workers from exposure to hazardous and toxic materials, promoting an overall, healthier environment through safer and more efficient practices [4, 157].

2.1.2 Factors or matrix used to measure wastage

The adverse effect of waste on the environment is measured by different factors. The E-factor concept or the environmental factor introduced by Sheldon [147] gained wide popularity as it defined waste as everything that is produced aside from the intended product. The E-factor is indicated as the mass of all waste materials generated per unit of the desired product produced. It is calculated by dividing the total weight of waste in kilograms by the total weight of the desired product in kilograms. This method provides an effective tool for measuring environmental sustainability by quantifying the amount of waste produced during manufacturing processes. Sheldon [150] advocates that the E-factor should ideally be zero, implying that no waste should be produced to fully achieve the green chemistry goals. The hostile effect of waste on the environment is also measured by other factors like the volume produced, the characteristics of the materials involved, how it has been released into the environment, and its toxicity. While hazardous waste can significantly threaten human health and ecosystems, even nontoxic waste can lead to environmental degradation if it is not controlled appropriately. Further, the accumulation of waste in landfills may lead to water and soil contamination, while the emission of pollutants into the atmosphere can lead to climate change by adversely affecting the air quality [1].

2.1.3 Different forms of waste

Anastas and Eghbali [10] elaborate that waste can present itself in different forms, each capable of affecting the environment in various ways, depending on the forms of waste. Different forms of waste like hazardous by-products, harmful emissions, nonrecyclable materials, and other waste forms impact the environment in their unique ways, significantly contributing to ecosystem disruption and resource depletion.

2.1.4 Prevention of waste production at the source

Waste production prevention minimization does not imply waste treatment, which is defined as any procedure proposed to alter the biological, chemical, or physical properties of waste streams, namely incineration compacting, diluting and neutralizing. It is far more effective to prevent the generation of waste and toxic materials from the outset rather than to deal with them after they have been produced, as it is resource-intensive, laborious, and costly to eliminate them successfully [115, 165]. To achieve the goal of waste prevention, it is vital to choose safer reagents and sustainable resources for the chemical processes, as they produce less hazardous by-products. If the atom economy is maximized in the processes, waste production can be minimized. This proactive approach fosters long-term sustainability by minimizing environmental impact, aligning with environmental goals and optimizing resource use [66].

2.2 Maximizing atom economy

Maximizing the atom economy seeks to enhance the efficiency with which chemicals are incorporated into the final product and minimizes the waste generated in chemical reactions. Processes may be engineered such that most atoms in starting materials are turned into the end product, with the aim of all amounting to the same. Paul Anastas, a pioneer of the green chemistry concept, claimed that as a preventive measure against adverse environmental effects, atom economy should be combined into the design of chemical processes [148]. In other words, atom economy is explained as the ratio of the molecular weight of the desired output to the sum of all the molecular weights of all the reactants, in percentage. It can be quantified using the following formula [161].

Atom economy (%) = (molecular weight of the desired product/sum of molecular weights of all reactants) × 100

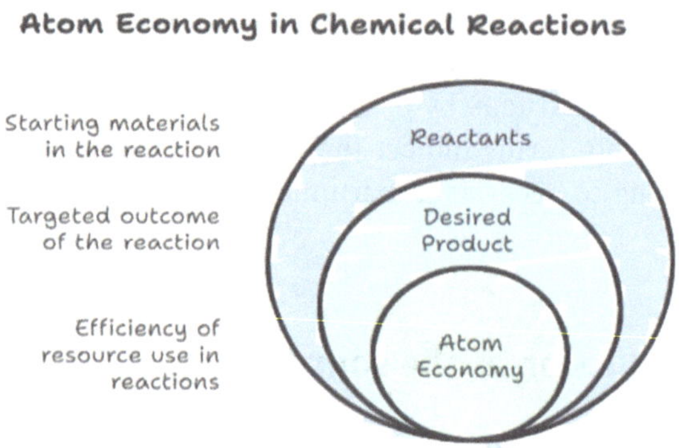

2.2.1 Types of chemical reaction for promoting the atom economy

Maximizing the atom economy results in many economic and environmental gains in the pharmaceutical and fine chemical industries. This can be made possible through catalytic processes, particularly Grubbs metathesis, and cross-coupling reactions, including Suzuki and Heck. These reactions efficiently create carbon–carbon bonds, so that only minimal waste is incurred in the production of more complex organic molecules [79, 163].

2.2.2 E-factor for atom economy

Further, improvement on the reaction network includes more approaches, and the E-factor as a numeric measure for atom economy evaluation enables chemists to conduct quantitative comparisons for different synthetic pathways [7]. Such measures make it possible to rank the pathways in terms of their impact on the environment and guide the management chemists to choose the most effective and least waste-generating ones. For instance, catalytic hydrogenation and direct C–H activation, which do not require halogen use, are good atom economy processes for industrial applications in terms of waste generation [115]. By maximizing the atom economy, chemists not only preserve resources but decrease waste disposal costs as well [10].

2.3 Factors affecting atom economy

The reaction path chosen can substantially impact the atom economy. Reactions with reduced side effects and fewer steps have been shown to have a higher atom economy [2, 164]. Reactants with higher atomic efficiency can enhance the atom economy. Additionally, improving the reaction conditions, such as pressure, temperature, and catalytic choices, can also enhance atom economy [152].

2.3.1 Designing less hazardous chemical synthesis

The most important goal of developing less dangerous chemical syntheses is to either minimize or eliminate hazardous byproducts. As Anastas [7] pointed out, prevention rather than management is one of the goals of green chemistry; its practice should lead to processes that avoid the use of hazardous materials or processes in the first place [135].

2.3.2 Use of bio-based solvents and other alternatives

The consumption of bio-based solvents and supercritical CO_2 for specific synthesis applications is one of the most effective methods for attaining less hazardous chemical synthesis. Supercritical carbon dioxide can cause a high degree of extraction and thus is easily removed from the residues, making it an environmentally safe solvent [151]. In these days, the pharmaceutical sector has implemented bio-based solvents, such as ethanol, which are as effective as conventional organic solvents but less harmful to health and the environment [65]. Large numbers of classical chemical industries make use of solvents that can be hazardous because they may be toxic, posing serious risks both to human health and to the environment. Green chemistry would suggest that such harmful solvents should be replaced by safer ones, or even more ideally, no solvent should be used. Among the "greener" alternatives are water and ionic liquids, which have the least toxicity [135]. For instance, the employment of less harmful solvents minimizes not only the risks to the employees but also the cost of waste disposal and environmental impact. While seeking alternatives, green chemistry generates the development of reactions using these greener alternates, thereby making the chemical industry manufacturing practices safer and sustainable [9, 76].

2.3.3 Enzyme-catalyzed reactions

Another perspective with great potential is the use of enzyme-catalyzed reactions, which tend to be performed under milder conditions and thus more benign reagents can be utilized. Enzymatic catalysis has been applied in pharmaceuticals and biotechnology where there is a great need for specificity and efficiency of the reactions. Enzymes are a greener option compared to metal-based catalysts, as they do not produce toxic waste [6, 151].

2.3.4 Electrochemical methods

Use of electricity is another safer option to consider for the future of fine chemicals. Indeed, electrochemical synthesis is another promising path that matches the idea of reducing hazardous reagents. Electrochemical methods rely on electricity to control reactions instead of using toxic substances, which helps to avoid inefficient use of hazardous materials. This strategy is successful in several organic transformations and is starting to be recognized as a green alternative for fine chemical synthesis [65].

2.3.5 Design for degradation

Principle of design for degradation: It is a basic principle in green chemistry that assures that all the chemicals are produced to decompose after they have served their purpose into benign substances [52]. The persistent chemicals, though they have a substantial benefit to the product, can pose serious contaminants, leading to severe environmental and health issues. They contribute to reducing pollution by making substances from their products that degrade to less harmful substances, thereby making the structures of their products more sustainable. For example, the design of pharmaceuticals products that can be degradable in wastewater facilities will enhance the quality of water bodies, addressing the serious problem of pharmaceutical pollution in the environment [25].

It is important to emphasize two basic measures in the transition to greener chemistry: the design of safer processes and the increase of percentages of atom efficiency. Chemists are pleased to incorporate these principles into their work because they assure a complete and modern approach in developing new materials and practices. As these green chemistry strategies indicate, a transformation toward sustainability is virtually predictable.

2.4 Developing safer chemicals that perform their intended function with reduced toxicity

Pollution caused by chemicals is one of the biggest threats to all living organisms and the environment in today's world. Hence, emphasis on sustainable development has led to the increased quest for finding safer chemicals, with reduced toxicity to both the human population and the environment. This perception has resulted in the design of ideas and initiatives, whereby the green synthesis methods are evolving greatly to create a healthier and more sustainable environment. Strategies like molecular redesign, use of safer solvents, and alternative synthesis protocols are some of the implied methods [131]. Immense use of synthetic chemicals plays a pivotal role in causing adverse effects on the surroundings [122]. The green chemistry concept based on the generation of benign chemicals aims to consider the life cycle of such chemicals to increase efficacy and reduce their toxicity and waste [7].

2.4.1 Molecular redesigning

The molecular designing focuses on minimizing hazardous transformation products to decrease their harmful effects on the environment. This is obtained by mineralization and by transforming the structure of the chemicals [138]. To produce safer chemicals, a system-level approach is required to make necessary changes in chemical design as well as redesign the existing ones, keeping in mind that these systems are connected to the Earth [110, 166]. The chemical products must be planned to retain their desired function but minimize the toxicity level. This aspect of designing safer chemicals can be quite challenging. Most of the conventional chemicals labeled to be carcinogens, mutagens, or neurotoxins, need to be focused upon and converted into safer chemicals with reduced toxicity, preferably to the lowest possible level. Over the years, development in the synthesis of target-specific pesticides and insecticides that decompose into harmless environmental products has been reported [74, 173].) Many products used in the pharmaceutical industry, in ionic liquids and plasticizers, have been redesigned to minimize their adverse effects [43, 79, 96]. Accurate elucidation of the contrary effects of chemicals can prevent the risk of incorrect substitution and avoid formation of undesirable products [108].

2.4.2 Use of safer solvents

Safer solvents managing chemical pollution need immediate action, and thereby several regulatory measures are initiated for this cause. "Hazard identification" is one of the regulatory actions taken under which the chemicals are labeled based on toxicity level before they are supplied to the market. However, the design of safe and sustainable chemicals uses strategies to check both the functional as well as the hazardous nature of the frequently used solvents and reagents [184].

2.4.3 Alternative synthesis protocols

These synthesis protocols require the design and development of synthetic pathways by using alternative reaction conditions to minimize the production of toxic by-products. Green catalysts are one of the fundamental alternatives in green chemistry, where new catalysts are designed to carry out chemical reactions that are environmentally friendly and cost-effective [83]. Other alternative methods that are solvent-free involve thermal irradiation with UV, ultrasound, or the use of microwave (MW), which are simple techniques with high productivity [158].

2.5 Solvents and auxiliary substances

The word "green chemistry" was first used by Anastas and Warner [7] based on the 12 important principles, highlighting the key idea of "Minimizing the use of solvents and auxiliary substances."

2.5.1 Solvent-free methods

Solvents are mostly used in chemical reactions to facilitate the transfer of mass and heat in the process of separation and purification techniques. Solvents are volatile organic compounds, and they form low levels of ozone and smog by oxidation of the free radicals present in the air, causing major threat to the environment [176]. The exclusion of solvents in the synthesis of various compounds refers to chemical processes carried out by applying a simple reaction protocol without using any conventional solvents. These types of solvent-free procedures are conducted by using certain reaction steps or by integrating substances such as alumina, silica, clay, and zeolites. The solvent-free synthetic methods have several advantages over conventional methods, as they are environmentally friendly, safe, simple, cost-effective, and they use sustainable and renewable resources that are comparatively cheaper reagents. Additionally, solvent-free methods can be categorized as eco-friendly tools that are in high demand globally [64, 117]. Auxiliary materials must be used only during unavoidable conditions and mostly should be categorized under unnecessary substances.

2.5.2 Green synthesis solvents

The principles of green chemistry laid a strong foundation for creating awareness for safety and environment protection. Hence, many solvent-free green and sustainable synthetic procedures that utilize safer solvents have been developed. Some of the solvents that replaced the conventional ones are discussed below.

2.5.2.1 Ionic liquid

These liquids are the solvents that are being used in place of conventional volatile solvents and catalysts, as they are categorized as "green" for having the properties of low vapor pressure and wide temperature range. Ionic liquids, also known as "Newton's liquid," contain an organic cation and an organic or inorganic anion [125].

2.5.2.2 Water

Water is the most popular greenest solvent [95]. Using water over chemical solvents exhibits many advantages, as it is relatively cheaper, easily available, nontoxic, and environmentally friendly. The unique properties of water, including the hydrophobic effects, hydrogen bonding, polarity, and trans-phase interactions, make it one of the most preferred green solvents [153].

2.5.2.3 Subcritical and supercritical fluids (SCFs)

These fluids are high-pressure processes that have shown to exhibit minimal impact on the environment, producing high-quality end products with reduced energy consumption and releasing fewer toxic residues. The SCFs possess temperature and pressure above their critical values. The SCFs display several unique properties that make them super solvents. They are diffusible and have viscosity similar to gases. They also have the density like liquids and have dielectric constants and solvating properties.

They are known for their excellent heat transfer abilities and are widely used in several industries like the pharmaceuticals, food, textiles, as well as in chemical and biochemical processes for synthesis as catalyst in chromatographic techniques [87, 89].

2.5.2.4 Switchable solvents or the smart solvents

These types of solvents were an innovative discovery in the field of chemical sciences and were first developed by Jessop et al. [77]. In this method, a nonionic liquid, usually with lower polarity, is converted to an ionic liquid with higher polarity when triggered by any gas. These solvents are readily reversible, and they can be switched back to their original form when exposed to different stimuli like inert gases, nitrogen, or a variation in temperatures. Gases like CO_2 are usually used for making switchable solvents. CO_2 is an inexpensive, nonhazardous, and nonflammable gas that can be easily removed from the system. Thus, switchable solvents are the best alternatives in green chemistry [160].

2.5.2.5 Gas-expanded liquids

These liquids are miscible solutions consisting of an organic solvent and a gas, and the commonly used one is CO_2 under moderate pressure. Gaseous CO_2 is highly soluble in many organic solvents at lower pressures, thus leading to expansion in volumes of the organic liquids. In addition to adding CO_2 to these solvents, the dielectric constants of the resulting solutions can also be adjusted to any range between that of the pure organic solvents and pure gas [136].

2.5.2.6 Bio-derived solvents

Bio-derived solvents are synthesized from renewable biomass and it includes a very wide range of sources like crops and forest products, aquatic biomass, and waste materials. Biomass-derived solvents offer several advantages over the conventional ones as they are biodegradable, environmentally benign, and lower in toxicity. Therefore, these solvents are very well categorized as green solvents [100].

2.5.2.7 Greener surfactants

These surfactants are amphiphilic solvents that contain both polar and nonpolar groups. Biosurfactants are categorized as greener surfactants because they are derived from bio-based renewable sources, developed through a sustainable and nonchemical process. They are usually made from eco-friendly organic compounds [121]. The solvent-free chemical processes have gained popularity over the years, as they enable the use of techniques and substances while minimizing the adverse effects on the environment.

Applying green synthesis techniques are highly recommended for their key qualities such as reduced reaction time, higher yields by converting all reactants into products, provision for safer reaction pathways, prevention of the formation of hazardous by-products or waste, and the ability to enable the synthesis of biodegradable materials through simpler methodologies [64].

2.6 Energy efficiency

The main green chemistry goal is to create environmentally pleasant and commercial production procedures. The following tips are essential to creating less waste, more efficient and greener processes.

2.6.1 Waste prevention

One of the green chemistry principles is that it is recommended to not create waste than to deal with it after the generation of it [7]. Green chemistry aims to create synthetic processes that produce little or no waste by selecting appropriate reaction chemicals and conditions, thus facilitating high process efficiency. For example, the prevention of waste from the atomic industry is an important issue, and the design process should ensure that all starting points are integrated into the finished product [146]. Furthermore, the pharmaceutical industry can reduce waste on site by focusing on reactions that produce fewer by-products, while also making the entire process more sustainable and well-functioning.

2.6.2 Renewable feedstock utilization

Another important principle of sustainable chemistry is to make use of renewable feedstocks [162]. Instead of using nonrenewable feedstocks such as petroleum, the green chemist seeks to utilize starting materials derived from renewable resources such as plants, algae, or waste biomass. This not only helps to decrease dependence upon finite raw materials but also contributes to the circular economy with the reuse and recycling of materials. Examples include the increased usage of renewable resource-based biopolymers, such as agricultural waste or plant origin, as an alternative to conventional plastics [13, 34]. The trend for the development of biodegradable and compostable materials fits into the objective of green chemistry through minimizing environmental impact and not causing long-term pollution.

2.6.3 Chemical processes with energy efficiency

Following pressing needs for energy efficiency, chemical processes should be designed to run at ambient pressure and temperature. By doing so, energy use is minimized and little or no negative environmental impact will result from the chemical production rooted on green chemistry principles [66]. Indeed, this objective is realized in catalysts, specifically enzymes in biocatalysis, since these allow for conducting reactions much more effectively under comparatively milder reaction conditions [140]. Besides this, many new chemical processes are in development, for which solar or wind power is used. It is such a transformation that will enable the chemical industry to further reduce its greenhouse gas (GHG) emissions and enhance the sustainability of its businesses [134].

2.6.4 Microwave (MW) irradiation

Guiguerre et al. [51] and Gedye et al. [50] implemented MW heating in 1986. In material synthesis, the benefits of MW heating over conventional heating are: i); enhanced control of the reaction parameters; ii) high rate of heating iii) selective heating in case of reaction mixture with compounds with varied MW absorbing properties; iv) reduced side reactions, hence offering better selectivity v) higher yields; vi) better reproducibility; vii) and good control of size and shape through high-throughput synthesis and automatization [36]. MW synthesis is used in field of nanomaterials [21], metal–organic frameworks [84], colloid nanocrystals [16], zeolites [113], graphene quantum dots [172], or polymers [185]. Compared to thermal methods, MW synthesis led to more dispersed and homogeneous catalysts, more economic gains, and generating less atmospheric pollutants [105]. MW heating creates environmental catalysts with similar crystal size, porous structure, desired interfacial properties, and morphology, which are significant for air pollution as well as catalytic remediation [17].

2.7 Renewable feedstock

The key objective of the seventh green chemistry principle is to substitute finite sources similar to fossil fuels with sustainable and renewable raw materials. To substantially reduce dependence and environmental impacts, renewable feedstocks should be used increasingly, as suggested by Anastas and Warner [8]. The use of renewable feedstocks will help reduce GHG emissions to reduce pollution and address challenges of climate change. This practice will encourage circular economy practices, enhance efficient recycling of useful resources, and reduce waste production [104]. Renewable feedstocks can be regenerated relatively quickly and easily as they are bio-based natural resources, while fossil resources like petroleum and coal take millions of years to form. Therefore, transitioning to renewable feedstocks plays a significant role in green chemistry, as they are eco-friendly and sustainable, in addition to assisting in reducing the depletion of finite fossil fuels [142].

2.7.1 Biomass feedstocks

Biomass feedstocks come from crops, agricultural sources, waste biomass, and by-products of forestry. During their growth, materials derived from biomass absorb carbon dioxide from the atmosphere via photosynthesis, store it in their structure, and release it out when they are being used or processed. However, this is much less harmful, in contrast to the fossil fuels that let out the carbon that had been stored underground for millions of years. Therefore, biomass resources like wood, waste and crops are considered carbon-neutral, provided they are utilized sustainably. Additionally, the carbon emissions while they are being processed are counterbalanced by the CO_2 absorbed from the atmosphere and retained by them during the phase of their growth [38].

2.7.2 CO_2 utilization

The idea of CO_2 utilization aligns with sustainable practices promoted by green chemistry principles, especially the utilization of renewable feedstocks because CO_2 is a low-impact raw material that is inexpensive and available in plenty. The CO_2 emissions from industrial processes can be captured and repurposed into valuable products like methanol, thereby redirecting waste CO_2 emissions to productive use, aiding in reducing the carbon footprint [12]. Carbon-intensive industries like petrochemicals, steel, and cement could also be impacted at a significant level if CO_2 is utilized on a large scale. It is highly challenging to decarbonize steel and cement production, as they rely heavily on high-temperature processes and a substantial amount of CO_2 emission happens in the chemical reactions involved in the process. This emission can be captured and put back to good use to produce valuable products like synthetic fuels and chemicals, thus turning waste into resources, which otherwise would have become an environmental burden [61].

2.7.3 Ethanol from corn starch

Corn-based ethanol is a biofuel that can replace gasoline. Although the process of producing ethanol from corn starch comes with some challenges, as it requires large amounts of water and a substantial land area, this process reduces the dependence on fossil fuel and aligns with green chemistry's goal of sustainability. In this procedure, corn starch is converted into sugars, which are fermented to ethanol. Ethanol burns cleaner than gasoline, resulting in less pollution [67].

Although renewable feedstock helps with waste minimization, contributes to energy security by reducing the dependency on fossil fuels, and lowers net carbon footprint, the transition to the use of renewable feedstock requires expensive modification in infrastructure, which can be complex [46]. The future of renewable feedstock seems to be promising, with continual advancements in technology. Policymakers are in support of the use of renewable feedstock to achieve environmental goals [12].

2.8 Reducing derivatization

In organic syntheses, reducing derivatization can be accomplished using various approaches and improvements in synthetic routes, tailored to definite reactions and anticipated outcomes. Here are some common approaches that can be adopted:

2.8.1 Non-covalent derivatives

These derivatives are used to control the physical and chemical properties of molecules in chemical reactions; the non-covalent interactions used are electrostatic, lipophilic, pi-stacking, and hydrogen bonding [8, 155]. In the early 1990s, Tayler and Warner employed the concept of non-covalent derivatization to explain the use of elevated pH for controlled diffusion and solubility of hydroquinones. Researchers at Polaroid developed an approach substituting the traditional approach of relying on covalent protecting base-labile for developing a non-covalent co-crystal between hydroquinones and bis-(*N*,*N*-dialkyl) as a protecting group [28]. The lowering of solubility of hydroquinone was made possible by the process of co-crystal formation [159, 169] of bis-[*N*, *N*-diethyl] terephthalamide and hydroquinone. This approach meets the requirements of using the original hydroquinone structures without modification, thereby minimizing energy and waste material. Hence, the process was termed viable and useful for industrial processes. Currently, in medicinal chemistry, the process of co-crystallization approach comprising two or more homogeneous crystalline materials in a fixed stoichiometric ratio enables the tuning of the physical stability of pharmaceuticals, bioavailability, or dissolution rate.

2.8.2 Supramolecules and molecular chaperones

Self-assembly or supramolecules is an extensively used approach for molecular assembly that takes place via two major steps; the first one is the rational design of building blocks and interactions developed between them and the second is the mixing of the building block in specific conditions, predominantly in solution, to assemble them to thermodynamically stable products [171]. Many of the functional materials used for electronics [30], drug delivery [68], light harvest [132], and sensing [31] use this approach. Another approach is "Catassembly", small molecule chemical chaperones (catassemblers) [97, 168] that aid in the formation molecular assembly called Catassembly (Figure 2.1). Catassemblers (i) lower the energy barrier, making reaction faster (ii) they select and guide specific subunits in multicomponent system (iii) support formation of the target product by choosing specific pathways in an assembly system with multiple pathways.

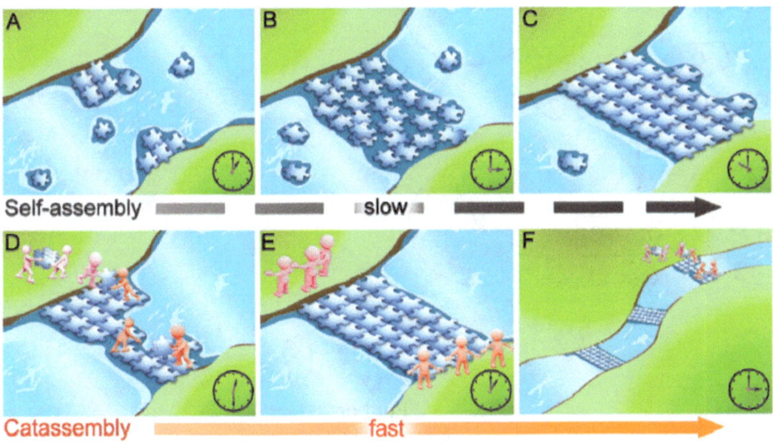

Figure 2.1: Self-assembly (A–C) process and catassemble (D–F) [168].

2.8.3 Flow chemistry

In syntheses, by applying the concept of flow chemistry, side reactions are minimized by using microfluids in flow reactors and maintaining precise thermal conditions and control mixing [174]. Different reagents are sent out as continuous streams using pumps, and mixed in reactors such as continuous-stirred tank reactor (CSTR) or a plug flow reactor (PFR) [32, 70]. When compared to conventional methods, such a practice enables analysis of reaction kinetics and real-time monitoring, trailed by increased reaction efficiency and scalability [58, 96]. Consequently, flow chemistry is adopted by various industries like fine chemicals, materials science, and pharmaceuticals [18, 72].

2.8.4 Click chemistry

Chemical reaction by nature joins small units to give larger molecules by forming thermodynamically favored carbon–heteroatom bonds over carbon–carbon bonds, hence termed "spring-loaded" reactions or "click chemistry" (Figure 2.2), as reported by Barry Sharpless [88]. The chemical reactions that benefit from the above principle are (i) cycloaddition [88]; (ii) nucleophilic substitution/ring-opening reactions [31, 179]; (iii) carbonyl reaction [47]; and (iv) addition reaction [123]. Most of these reactions give high yields, create biodegradable products that can be separated without chromatographic techniques, and are conducted in aqueous medium [93]. Energetically favorable linking reaction establishes a carbon framework with carbon–heteroatom bonds, where new groups are attached [88].

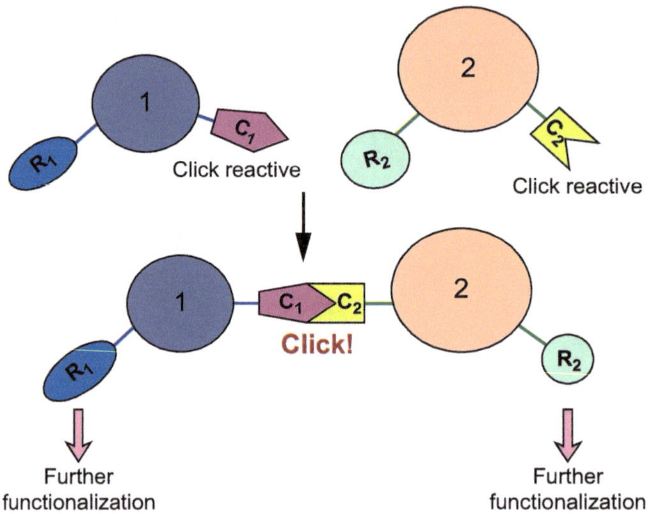

Figure 2.2: Multifunctional design of materials using "click chemistry" (from [124]).

2.8.5 Electrosynthesis

Functional group interconversions or C–C bond formation in organic synthesis eliminate intermediates formation, materialized by electrochemical synthesis, using the principle of selective electron transfer. The selective control (Figure 2.3) could be achieved by (i) paired electrolysis, where a combination of two half-cell reactions is used to give the desired product [24]; (ii) the synergism of electrochemistry and photochemistry, which is aimed at achieving extreme redox potentials for molecular transformations [177]; (iii) the conversion of small molecules (CO_2), in conjunction with electrochemical methods, establishing a green method for synthetic organic chemistry

[181]; and (iv) asymmetric electrochemical transformations, leading to organic synthesis with high enantioselectivity and good functional group tolerance [29].

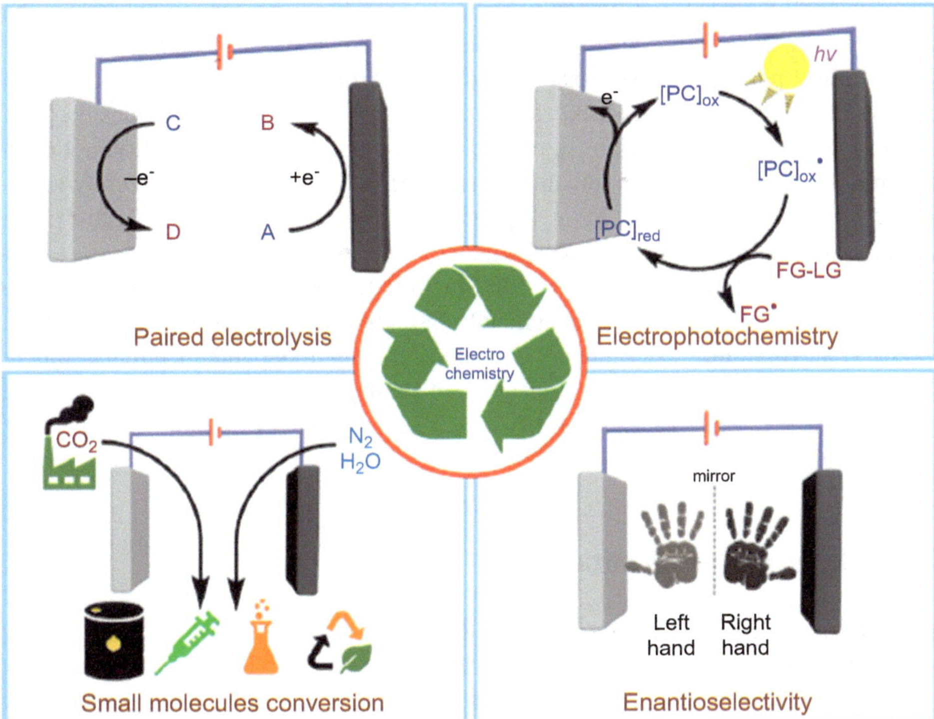

Figure 2.3: Electrochemical methods for synthetic organic chemistry (from [183]).

2.9 Catalysis

Catalytic reagents improve the atom economy, lower energy, increase selectivity, and reduce waste thereby, making them a versatile tool for green synthesis [5]. This part will discuss the improvements and strategies adopted in homogeneous, heterogeneous, and biocatalysts to meet the goals of green chemistry.

2.9.1 Homogeneous catalysts

Homogeneous catalysts, with their uniform distribution, have contributed to high selectivity and activity along with ease of characterization, in terms of intermediate species and kinetics, compared to heterogeneous catalysts [69]. Since 1980, palladacycles have been one of the most effective systems used for C–C bond formation [5]. The recent trend in green chemistry focuses on the usage of non-precious metals in the first row of transition metals such as copper, manganese, and iron; examples are oxidative cleavage of 1,2-diols [44] performed using Mn/Na catalyst. Another major trend is the use of biomimetic approaches like enzymes to achieve efficiency and reactivity similar to natural phenomena. For example, in green oxidation processes, Fe-tetraamido macrocyclic ligand (TAML) is used as an activator. Various other techniques incorporated to enhance the performance of homogeneous catalysis are simplifying downstream operation, immobilization, and workup [112]. This concept was recently established with solid supports, ionic liquids, and supramolecular architectures, in combination with supercritical CO_2 [37]. Significant developments that lead to economic and toxicological advantages [107] include the usage of organocatalysts like nonhazardous amino acids replacing metals [39]. With the Nobel Prize being awarded to List et al. [102] and MacMillan's [39] group for their independent publications in the year 2000, it is now a well-established and fast-developing field [49]. The existing challenges are catalyst recycling [145], toxicity assessments [120], and high catalyst loading [39].

2.9.2 Heterogeneous catalysts

The heterogeneous catalysts have more stability and ease of separation and handling as well as recycling ability [101]. The drawback is its structural instability along with toxicity awareness of non-precious abundant metal [40]. Abundant materials like clay [83], zeolites [106], and solid base and acid catalysts with increased safety have replaced liquid systems, in agreement with Principles 3 and 12 of green synthesis [54, 81]. Currently, nanoparticles from non-precious metals like copper and nickel [71, 137] have gained attention, but their environmental and human health concerns are debatable. Using advanced knowledge, it has been suggested to use immobilized solid catalysts in continuous flow conditions [139] to generate multiple functional products, which are inevitable for developing economies. The fine chemical industry [129] has started using modularized production units with a network of smaller decentralized units, where fine chemicals and active pharmaceutical ingredients (APIs) are flexibly manufactured under mild reaction conditions using microreactor technology coupled with new-generation immobilized solid catalyst with no or little waste.

2.9.3 Biocatalysts

Biocatalysts paved the way for designing a chemical reaction in ambient conditions and in aqueous medium. The process conducted culminates with catalysts that are biocompatible, biodegradable, and renewable [98, 148]. The different stages involved in the development of biocatalysts for industry include designing of reaction, choice of biocatalyst, and its optimization, followed by development to suit each reaction type [19].

Biocatalytic cascades [111, 126] consist of two or more steps (bond forming or interconversions of the functional group), of which at least one step of transformation uses enzymes and without isolations of intermediate (Figure 2.4) [19]. Biocatalysis is appropriate to the cascade processes, with its inherent properties like regioselectivity, stereoselectivity, and chemoselectivity, and is functional in aqueous media [19]. Recent reviews published on Biocatalytic cascades throw insight into its scope and potential [98, 141, 154]. Biocatalytic cascade processes streamline the overall synthesis [26] from an industrial standpoint, strikingly eliminating the requirement of isolation of intermediate, saving time, and reducing waste and costs. The synthesis of direct-acting COVID-19 antiviral molnupiravir [20] is a recent example of the three-step route with two enzymatic steps developed within six months. The use of machine learning [174] and ultra-high-throughput screening techniques [109] in protein engineering would open a futuristic way to synthesize efficient biocatalysts.

2.10 Design for degradation

Degradation processes occur via three major pathways: (a) hydrolysis (b) biodegradation and (c) atmospheric oxidation. While designing a chemical reaction, degradability of the product to sustainable materials must be ensured.

2.10.1 Designing reactions using database

The design stage of green organic synthesis emphasizes the usage of databases and programs like the University of Minnesota Biocatalysis/Biodegradation Database (UMBBD) [42, 48], Pathway Prediction System (PPS) [22], in association with CATABOL program [74], and QSAR/SAR in silico models [113, 128]. The US EPA (EpiSuite^{TM} software), together with "rules of thumb," is currently supporting synthetic chemists to improve the degradability of molecules [23, 133] in the design stage.

2.10.2 Chemical modification

The degradation is promoted by using the principle of non-covalent derivatives [155] to develop a "molecular switch," a phenomenon by which degradation would be activated under specific environmental conditions [86]. Lately, efforts are on for the chemical modification of "big 6" polymers by alternative biodegradable monomers.

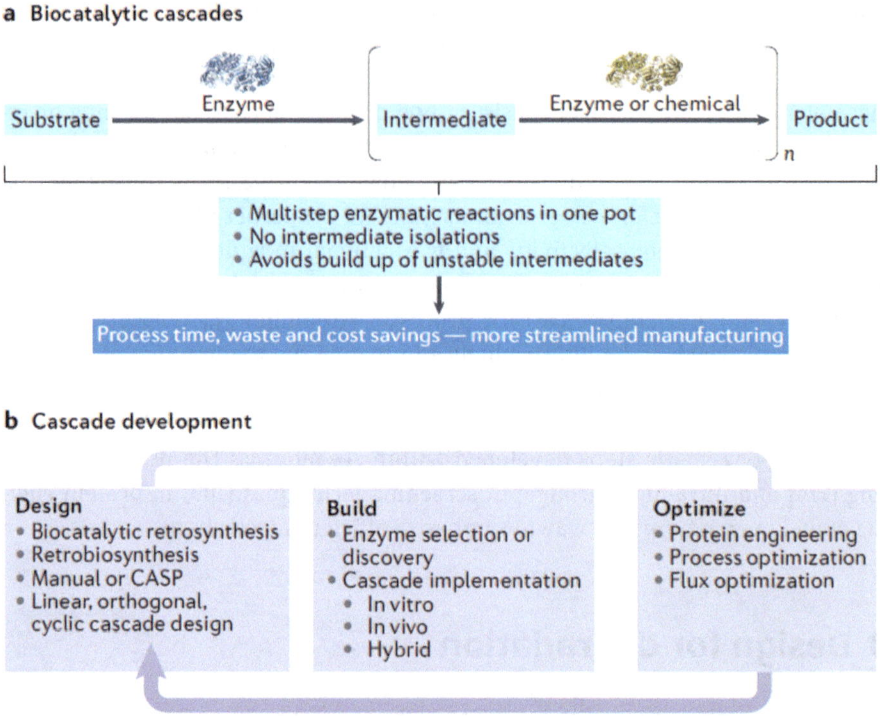

Figure 2.4: (a) Biocatalytic cascades are reaction sequences in which at least one chemical step is catalyzed by an enzyme. Reaction conditions for enzymatic reactions are often compatible with multistep one-pot protocols. (b) Steps to generate optimized biocatalytic cascades (design–build–optimize cycle) [19].

About 76% of the plastics used currently are made from the "big 6" polymers and are the most evident source of pollution [156]. The alternative monomers used include cellulose derivatives and starch [14], poly(lactic acid) (PLA) [63], poly(butylene succinate) (PBS) [143], poly(vinyl alcohol) (PVA) [33], poly(hydroxyalkanoates) (PHAs) [11], and poly(caprolactone) (PCL) [143]. Although their commercial applications are wide in range, they are yet to replace petroleum-based polymers on a large scale.

2.10.3 Practice in pharmaceutical

Around 3,000 APIs used in the pharmaceutical industry [92] have been identified. API metabolites with longer shelf lives [75] in design, when released in the environment, offer challenges and pose significant threats, for example, immense deterioration in the vulture population due to acquaintance to diclofenac [126] and with antibiotic resistance [91, 105]. Several countries [35, 60] have implemented systems that incorporate the possible ecological threats of prescribed APIs to propose environmentally friendly and functionally suitable medications to both doctors and patients [59, 91].

2.11 Real-time analysis

Pollution caused by chemical industries, the leading source of production of hazardous chemical wastes, is one of the biggest threats to the modern world, causing harm to all living organisms and the environment. Thus, one major area of emphasis in green chemistry is the development of systematic tools for the real-time observation of the chemical processes. As such the main goal of real-time monitoring is to obtain analytical data close to the reaction process, which leads to an improved control over the overall process as well as minimizes the impact on the environment [167]. Many factors determine the pace of a chemical reaction and every reaction takes its own time to happen. Some reactions are very slow, while some can occur rapidly [182]. Hence, it is important to monitor chemical reactions and simultaneously take care of the concentrations of different products formed in real time under the reaction conditions. These monitoring tools help to determine the reaction kinetics as well as elucidate the reaction mechanism altogether. The advanced monitoring tools can provide information about the components present within the reaction mixtures. They also give valuable information related to the structure of the products and help in the quantitative and qualitative analysis of the end products [55]. Chemical reactions can be monitored in different ways. It can be coupled directly into the reaction system by inserting a probe into the reaction vessel, and this type of monitoring that is usually done is known as in line or in situ monitoring. In the other type of observation, the instrument is connected directly to the reaction system, and is known as online monitoring. There is a third type known as at-line or off-line monitoring, where monitoring is done remotely from the reaction system by sampling.

2.11.1 Continuous flow analysis using spectroscopic instruments

The continuous flow analysis serves as an integral and important tool in the field of scientific research as well as in industrial scenarios. The advantages of using in-line monitoring tools are that they help in the optimization of reactions, keep a check over quality control, and also detect sample conditions. These instruments are faster and highly reliable for data analysis. The most important spectroscopic instruments that are commonly used are infrared (IR), Raman, ultraviolet-visible (UV-vis), and nuclear magnetic resonance (NMR), along with methods like calorimetry, mass spectrometry, electrochemistry, and various chromatographic techniques [56].

2.11.2 Electrochemical biosensors

These biosensors have also evolved in recent times as an analytical tool based on the principles of green chemistry. They are equipped with the methodologies for real-time and in situ monitoring, managing, and preventing the formation of hazardous substances by performing direct sample analysis. In the electrochemical detection technique, portable mini-sized devices are used like the in situ analysis of redox reaction, without using any reagents. The biosensors are prepared following the green chemistry and sustainability standards. Environmentally friendly materials are used to construct such devices. These tools are also being used for the in situ detection of polluting substances like pesticides [130]. Real-time bioluminescence was used in one of the studies to detect the detailed mechanism of a protein called KaiABC-ATP, related to the circadian rhythm of the human body [119].

2.11.3 Chromatography

Chromatography is one of the most commonly used tools for real-time monitoring and for separation and purification of samples as well as for data analysis. There are different types of chromatography techniques used in research and industries. Column chromatography is used in the pharmaceutical, food, and petroleum industries [180]. High-performance liquid chromatography (HPLC) coupled to UV-Vis or fluorescence spectrophotometer is very efficiently used for off-line sample analysis. Other frequently used column techniques are ion exchange, reversed-phase liquid chromatography, and hydrophobic interaction liquid chromatography are the other types of columns used effectively for fast and reliable data analysis [53].

2.11.4 Computational methods

The methods used in quantum mechanics help in the study and identification of interactions in interatomic materials, while data-driven techniques are helpful in understanding the structure and functions of different substances [57]. These methods save time and are cost-effective. Large amounts of datasets of different materials have been generated for analysis and information [178]. Recently, biosensors, very sensitive devices, have been used in the detection of pathogens like viruses [3]. Microfluid-integrated biosensors were designed with the aim of detecting coronavirus, during the COVID pandemic. These biosensors, containing microfluidic channels, are connected to electrical chips for analysis. Different types of software tools and applications have been generated to study the detailed mechanism of chemical reactions [143].

2.12 Hazard and accident prevention

Green chemistry advocates eradicating intrinsic hazards at the source rather than minimizing risks due to exposure to harmful toxic substances [170]. Most traditional chemical processes lead to health hazards and environmental pollution that affect the whole community, including the workers involved in the process due to the exposure to the chemical emissions and waste produced.

2.12.1 Safety in green chemistry

Traditional methods often rely on controls from outside and the use of protective gear to ensure safety. However, this principle of green chemistry seeks to remodel chemical processes, materials, and products to prevent dangers from arising in the first place. Inherent safety means making the process safer from the outset so that safety is built into the design of the process strategically by being proactive rather than reactive [27]. In simple terms, safety in green chemistry is controlling the known hazards by minimizing risks to an acceptable level by designing eco-friendly chemical processes to reduce the production of hazardous substances [8, 41]. It also promotes designing strategies and control measures that can be achieved by minimizing the need for personal protective equipment [1].

2.12.2 Risk

The term "risk" is not to be confused with hazard, as chemists and chemistry students often use these terms interchangeably. Risk is considered as the likelihood of harm occurring due to the exposure to anything potentially hazardous, while hazard is something that can cause harm. It is represented in the following equation:

Risk = (probability of exposure) × (severity of the consequence of the hazard)

This kind of quantification of risk aids in proactively identifying hazards [45].

2.12.3 RAMP

The RAMP is a proactive risk management strategy used in green chemistry to improve safety. RAMP stands for recognize hazards, assess risks, minimize risks, and prepare for emergencies. This promotes a protocol that encourages safer and more sustainable chemical practices directly integrated into the design [62]. The first step of this protocol is to recognize or identify hazardous chemicals, processes, or conditions that might be present. By evaluating these risks from the beginning, green chemists can take steps to avoid or reduce hazards within the process, although hazards are associated with all chemicals. The next stage is to assess the probability of accidents and the potential severity of the impact, which is considered to be the most important step as per Hill and Finster [62]. Such proactive risk assessment aids in making crucial decisions on which type of hazards need the most attention based on the risk levels. Once the hazards are identified and risks are evaluated, it is important to devise strategies to minimize the risks. This could be achieved by careful selection of safer solvents, and minimal use of hazardous materials and catalytic reactions. Despite all these precautionary measures to prevent emergencies, unforeseen incidents and accidents might happen. Therefore, the RAMP framework strongly emphasizes the importance of being well prepared with an effective emergency response plan to minimize personal injury and potential damage in case of unforeseen incidents [45].

2.13 Conclusion

To conclude, these techniques establish that greener syntheses with reduced environmental pollution and fewer workplace hazards are in line with the philosophy of green chemistry in achieving sustainable development [78]. For the development of safer and more sustainable chemicals and solvents, many initiatives and regulations have been planned and executed over the years. Regarding the concerns about using harmful volatile chemicals, green chemistry focuses on chemical processes that are gentle on the environment and are designed with methods that cause the least pollution to the ecosystem [64]. Real-time monitoring critically emphasizes the implication of monitoring chemical reactions to avoid the formation of hazardous by-products causing pollution and to use safer reaction processes. These in-process monitoring systems can be used to prevent waste products, improve synthesis efficacy, enhance the use of catalysts, and support unconventional methods like solvent-free chemistry [43]. By using real-time monitoring with continuous flow processing techniques, it is easy for the researchers to perform a wide range of chemical reactions safely, effectively, and easily [56].

References

[1] Abdussalam-Mohammed, W., Qasem Ali, A., & Errayes, A. O. (2020). Green chemistry: Principles, applications, and disadvantages. *Chemical Methodologies*, 4(4), 408–423. https://doi.org/10.33945/sami/chemm.2020.4.4

[2] Amesho, K. T. T., Kadhila, T., Shangdiar, S., Shihomeka, S. P., Shopati, A. K., Edoun, E. I., & Mohan, C. (2024). Bioeconomy for sustainable Environment management practices. In: *Book: Bioeconomy for Sustainability*, (1 ed.). *Springer Singapore*, pp. 41–63. https://doi.org/10.1007/978-981-97-1837-5_2

[3] Amritsar, J., Foroughi, S., Raju, D., Pakkiriswami, S., & Packirisamy, M. (2020). Conformational detection of heat shock protein through bio-interactions with microstructures. *Research on Biomedical Engineering*, 36(1), 89–98. https://doi.org/10.1007/s42600-019-00038-7
[4] Anastas, P. T. (1999). Green chemistry and the role of analytical methodology development. *Critical Reviews in Analytical Chemistry*, 29(3), 167–175. https://doi.org/10.1080/10408349891199356
[5] Anastas, P. T., & Crabtree, R. H. (2009). Chapter 10. In: *Green Catalysis: Volume 1: Homogeneous Catalysis*, Wiley-VCH, pp. 319–338.
[6] Anastas, P. T., & Crabtree, R. H., (Eds.). (2009). *Green Catalysis: Volume 1: Homogeneous Catalysis*, (vol. 1). Wiley-VCH.
[7] Anastas, P. T., & Warner, J. C. (1998). *Green Chemistry: Theory and Practice*. Oxford University Press.
[8] Anastas, P. T., & Warner, J. C. (2000). Chapter 19. In: *Green Chemistry: Theory and Practice*, Oxford University Press, pp. 336–346.
[9] Anastas, P. T., & Zimmerman, J. B. (2003). Design through the 12 principles of green engineering. *Environmental Science & Technology*, 37(5), 94A–101A. https://doi.org/10.1021/es032373g
[10] Anastas, P., & Eghbali, N. (2010). Green chemistry: Principles and practice. *Chem. Soc. Rev*, 39(1), 301–312. https://doi.org/10.1039/b918763b
[11] Anbukarasu, P., Sauvageau, D., & Elias, A. (2015). Tuning the properties of polyhydroxybutyrate films using acetic acid via solvent casting. *Scientific Reports*, 5(1), https://doi.org/10.1038/srep17884
[12] Artz, J., Müller, T. E., Thenert, K., Kleinekorte, J., Meys, R., Sternberg, A., Bardow, A., & Leitner, W. (2017). Sustainable conversion of carbon dioxide: An integrated review of catalysis and life cycle assessment. *Chemical Reviews*, 118(2), 434–504. https://doi.org/10.1021/acs.chemrev.7b00435
[13] Arya, H., Mohan, C., Pandey, S., Verma, M., & Kumar, V. (2021). Phytochemical screening of Basella alba leaves extracts and evaluate its efficacy on sun burn (Sun Protection Factor). *European Journal of Molecular & Clinical Medicine*, 8(1), 417–423.
[14] Avérous, L. (2004). Biodegradable multiphase systems based on plasticized starch: A Review. *Journal of Macromolecular Science, Part C: Polymer Reviews*, 44(3), 231–274. https://doi.org/10.1081/mc-200029326
[15] Babita, S. V., & Mohan, C. (2023). Plant-derived compounds and their green synthesis in pharmaceuticals and nutraceuticals. In: *Book: Green Chemistry Approaches to Environmental Sustainability*. Elsevier, pp. 149–163. https://doi.org/10.1016/B978-0-443-18959-3.00001-X.
[16] Baghbanzadeh, M., Carbone, L., Cozzoli, P. D., & Kappe, C. O. (2011). Microwave-assisted synthesis of colloidal inorganic nanocrystals. *Angewandte Chemie International Edition*, 50(48), 11312–11359. https://doi.org/10.1002/anie.201101274
[17] Bao, C., Serrano-Lotina, A., Niu, M., Portela, R., Li, Y., Lim, K. H., Liu, P., Wang, W. J., Bañares, M. A., & Wang, Q. (2023). Microwave-associated chemistry in environmental catalysis for air pollution remediation: A review. *Chemical Engineering Journal*, 466, 142902. https://doi.org/10.1016/j.cej.2023.142902
[18] Baumann, M., Moody, T. S., Smyth, M., & Wharry, S. (2020). A perspective on continuous flow chemistry in the pharmaceutical industry. *Organic Process Research & Development*, 24(10), 1802–1813.
[19] Bell, E. L., Finnigan, W., France, S. P., Green, A. P., Hayes, M. A., Hepworth, L. J., Lovelock, S. L., Niikura, H., Osuna, S., Romero, E., Ryan, K. S., Turner, N. J., & Flitsch, S. L. (2021). Biocatalysis. *Nature Reviews Methods Primers*, 1(1), 1–21. https://doi.org/10.1038/s43586-021-00044-z
[20] Benkovics, T., McIntosh, J., Silverman, S., Kong, J., Maligres, P., Itoh, T., Yang, H., Huffman, M., Verma, D., Pan, W., Ho, H.-I., Vroom, J., Knight, A., Hurtak, J., Morris, W., Strotman, N., Murphy, G., Maloney, K., & Fier, P. (2020). Evolving to an ideal synthesis of molnupiravir, an investigational treatment for COVID-19. *Chemrxiv.org*, https://doi.org/10.26434/chemrxiv.13472373.v1
[21] Bilecka, I., & Niederberger, M. (2010). Microwave chemistry for inorganic nanomaterials synthesis. *Nanoscale*, 2(7), 1358–1374. https://doi.org/10.1039/B9NR00377K

[22] Lynda, B. K. H., & Wackett, L. P. (2004). Encoding microbial metabolic logic: Predicting biodegradation. *Journal of Industrial Microbiology & Biotechnology*, 31(6), 261–272. https://doi.org/10.1007/s10295-004-0144-7
[23] Boethling, R. S., Sommer, E., & DiFiore, D. (2007). Designing small molecules for biodegradability. *Chemical Reviews*, 107(6), 2207–2227. https://doi.org/10.1021/cr050952t
[24] Bortnikov, E. O., & Semenov, S. N. (2020). Coupling of alternating current to transition-metal catalysis: Examples of nickel-catalyzed cross-coupling. *The Journal of Organic Chemistry*, 86(1), 782–793.
[25] Boxall, A. B., & Rudd, M. A. (2007). Pharmaceuticals and personal care products in the environment: What are the big questions?. *Environmental Health Perspectives*, 115(10), 1229–1235. https://doi.org/10.1289/ehp.10269
[26] Bruggink, A., Schoevaart, R., & Kieboom, T. (2003). Concepts of nature in organic synthesis: cascade catalysis and multistep conversions in concert. *Organic Process Research & Development*, 7(5), 622–640. https://doi.org/10.1021/op0340311
[27] Caldeira, C., Abbate, E., Moretti, C., Mancini, L., & Sala, S. (2024). Safe and sustainable chemicals and materials: A review of sustainability assessment frameworks. *Green Chemistry*, 26(26), 7456–7477. https://doi.org/10.1039/D3GC04598F
[28] Cannon, A. S., & Warner, J. C. (2002). Noncovalent derivatization: Green chemistry applications of crystal engineering. *Crystal Growth & Design*, 2(4), 255–257.
[29] Chang, X., Zhang, Q., & Guo, C. (2020). Asymmetric electrochemical transformations. *Angewandte Chemie*, 132(31), 12712–12722.
[30] Chen, H., & Fraser Stoddart, J. (2021). From molecular to supramolecular electronics. *Nature Reviews Materials*, 6, 804–828. https://doi.org/10.1038/s41578-021-00302-2
[31] Chen, Y., & Tong, Z. R. (2017). *Click Chemistry: Approaches, Applications and Challenges*. Nova Science Publishers, Inc.
[32] Cherkasov, N., Adams, S. J., Bainbridge, E. G., & Thornton, J. A. (2023). Continuous stirred tank reactors in fine chemical synthesis for efficient mixing, solids-handling, and rapid scale-up. *Reaction Chemistry & Engineering*, 8(2), 266–277.
[33] Chiellini, E., Corti, A., D'Antone, S., & Solaro, R. (2003). Biodegradation of poly (vinyl alcohol) based materials. *Progress in Polymer Science*, 28(6), 963–1014. https://doi.org/10.1016/s0079-6700(02)00149-1
[34] Clark, J. H., & Deswarte, F. E. I. (2015). *Introduction to Chemicals from Biomass*. John Wiley & Sons.
[35] Cussans, A., Harvey, G., Kemple, T., & Tomson, M. (2021). Interventions to reduce the environmental impact of medicines: A UK perspective☆. *The Journal of Climate Change and Health*, 4, 100079. https://doi.org/10.1016/j.joclim.2021.100079
[36] Dahl, J. A., Maddux, B. L. S., & Hutchison, J. E. (2007). Toward greener nanosynthesis. *Chemical Reviews*, 107(6), 2228–2269. https://doi.org/10.1021/cr050943k
[37] Dewaele, A., Verpoort, F., & Sels, B. (2016). Opportunities of immobilized homogeneous metathesis complexes as prominent heterogeneous catalysts. *ChemCatChem*, 8(19), 3010–3030. https://doi.org/10.1002/cctc.201600591
[38] Dion, L.-M., Lefsrud, M., & Orsat, V. (2011). Review of CO2 recovery methods from the exhaust gas of biomass heating systems for safe enrichment in greenhouses. *Biomass and Bioenergy*, 35(8), 3422–3432. https://doi.org/10.1016/j.biombioe.2011.06.013
[39] Dondoni, A., & Massi, A. (2008). Asymmetric organocatalysis: From infancy to adolescence. *Angewandte Chemie International Edition*, 47(25), 4638–4660.
[40] Egorova, K. S., & Ananikov, V. P. (2016). Which metals are green for catalysis? comparison of the toxicities of Ni, Cu, Fe, Pd, Pt, Rh, and Au Salts. *Angewandte Chemie International Edition*, 55(40), 12150–12162. https://doi.org/10.1002/anie.201603777

[41] Egorova, K. S., Kolesnikov, A. E., Posvyatenko, A. V., Galushko, A. S., Shaydullin, R. R., & Ananikov, V. P. (2024). Establishing the main determinants of the environmental safety of catalytic fine chemical synthesis with catalytic cross-coupling reactions. *Green Chemistry*, 26(5), 2825–2841. https://doi.org/10.1039/d3gc04572b

[42] Ellis, L. B. M. (2006). The University of Minnesota Biocatalysis/Biodegradation database: The first decade. *Nucleic Acids Research*, 34(90001), D517–D521. https://doi.org/10.1093/nar/gkj076

[43] Erythropel, H. C., Zimmerman, J. B., Winter, T. M. D., Petitjean, L., Melnikov, F., Lam, C. H., Lounsbury, A. W., Mellor, K. E., Janković, N. Z., Tu, Q., Pincus, L. N., Falinski, M. M., Shi, W., Coish, P., Plata, D. L., & Anastas, P. T. (2018). The Green ChemisTREE: 20 years after taking root with the 12 principles. *Green Chemistry*, 20(9), 1929–1961. https://doi.org/10.1039/C8GC00482J

[44] Escande, V., Lam, C. H., Coish, P., & Anastas, P. T. (2017). Heterogeneous Sodium-Manganese Oxide Catalyzed Aerobic Oxidative Cleavage of 1,2-diols. *Angewandte Chemie International Edition*, 56(32), 9561–9565. https://doi.org/10.1002/anie.201705934

[45] Finster, D. C. (2020). RAMP: A safety tool for chemists and chemistry students. *Journal of Chemical Education*, 98(1), 19–24. https://doi.org/10.1021/acs.jchemed.0c00142

[46] Gallezot, P. (2012). Conversion of biomass to selected chemical products. *Chem. Soc. Rev*, 41(4), 1538–1558. https://doi.org/10.1039/c1cs15147a

[47] Gandhi, P. S., Gaikwad, P. L., Jagdale, D. M., & Kadam, V. J. (2012). Growing advances and applications of click reactions. *American Journal of PharmTech Research*, 2, 83–103.

[48] Gao, J., Ellis, L. B. M., & Wackett, L. P. (2010). The University of Minnesota Biocatalysis/ Biodegradation database: Improving public access. *Nucleic Acids Research*, 38(Suppl_1), D488–D491. https://doi.org/10.1093/nar/gkp771

[49] García Mancheño, O., & Waser, M. (2023). Recent developments and trends in asymmetric organocatalysis. *European Journal of Organic Chemistry*, 26(1), e202200950.

[50] Gedye, R., Smith, F., Westaway, K., Ali, H., Baldisera, L., Laberge, L., & Rousell, J. (1986). The use of microwave ovens for rapid organic synthesis. *Tetrahedron Letters*, 27(3), 279–282. https://doi.org/10.1016/S0040-4039(00)83996-9

[51] Giguere, R. J., Bray, T. L., Duncan, S. M., & Majetich, G. (1986). Application of commercial microwave ovens to organic synthesis. *Tetrahedron Letters*, 27(45), 4945–4948. https://doi.org/10.1016/S0040-4039(00)85103-5

[52] Gleiter, C. H., Schuchmann, H. P., & Rasper, J. (1995). Degradable plastics and their implications on green chemistry. *Journal of Polymers and the Environment*, 3(4), 215–226. https://doi.org/10.1007/BF01457977

[53] Graf, T., Naumann, L., Bonnington, L., Heckel, J., Spensberger, B., Klein, S., Brey, C., Nachtigall, R., Mroz, M., Hogg, T. V., McHardy, C., Martinez, A., Braaz, R., & Leiss, M. (2024). Expediting online liquid chromatography for real-time monitoring of product attributes to advance process analytical technology in downstream processing of biopharmaceuticals. *Journal of Chromatography A*, 1729, 465013–465013. https://doi.org/10.1016/j.chroma.2024.465013

[54] Gupta, P., & Paul, S. (2014). Solid acids: Green alternatives for acid catalysis. *Catalysis Today*, 236, 153–170. https://doi.org/10.1016/j.cattod.2014.04.010

[55] Hall, A. M. R., Chouler, J. C., Codina, A., Gierth, P. T., Lowe, J. P., & Hintermair, U. (2016). Practical aspects of real-time reaction monitoring using multi-nuclear high resolution FlowNMR spectroscopy. *Catalysis Science & Technology*, 6(24), 8406–8417. https://doi.org/10.1039/c6cy01754a

[56] Hamlin, T. A., & Leadbeater, N. E. (2015). Real-time monitoring of reactions performed using continuous-flow processing: The preparation of 3-Acetylcoumarin as an example. *Journal of Visualized Experiments: JoVE*, 105, 52393. https://doi.org/10.3791/52393

[57] Harada, J. K., Charles, N., Poeppelmeier, K. R., & Rondinelli, J. M. (2019). Heteroanionic materials by design: Progress toward targeted properties. *Advanced Materials*, 31(19), 1805295–1805295. https://doi.org/10.1002/adma.201805295

[58] Hartman, R. L., McMullen, J. P., & Jensen, K. F. (2011). Deciding whether to go with the flow: Evaluating the merits of flow reactors for synthesis. *Angewandte Chemie International Edition*, 50(33), 7502–7519.

[59] Heidari, M., Kumar, A., Kumar, A., & Kumar, A. (2024). Navigating the landscape: Cutting-edge biomedical manufacturing techniques. In: Kumar, A., Kumar, A., & Kumar, A., (eds.). *Applications of Biotribology in Biomedical Systems*. Cham: Springer, (pp. [page numbers if known]). https://doi.org/10.1007/978-3-031-58327-8_7

[60] Helwig, K., Niemi, L., Stenuick, J. -. Y., Alejandre, J. C., Pfleger, S., Roberts, J., Harrower, J., Nafo, I., & Pahl, O. (2024). Broadening the perspective on reducing pharmaceutical residues in the environment. *Environmental Toxicology and Chemistry*, https://doi.org/10.1002/etc.5563

[61] Hepburn, C., Adlen, E., Beddington, J., Carter, E. A., Fuss, S., Mac Dowell, N., Minx, J. C., Smith, P., & Williams, C. K. (2019). The technological and economic prospects for CO2 utilization and removal. *Nature*, 575(7781), 87–97. https://doi.org/10.1038/s41586-019-1681-6

[62] Hill, R. H. Jr, & Finster, D. C. (2010). *Laboratory Safety for Chemistry Students*. Wiley.

[63] Hillmyer, M. A., & Tolman, W. B. (2014). Aliphatic polyester block polymers: renewable, degradable, and sustainable. *Accounts of Chemical Research*, 47(8), 2390–2396. https://doi.org/10.1021/ar500121d

[64] Himaja, M., Rao, B. S., & Kumar, K. P. (2011). Green technique-solvent free synthesis and its advantages. *International Journal of Research in Ayurveda and Pharmacy*, 2(4), 1079–1086. https://www.researchgate.net/publication/232710079

[65] Horn, E. J., Rosen, B. R., & Baran, P. S. (2016). Synthetic organic electrochemistry: An enabling and sustainable technology. *ACS Central Science*, 2(5), 302–308. https://doi.org/10.1021/acscentsci.6b00091

[66] Horváth, I. T., & Anastas, P. T. (2007). Innovations and green chemistry. *Chemical Reviews*, 107(6), 2169–2173. https://doi.org/10.1021/cr0783877

[67] Shapouri, H., Duffield, J. A., & Wang, M. (2002). The energy balance of corn ethanol: An update. https://doi.org/10.22004/ag.econ.34075

[68] Hou, X., Zaks, T., Langer, R., et al. (2021). Lipid nanoparticles for mRNA delivery. *Nature Reviews Materials*, 6, 1078–1094. https://doi.org/10.1038/s41578-021-00358-0

[69] Housecroft, C. E., & Sharpe, A. G. (2001). *Inorganic Chemistry*, (4th ed.). Pearson Education.

[70] Hu, C. (2021). Reactor design and selection for effective continuous manufacturing of pharmaceuticals. *Journal of Flow Chemistry*, 11(3), 243–263.

[71] Hudson, R., Li, C.-J., & Moores, A. (2012). Magnetic copper–iron nanoparticles as simple heterogeneous catalysts for the azide–alkyne click reaction in water. *Green Chemistry*, 14(3), 622. https://doi.org/10.1039/c2gc16421c

[72] Hughes, D. L. (2020). Applications of flow chemistry in the pharmaceutical industry – Highlights of the recent patent literature. *Organic Process Research & Development*, 24(10), 1850–1860.

[73] Ivanković, A. (2017). Review of 12 principles of green chemistry in practice. *International Journal of Sustainable and Green Energy*, 6(3), 39. https://doi.org/10.11648/j.ijrse.20170603.12

[74] Jaworska, J., Dimitrov, S., Nikolova, N., & Mekenyan, O. (2002). Probabilistic assessment of biodegradability based on metabolic pathways: CATABOL System. *SAR and QSAR in Environmental Research*, 13(2), 307–323. https://doi.org/10.1080/10629360290002794

[75] Jagiello, K., Mostrag-Szlichtyng, A., Gajewicz, A., Kawai, T., Imaizumi, Y., Sakurai, T., Yamamoto, H., Tatarazako, N., Mizukawa, K., Aoki, Y., Suzuki, N., Watanabe, H., & Puzyn, T. (2015). Towards modelling of the environmental fate of pharmaceuticals using the QSPR-MM scheme. *Environmental Modelling and Software*, 72, 147–154. https://doi.org/10.1016/j.envsoft.2015.06.013

[76] Jessop, P. G. (2011). Searching for green solvents. *Green Chemistry*, 13(6), 1391–1398. https://doi.org/10.1039/C0GC00962K

[77] Jessop, P. G., Mercer, S. M., & Heldebrant, D. J. (2012). CO2-triggered switchable solvents, surfactants, and other materials. *Energy and Environmental Science*, 5(6), 7240–7240. https://doi.org/10.1039/c2ee02912j

[78] Johnson, M. P., & Wilson, L. K. (2018). Advances in green solvents: Moving towards a safer synthesis. *Environmental Chemistry Reviews*, 29(5), 455–468.

[79] Jordan, A., Haiß, A., Spulak, M., Karpichev, Y., Kümmerer, K., & Gathergood, N. (2016). Synthesis of a series of amino acid derived ionic liquids and tertiary amines: Green chemistry metrics including microbial toxicity and preliminary biodegradation data analysis. *Green Chemistry*, 18(16), 4374–4392. https://doi.org/10.1039/c6gc00415f

[80] Joshi, P. B., Durve, N. C., & Mohan, C. (2023). Full blown green metrics. In: *Book: Green Chemistry Approaches to Environmental Sustainability*. Elsevier, pp. 109–129. https://doi.org/10.1016/B978-0-443-18959-3.00013-6.

[81] Jothiramalingam, R., & Wang, M. K. (2009). Review of recent developments in solid acid, base, and enzyme catalysts (Heterogeneous) for biodiesel production via transesterification. *Industrial & Engineering Chemistry Research*, 48(13), 6162–6172. https://doi.org/10.1021/ie801872t

[82] Kate, A., Sahu, L. K., Pandey, J., Mishra, M., & Sharma, P. K. (2022). Green catalysis for chemical transformation: The need for the sustainable development. *Current Research in Green and Sustainable Chemistry*, 5, 100248. https://doi.org/10.1016/j.crgsc.2021.100248

[83] Kaur, N., & Kishore, D. (2012). Montmorillonite: An efficient, heterogeneous, and green catalyst for organic synthesis. *Journal of Chemical and Pharmaceutical Research*, 4(2), 991–1015.

[84] Khan, N. A., & Jhung, S. H. (2015). Synthesis of metal-organic frameworks (MOFs) with microwave or ultrasound: Rapid reaction, phase-selectivity, and size reduction. *Coordination Chemistry Reviews*, 285, 11–23. https://doi.org/10.1016/j.ccr.2014.10.008

[85] Kharissova, O. V., Kharisov, B. I., Oliva González, C. M., Méndez, Y. P., & López, I. (2019). Greener synthesis of chemical compounds and materials. *Royal Society Open Science*, 6(11), 191378. https://doi.org/10.1098/rsos.191378

[86] Khetan, S. K., & Collins, T. J. (2007). Human pharmaceuticals in the aquatic environment: A challenge to green chemistry. *Chemical Reviews*, 107(6), 2319–2364. https://doi.org/10.1021/cr020441w

[87] Knez, Ž., Pantić, M., Cör, D., Novak, Z., & Knez-Hrnčič, M. (2019). Are supercritical fluids solvents for the future?. *Chemical Engineering and Processing – Process Intensification*, 141, 107532. https://doi.org/10.1016/j.cep.2019.107532

[88] Kolb, H. C., & Sharpless, K. B. (2003). The growing impact of click chemistry on drug discovery. *Drug Discovery Today*, 8(24), 1128–1137.

[89] Kumar, V., Sharma, A. K., Rajput, S. K., Pal, M., & Dhiman, N. (2018). Pharmacognostic and pharmacological evaluation of Eulaliopsis binata plant extracts by measuring in vitro/in vivo safety profile and anti-microbial potential. *Toxicology Research*, 7(3), 454–464.

[90] Kümmerer, K. (2008). Pharmaceuticals in the environment – A brief summary. In Kümmerer, K., (ed.). *Pharmaceuticals in the Environment*. Springer, (pp. 3–21). https://doi.org/10.1007/978-3-540-74664-5_1

[91] Kümmerer, K. (2009). Antibiotics in the aquatic environment – A review – Part I. *Chemosphere*, 75(4), 417–434. https://doi.org/10.1016/j.chemosphere.2008.11.086

[92] Kümmerer, K. (2010). Pharmaceuticals in the Environment. *Annual Review of Environment and Resources*, 35(1), 57–75. https://doi.org/10.1146/annurev-environ-052809-161223

[93] Kwan, T. A., Tu, Q., & Zimmerman, J. B. (2016). Simultaneous extraction, fractionation, and enrichment of microalgal triacylglycerides by exploiting the tunability of neat supercritical carbon dioxide. *ACS Sustainable Chemistry & Engineering*, 4(11), 6222–6230.

[94] Lajoie, L., Paquin, C., Morin-Crini, N., & Crini, G. (2022). Water as a green solvent: Methods of solubilisation and extraction of natural products – Past, present and future solutions. *Pharmaceuticals*, 15(12), 1507. https://doi.org/10.3390/ph15121507

[95] Laybourn, A., Robertson, K., & Slater, A. G. (2023). Quid pro flow. *Journal of the American Chemical Society*, 145(8), 4355–4365.

[96] Leder, C., Rastogi, T., & Kümmerer, K. (2015). Putting benign by design into practice-novel concepts for green and sustainable pharmacy: Designing green drug derivatives by non-targeted synthesis and screening for biodegradability. *Sustainable Chemistry and Pharmacy*, 2, 31–36. https://doi.org/10.1016/j.scp.2015.07.001

[97] Lei, Z. C., Wang, X., Yang, L., Qu, H., Sun, Y., Yang, Y., Li, W., Zhang, W. B., Cao, X. Y., Fan, C., & Li, G. (2024). What can molecular assembly learn from catalysed assembly in living organisms?. *Chemical Society Reviews*, 53(4), 1892–1914.

[98] Lenz, M., Borlinghaus, N., Weinmann, L., & Nestl, B. M. (2017). Recent advances in imine reductase-catalyzed reactions. *World Journal of Microbiology and Biotechnology*, 33(11), https://doi.org/10.1007/s11274-017-2365-8

[99] Li, C. J., & Trost, B. M. (2008). Green chemistry for chemical synthesis. *Proceedings of the National Academy of Sciences of the United States of America*, 105(36), 13197–13202. https://doi.org/10.1073/pnas.0804348105

[100] Li, Z., Dong, W., Li, X., & Yang, Y. (2016). The use of environmentally sustainable bio-derived solvents in solvent extraction applications – A review. *Chinese Journal of Chemical Engineering*, 24(2), 215–220. https://doi.org/10.1016/j.cjche.2015.07.021

[101] Lin, G. Q., Li, Y. M., & Chan, A. S. (2001). *Principles and Applications of Asymmetric Synthesis*. John Wiley & Sons.

[102] List, B., Lerner, R. A., & Barbas, C. F. (2000). Proline-catalyzed direct asymmetric aldol reactions. *Journal of the American Chemical Society*, 122(10), 2395–2396.

[103] Lopez, G., Keiner, D., Fasihi, M., Koiranen, T., & Breyer, C. (2023). From fossil to green chemicals: Sustainable pathways and new carbon feedstocks for the global chemical industry. *Energy & Environmental Science*, 16(7), 2879–2909. https://doi.org/10.1039/D3EE00478C

[104] Lu, Z.-N., Chen, H., Hao, Y., Wang, J., Song, X., & Mok, T. M. (2017). The dynamic relationship between environmental pollution, economic development and public health: Evidence from China. *Journal of Cleaner Production*, 166, 134–147. https://doi.org/10.1016/j.jclepro.2017.08.010

[105] Lübbert, C., Baars, C., Dayakar, A., Lippmann, N., Rodloff, A. C., Kinzig, M., & Sörgel, F. (2017). Environmental pollution with antimicrobial agents from bulk drug manufacturing industries in Hyderabad, South India, is associated with dissemination of extended-spectrum beta-lactamase and carbapenemase-producing pathogens. *Infection*, 45(4), 479–491. https://doi.org/10.1007/s15010-017-1007-2

[106] Ma, Y., Tong, W., Zhou, H., & Suib, S. L. (2000). *A Review of Zeolite-like Porous Materials*, 37(1–2), 243–252. https://doi.org/10.1016/s1387-1811(99)00199-7

[107] MacMillan, D. W. C. (2008). The advent and development of organocatalysis. *Nature*, 455(7211), 304–308. https://doi.org/10.1038/nature07367

[108] Maertens, A., Golden, E., & Hartung, T. (2021). Avoiding regrettable substitutions: green toxicology for sustainable chemistry. *ACS Sustainable Chemistry & Engineering*, 9(23), 7749–7758. https://doi.org/10.1021/acssuschemeng.0c09435

[109] Markel, U., Essani, K. D., Besirlioglu, V., Schiffels, J., Streit, W. R., & Schwaneberg, U. (2020). Advances in ultrahigh-throughput screening for directed enzyme evolution. *Chemical Society Reviews*, 49(1), 233–262. https://doi.org/10.1039/c8cs00981c

[110] Matlin, S. A., Mehta, G., Hopf, H., & Krief, A. (2016). One-world chemistry and systems thinking. *Nature Chemistry*, 8(5), 393–398. https://doi.org/10.1038/nchem.2498

[111] McIntosh, J. A., & Owens, A. E. (2021). Enzyme engineering for biosynthetic cascades. *Current Opinion in Green and Sustainable Chemistry*, 29, 100448. https://doi.org/10.1016/j.cogsc.2021.100448

[112] McMorn, P., & Hutchings, G. J. (2004). Heterogeneous enantioselective catalysts: Strategies for the immobilisation of homogeneous catalysts. *Chemical Society Reviews*, 33(2), 108–108. https://doi.org/10.1039/b200387m
[113] Meng, X., & Xiao, F.-S. (2014). Green routes for synthesis of zeolites. *Chemical Reviews*, 114(2), 1521–1543. https://doi.org/10.1021/cr4001513
[114] Meylan, W. M., & Howard, P. H. (2003). A review of quantitative structure–activity relationship methods for the prediction of atmospheric oxidation of organic chemicals. *Environmental Toxicology and Chemistry*, 22(8), 1724. https://doi.org/10.1897/01-275
[115] Koel, M., & Rikhovich Kal'iūrand, M. (2019). *Green Analytical Chemistry*. Royal Society Of Chemistry.
[116] Miyaura, N., & Suzuki, A. (1995). Palladium-catalyzed cross-coupling reactions of organoboron compounds. *Chemical Reviews*, 95(7), 2457–2483. https://doi.org/10.1021/cr00039a007
[117] Mohan, C., Kumar, V., & Kumari, S. (2018). synthesis, characterization, and antibacterial activity of Schiff bases derived from thiosemicarbazide, 2-acetylthiophene and thiophene-2-aldehyde. *Int. Res. J. Pharm*, 9, 153–158.
[118] Mondal, A., & Mukhopadhyay, C. (2024). Solvent-free microwave reactions towards significant organic transformations: A green approach. *Tetrahedron Green Chemistry*, 4, 100054–100054. https://doi.org/10.1016/j.tgchem.2024.100054
[119] Mutoh, R., Iwata, K., Iida, T., Ishiura, M., & Onai, K. (2021). Rhythmic adenosine triphosphate release from the cyanobacterial circadian clock protein KaiC revealed by real-time monitoring of bioluminescence using firefly luciferase. *Genes to Cells*, 26(2), 83–93. https://doi.org/10.1111/gtc.12825
[120] Nachtergael, A., Coulembier, O., Dubois, P., Helvenstein, M., Duez, P., Blankert, B., & Mespouille, L. (2015). Organocatalysis paradigm revisited: Are metal-free catalysts really harmless?. *Biomacromolecules*, 16(2), 507–514.
[121] Nagtode, V. S., Kale, S., & Sudeep, P. (2023). Green surfactants (biosurfactants): A petroleum-free substitute for sustainability—comparison, applications, market, and future prospects. *ACS Omega*, 8(13), https://doi.org/10.1021/acsomega.3c00591
[122] Naidu, R. (2021). Chemical pollution: A growing peril and potential catastrophic risk to humanity. *Environment International*, 156, 106616. https://doi.org/10.1016/j.envint.2021.106616
[123] Nair, D. P., Podgorski, M., Chatani, S., Gong, T., Xi, W., Fenoli, C. R., & Bowman, C. N. (2014). The thiol-Michael addition click reaction: A powerful and widely used tool in materials chemistry. *Chemistry of Materials*, 26(1), 724–744.
[124] Nandivada, H., Jiang, X., & Lahann, J. (2007). Click chemistry: Versatility and control in the hands of materials scientists. *Advanced Materials*, 19(17), 2197–2208. https://doi.org/10.1002/adma.200602739
[125] Nasirpour, N., Ebrahimi, H., & Mahmood, H. (2020). Ionic liquids: Promising compounds for sustainable chemical processes and applications. *Chemical Engineering Research and Design*, 160, 264–300. https://doi.org/10.1016/j.cherd.2020.06.006
[126] Nazor, J., Liu, J., & Huisman, G. (2021). Enzyme evolution for industrial biocatalytic cascades. *Current Opinion in Biotechnology*, 69, 182–190. https://doi.org/10.1016/j.copbio.2020.12.013
[127] Oaks, J. L., Gilbert, M., Virani, M. Z., Watson, R. T., Meteyer, C. U., Rideout, B. A., Shivaprasad, H. L., Ahmed, S., Iqbal Chaudhry, M. J., Arshad, M., Mahmood, S., Ali, A., & Ahmed Khan, A. (2004). Diclofenac residues as the cause of vulture population decline in Pakistan. *Nature*, 427(6975), 630–633. https://doi.org/10.1038/nature02317
[128] Öberg, T. (2005). A QSAR for the hydroxyl radical reaction rate constant: Validation, domain of application, and prediction. *Atmospheric Environment*, 39(12), 2189–2200. https://doi.org/10.1016/j.atmosenv.2005.01.007
[129] Pagliaro, M. (2019). An industry in transition: The chemical industry and the megatrends driving its forthcoming transformation. *Angewandte Chemie International Edition*, 58(33), 11154–11159. https://doi.org/10.1002/anie.201905032

[130] Yáñez-Sedeño, P., Campuzano, S., & Pingarrón, J. M. (2019). Electrochemical (bio)sensors: Promising tools for green analytical chemistry. *Current Opinion in Green and Sustainable Chemistry*, 19, 1–7. https://doi.org/10.1016/j.cogsc.2019.01.004

[131] Parker, O. (2024). Designing safer chemicals: reducing toxicity and environmental impact. *TO Chemistry Journal*, 11, 2581–7507.

[132] Peng, H. Q., Niu, L. Y., Chen, Y. Z., Wu, L. Z., Tung, C. H., & Yang, Q. Z. (2015). Biological applications of supramolecular assemblies designed for excitation energy transfer. *Chemical Reviews*, 115(15), 7502–7542.

[133] Petrovic, P. V., & Anastas, P. T. (2023). *First Do No Harm: A Chemist's Guide to Molecular Design for Reduced Hazard*, (1st ed.). Jenny Stanford Publishing, https://doi.org/10.1201/9781003359647

[134] Poliakoff, M., & Licence, P. (2007). Sustainable technology: Green chemistry for the environment. *Chemical Society Reviews*, 36(8), 1326–1350. https://doi.org/10.1039/B702669M

[135] Poliakoff, M., Fitzpatrick, J. M., Farren, T. R., & Anastas, P. T. (2002). Green chemistry: Science and politics of change. *Science*, 297(5582), 807–810. https://doi.org/10.1126/science.297.5582.807

[136] Pollet, P., Dave, R. N., & Jessop, P. G. (2014). Solvents for sustainable chemical processes. *Green Chemistry*, 16(3), 1034–1055. https://doi.org/10.1039/c3gc42302f

[137] Polshettiwar, V., & Varma, R. S. (2010). Green chemistry by nano-catalysis. *Green Chemistry*, 12(5), 743. https://doi.org/10.1039/b921171c

[138] Puhlmann, N., Mols, R., Olsson, O., Slootweg, J. C., & Kümmerer, K. (2021). Towards the design of active pharmaceutical ingredients mineralizing readily in the environment. *Green Chemistry*, 23(14), 5006–5023. https://doi.org/10.1039/d1gc01048d

[139] Samburski, G. (2015, February 17–18). Anelli's (TEMPO) oxidation of alcohol by flow chemistry, industrial scale. In: *Paper Presented at the 80th Annual Meeting of the Israel Chemical Society*. Tel Aviv, Israel.

[140] Schmidt, W. M., & Hutchison, J. E. (2000). Green chemistry in the curriculum: Challenging convention through renewable feedstocks and energy efficiency. *Journal of Chemical Education*, 77(12), 1627–1632. https://doi.org/10.1021/ed077p1627

[141] Schrittwieser, J. H., Velikogne, S., Hall, M., & Kroutil, W. (2017). Artificial biocatalytic linear cascades for preparation of organic molecules. *Chemical Reviews*, 118(1), 270–348. https://doi.org/10.1021/acs.chemrev.7b00033

[142] Serrano-Ruiz, J. C. (2020). Biomass: A renewable source of fuels, chemicals and carbon materials. *Molecules*, 25(21), 5217. https://doi.org/10.3390/molecules25215217

[143] Shah, A. A., Kato, S., Shintani, N., Kamini, N. R., & Nakajima-Kambe, T. (2014). Microbial degradation of aliphatic and aliphatic-aromatic co-polyesters. *Applied Microbiology and Biotechnology*, 98(8), 3437–3447. https://doi.org/10.1007/s00253-014-5558-1

[144] Shahbazi, F., Jabbari, M., Esfahani, M. N., & Keshmiri, A. (2021). A computational simulation platform for designing real-time monitoring systems with application to COVID-19. *Biosensors and Bioelectronics*, 171, 112716. https://doi.org/10.1016/j.bios.2020.112716

[145] Shaikh, I. R. (2014). Organocatalysis: Key trends in green synthetic chemistry, challenges, scope towards heterogenization, and importance from research and industrial point of view. *Journal of Catalysts*, 2014(1), 402860.

[146] Sheldon, R. A. (1992). Organic synthesis: Past, present, and future: A cost-effective, sustainable route to chemical products. *Chemical Society Reviews*, 21(5), 503–512. https://doi.org/10.1039/CS9922100503

[147] Sheldon, R. A. (1997). Catalysis: The key to waste minimization. *Journal of Chemical Technology & Biotechnology*, 68(4), 381–388. https://doi.org/10.1002/(sici)1097-4660(199704)68:4%3C381::aid-jctb620%3E3.0.co;2-3

[148] Sheldon, R. A. (2000). Atom efficiency and catalysis in organic synthesis. *Green Chemistry*, 2(6), 363–368. http://dx.doi.org/10.1351/pac200072071233

[149] Sheldon, R. A. (2012). Fundamentals of green chemistry: Efficiency in reaction design. *Chem. Soc. Rev*, 41(4), 1437–1451. https://doi.org/10.1039/c1cs15219j

[150] Sheldon, R. A. (2017). The E factor 25 years on: The rise of green chemistry and sustainability. *Green Chemistry*, 19(1), 18–43. https://doi.org/10.1039/c6gc02157c

[151] Sheldon, R. A., & van Pelt, S. (2013). Enzyme immobilization in biocatalysis: Why, what, and how. *Chemical Society Reviews*, 42(15), 6223–6235. https://doi.org/10.1039/C3CS60075K

[152] Sheldon, R. A., Arends, I., & Hanefeld, U. (2007). *Green Chemistry and Catalysis*. Wiley-VCH.

[153] Simon, M.-O., & Li, C.-J. (2012). Green chemistry oriented organic synthesis in water. *Chemical Society Reviews*, 41(4), 1415–1427. https://doi.org/10.1039/c1cs15222j

[154] Sperl, J. M., & Sieber, V. (2018). Multienzyme cascade reactions – status and recent advances. *ACS Catalysis*, 8(3), 2385–2396. https://doi.org/10.1021/acscatal.7b03440

[155] Stoler, E., & Warner, J. (2015). Non-covalent derivatives: cocrystals and eutectics. *Molecules*, 20(8), 14833–14848. https://doi.org/10.3390/molecules200814833

[156] Thompson, R. C. (2004). Lost at Sea: Where is all the plastic?. *Science*, 304(5672), 838–838. https://doi.org/10.1126/science.1094559

[157] Tiwari, A., Niveita, S. V., Mohan, C., & Bendi, A. (2024). Waste management from agricultural field: Sustainable practices and framework. *Book: Exploring Waste Management in Sustainable Development Contexts, IGI Global, USA*, 154–169. 10.4018/979-8-3693-4264-0.ch011

[158] Tiwari, S., & Talreja, S. (2022). Green chemistry and microwave irradiation technique: A review. *Journal of Pharmaceutical Research International*, 74–79. https://doi.org/10.9734/jpri/2022/v34i39a36240

[159] Trakhtenberg, S., & Warner, J. C. (2007). Green chemistry considerations in entropic control of materials and processes. *Chemical Reviews*, 107(6), 2174–2182.

[160] Tripathy, D. B., Singh, R., & Gupta, N. (2021). Switchable solvents as alternative solvents for green chemistry. *Elsevier EBooks*, 109–131. https://doi.org/10.1016/b978-0-12-819850-6.00011-5

[161] Trost, B. (1991a). The atom economy–a search for synthetic efficiency. *Science*, 254(5037), 1471–1477. https://doi.org/10.1126/science.1962206

[162] Trost, B. M. (1991b). Atom economy: A challenge for organic synthesis: Homogeneous catalysis leads the way. *Angewandte Chemie International Edition*, 34(3), 259–281. https://doi.org/10.1002/anie.1991000259

[163] Trost, B. M. (1995). Atom economy: A challenge for organic synthesis: Homogeneous catalysis leads the way. *Angewandte Chemie International Edition*, 34(3), 259–281. https://doi.org/10.1002/anie.199502591

[164] Trost, B. M. (2002). On inventing reactions for atom economy. *Accounts of Chemical Research*, 35(9), 695–705. https://doi.org/10.1021/ar010068z

[165] Vatin, N. I., Negi, G. S., Yellanki, S. V., Mohan, C., & Singla, N. (2024). Sustainability measures: An experimental analysis of AI and big data insights in Industry 5.0. *BIO Web of Conferences*, 86, 01072. https://doi.org/10.1051/bioconf/20248601072

[166] Verma, V., Joshi, A., Kalra, S., Singh, A., Singh, A., & Mohan, C. (2024). Novel nanocatalysts for sustainable hydrogen production from renewable resources. *E3S Web of Conferences*, 511, 01006. https://doi.org/10.1051/e3sconf/202451101006

[167] Wang, J. (2002). Real-time electrochemical monitoring: Toward green analytical chemistry. *Accounts of Chemical Research*, 35(9), 811–816. https://doi.org/10.1021/ar010066e

[168] Wang, Y., Lin, H.-X., Chen, L., Ding, S.-Y., Lei, Z.-C., Liu, D.-Y., Cao, X.-Y., Liang, H.-J., Jiang, Y.-B., & Tian, Z.-Q. (2014). What molecular assembly can learn from catalytic chemistry. *Chem. Soc. Rev*, 43(1), 399–411. https://doi.org/10.1039/c3cs60212e

[169] Warner, J. C. (2006). Entropic control in green chemistry and materials design. *Pure and Applied Chemistry*, 78(11), 2035–2041.

[170] Warner, J. C., Cannon, A. S., & Dye, K. M. (2004). Green chemistry. *Environmental Impact Assessment Review*, 24(7), 775–799. https://doi.org/10.1016/j.eiar.2004.06.006
[171] Whitesides, G. M., & Grzybowski, B. (2002). Self-assembly at all scales. *Science*, 295(5564), 2418–2421.
[172] Xue, C., Mao, Y., Wang, W., Song, Z., Zhao, X., Sun, J., & Wang, Y. (2019). Current status of applying microwave-associated catalysis for the degradation of organics in aqueous phase – A review. *Journal of Environmental Sciences*, 81, 119–135. https://doi.org/10.1016/j.jes.2019.01.019
[173] Yadav, R., Sanduja, M., Kumar, V., Sharma, K., Khan, S., & Kumar, K. (2024). Microwave-assisted synthetic pathways of pyrrole: A comprehensive review. *Asian Journal of Organic Chemistry*, 13(12), e202400401.
[174] Yang, K. K., Wu, Z., & Arnold, F. H. (2019). Machine-learning-guided directed evolution for protein engineering. *Nature Methods*, 16(8), 687–694. https://doi.org/10.1038/s41592-019-0496-6
[175] Ye, R., Hurlburt, T. J., Sabyrov, K., Alayoglu, S., & Somorjai, G. A. (2016). Molecular catalysis science: Perspective on unifying the fields of catalysis. *Proceedings of the National Academy of Sciences*, 113(19), 5159–5166.
[176] Younïs, A., & Osman, A. (2023). Solvent-free organic reaction techniques as an approach for green chemistry. *Journal of the Turkish Chemical Society Section A: Chemistry*, 10(2), 549–576. https://doi.org/10.18596/jotcsa.1188983
[177] Yu, Y., Guo, P., Zhong, J. S., Yuan, Y., & Ye, K. Y. (2020). Merging photochemistry with electrochemistry in organic synthesis. *Organic Chemistry Frontiers*, 7(1), 131–135.
[178] Gunji, Y. P., & Adamatzky, A. (2024). Computation implemented by the interaction of chemical reaction, clustering, and de-clustering of molecules. *Biomimetics*, 9(7), 432–432. https://doi.org/10.3390/biomimetics9070432
[179] Zhang, M., June, S. M., & Long, T. E. (2012). Principles of step-growth polymerization (polycondensation and polyaddition). In: *Polymer Science: A Comprehensive Reference, 10 Volume Set*, Elsevier, pp. 7–47.
[180] Zhang, S., Zhang, S., Gong, X., & Qu, H. (2024). Real-time monitoring of the column chromatography process of Ginkgo biloba using near-infrared and Raman spectroscopy combined with spectral fusion strategy. *Process Biochemistry*, 145, 50–62. https://doi.org/10.1016/j.procbio.2024.06.012
[181] Zhang, W., Lai, W., & Cao, R. (2017). Energy-related small molecule activation reactions: Oxygen reduction and hydrogen and oxygen evolution reactions catalyzed by porphyrin- and corrole-based systems. *Chemical Reviews*, 117(4), 3717–3797.
[182] Zhang, Y., Wang, J., Wang, Y., & Zhao, K. (2022). In-situ real-time monitoring of chemical kinetics by a self-designed automatic micro-reaction device. *Social Science Research Network*, https://doi.org/10.2139/ssrn.4292578
[183] Zhu, C., Ang, N. W. J., Meyer, T. H., Qiu, Y., & Ackermann, L. (2021). Organic electrochemistry: Molecular syntheses with potential. *ACS Central Science*, 7(3), 415–431. https://doi.org/10.1021/acscentsci.0c01532
[184] Zimmerman, J. B., Anastas, P. T., Erythropel, H. C., & Leitner, W. (2020). Designing for a green chemistry future. *Science*, 367(6476), 397–400. https://doi.org/10.1126/science.aay3060
[185] Zong, L., Zhou, S., Sgriccia, N., Hawley, M. C., & Kempel, L. C. (2003). A review of microwave-assist polymer chemistry (MAPC). *Journal of Microwave Power and Electromagnetic Energy*, 38(1), 49–74. https://doi.org/10.1080/08327823.2003.11688487

Anita Singh*, Chandra Mohan, Chhagan Lal, and Jaspal Singh

Chapter 3
The role of solvents and catalysts in green chemistry

Abstract: In the pursuit of sustainable practices, green chemistry emerges as a pivotal field aimed at reducing the environmental impact of chemical processes. This chapter delves into the critical roles that solvents and catalysts play in the advancement of green chemistry, highlighting the need for safer, more efficient alternatives. Solvents, traditionally viewed as necessary but often hazardous components of chemical reactions, are evaluated through the lens of their environmental footprints, toxicity profiles, and recyclability. The chapter presents various green solvent alternatives, such as water, ionic liquids, supercritical fluids, and bio-based solvents, elucidating their advantages in minimizing waste and energy consumption.

In parallel, the discussion on catalysts emphasizes their role in enhancing reaction efficiency and selectivity, thereby reducing by-products and energy requirements. Traditional catalysts are contrasted with novel and environmentally benign options, such as biocatalysts and green catalytic processes that employ Earth-abundant materials. The synergistic relationship between solvents and catalysts is also explored, demonstrating how their careful selection and integration can lead to innovative processes that align with the principles of green chemistry.

Ultimately, this chapter underscores the transformative potential of optimizing solvents and catalysts in industrial applications, promoting a paradigm shift toward sustainability in chemical manufacturing. By detailing the advancements and ongoing research in these areas, the chapter contributes to the broader discourse on achieving an eco-friendlier chemical industry, advocating for continued innovation and regulatory support in the development of green alternatives.

Keywords: Green chemistry, solvent, catalyst, industrial applications, ionic liquid, supercritical fluid

***Corresponding author: Anita Singh**, Department of Basic Sciences, IES College of Technology, Bhopal, Madhya Pradesh, India, e-mail: a4anitasingh@gmail.com
Chandra Mohan, School of Basic and Applied Sciences, K. R. Mangalam University, Gurugram, India
Chhagan Lal, Department of Chemistry, Swami Shraddhanand College, University of Delhi, Delhi, India
Jaspal Singh, Laboratory of Advanced Materials for Energy and Environment, Université Du Québec à Trois-Rivières, Québec, Canada G9A 5H7

https://doi.org/10.1515/9783111618906-003

3.1 Introduction to green chemistry

3.1.1 Definition of green chemistry

Green chemistry is a scientific discipline that seeks to design chemical products and processes that reduce or eliminate the generation of hazardous substances. By focusing on sustainability, green chemistry aims to minimize the negative impacts of chemical production on both human health and the environment [56]. Pioneered in the 1990s, green chemistry encompasses a set of principles and practices that seek to encourage the development of more sustainable chemical methodologies. According to the U.S. Environmental Protection Agency (EPA), green chemistry is fundamentally about ensuring that chemical processes are not only efficient and effective but also environmentally benign [2, 90].

3.1.1.1 Principles of green chemistry

Green chemistry is built upon 12 fundamental principles that guide researchers and practitioners in their pursuit of sustainable chemistry. These principles were outlined by Paul Anastas and John Warner in their seminal book *Green Chemistry: Theory and Practice* [3]. The 12 principles are as follows:

1. **Prevention**: It is better to prevent waste than to treat or clean up waste after it has been created. This principle underscores the importance of designing processes that minimize or prevent waste from the outset.
2. **Atom economy**: Synthetic methods should be designed to maximize the incorporation of all materials used in the process into the final product. This means that more efficient reactions, which use fewer reagents and generate less by-products, should be prioritized.
3. **Less hazardous chemical syntheses**: Whenever possible, synthetic methods should be designed to use and generate substances that possess little or no toxicity to human health and the environment.
4. **Designing safer chemicals**: Chemical products should be designed to preserve efficacy while reducing toxicity. The aim is to create chemicals that are effective for their intended use yet carry minimal risk to health and the environment.
5. **Safer solvents and auxiliaries**: The use of auxiliary substances (solvents, separation agents, etc.) should be made unnecessary wherever possible and innocuous when used. The reduction or elimination of solvents is pivotal, as many solvents can be hazardous.
6. **Energy efficiency**: Energy requirements should be recognized for their environmental and economic impacts and methods should be designed to minimize energy consumption. When possible, synthetic methods should be conducted at room temperature.

7. **Use of renewable feedstocks**: A raw material feedstock should be renewable rather than depleting whenever technically and economically practicable. This encourages the use of agricultural products, biologically derived materials, and other sustainable resources.
8. **Reduce derivatives**: Unnecessary derivatization (blocking a functional group, protection/deprotection steps) should be minimized or avoided if possible because such steps require additional reagents and can generate waste.
9. **Catalysis**: Catalytic reagents (as selective as possible) are superior to stoichiometric reagents. Catalysts can greatly enhance reaction efficiency and often reduce waste.
10. **Design for degradation**: Chemical products should be designed so that at the end of their function they break down into innocuous degradation products and do not persist in the environment.
11. **Real-time analysis for pollution prevention**: Analytical methodologies need to be further developed to allow for real-time, in-process monitoring and control prior to the formation of hazardous substances.
12. **Inherently safer chemistry for accident prevention**: Substances and the forms of substances used in a chemical process should be chosen to minimize the potential for chemical accidents, including releases, explosions, and fires.

3.1.2 Importance of sustainability and the role of solvents and catalysts in chemical processes

Sustainability refers to the ability to meet present needs without compromising the ability of future generations to meet their own needs [58]. The concept is deeply rooted in environmental science, economics, and social equity, recognizing that the health of our planet, economy, and societies are interlinked. The significance of sustainability can be analyzed from various perspectives:

3.1.2.1 Environmental impact

The ever-increasing human footprint has led to severe environmental consequences, including climate change, resource depletion, and biodiversity loss. Sustainable practices are essential to minimize these negative impacts. According to the United Nations Sustainable Development Goals [91], sustainability entails responsible management of resources, which is critical for mitigating climate change and protecting ecosystems. For instance, adopting sustainable practices in agriculture and industry can reduce the emission of greenhouse gases and toxic waste.

3.1.2.2 Economic viability

Sustainable development promotes economic viability by fostering green technologies and renewable resources. Businesses that integrate sustainability into their operations often see long-term benefits such as cost reductions, enhanced brand reputation, and increased consumer loyalty [47, 57]. For example, companies that adopt energy-efficient technologies can lower operational costs and enhance competitiveness in an increasingly eco-conscious market.

3.1.2.3 Social equity

Ensuring sustainability also addresses social equity issues. It emphasizes the need for policies that promote fair labor practices, equal access to resources, and participation in decision-making processes. The International Labour Organization [49] states that sustainable economic practices can lead to job creation and better working conditions, contributing to social stability and community well-being.

3.1.2.4 Global collaboration

Sustainability fosters international collaboration. Global issues such as climate change and resource scarcity require cohesive action across borders. The Paris Agreement, for instance, exemplifies how countries can work together to combat climate change, showing that sustainability transcends national interests and promotes collective responsibility [93].

3.1.3 Overview of solvents and catalysts in chemical processes

Chemical processes are central to various industries, including pharmaceuticals, plastics, and petrochemicals. Solvents and catalysts play crucial roles in these processes, influencing efficiency, selectivity, and environmental impact.

3.1.3.1 Solvents

Solvents are substances, typically liquids, used to dissolve solutes, facilitating chemical reactions or processes. They are critical in a myriad of applications, including extraction, purification, and synthesis. However, the choice of solvent can significantly affect the sustainability of chemical processes.

3.1.3.1.1 Types of solvents

1. **Conventional solvents:** These are often organic solvents like benzene, toluene, and acetone, which can be hazardous to health and the environment [9]. Many of these solvents are volatile organic compounds (VOCs) that contribute to air pollution and pose health risks.
2. **Green solvents:** The push toward sustainability has led to the development of green solvents, which are environmentally benign alternatives. Examples include water, ionic liquids (ILs), and supercritical carbon dioxide ($scCO_2$) [8]. These solvents offer advantages such as lower toxicity, reduced environmental impact, and often improved efficiency in chemical processes.

3.1.3.2 Catalysts

Catalysts are substances that increase the rate of a chemical reaction without being consumed in the process. They are vital in industrial processes, enabling reactions to occur at lower temperatures and pressures, thereby saving energy and resources [1].

3.1.3.2.1 Types of catalysts

1. **Homogeneous catalysts:** These are in the same phase as the reactants and often lead to high selectivity. However, they can be challenging to separate from products, potentially leading to waste and inefficiency.
2. **Heterogeneous catalysts:** Present in a different phase than the reactants, these catalysts can be easily separated and reused, making them more sustainable. Common examples include metal catalysts employed in hydrogenation reactions [14].

3.1.3.3 The role of solvents and catalysts in sustainability

The integration of green solvents and efficient catalysts aligns with the principles of sustainable chemistry. By optimizing solvent use and employing catalysts effectively, chemical processes can reduce waste and energy consumption. For instance, the shift to using renewable feedstocks and employing biocatalysis is a growing trend that enhances sustainability in the chemical industry.

3.1.3.4 Future perspectives

As industries continue to prioritize sustainability, the development of novel solvents and catalysts is critical. Ongoing research in nanotechnology and biomimicry presents opportunities for creating more efficient and environmentally friendly chemical processes. The implementation of sustainable practices not only reduces environmental impact but also encourages innovation in the development of alternative technologies and processes.

3.2 Understanding solvents

3.2.1 Definition and classification of solvents

3.2.1.1 Introduction to solvents

A solvent can be defined as a substance, usually in the liquid state, that has the ability to dissolve a solute, leading to the formation of a solution. In a solution, the solvent is present in the largest amount. The solute is the substance that is being dissolved. The resulting liquid mixture is typically uniform and homogeneous. Solvents are a fundamental aspect of chemistry, playing crucial roles in various applications ranging from laboratory research to industrial processes and everyday products. A solvent is typically defined as a substance in which solutes dissolve, resulting in a solution [58]. Solvents can be gases, liquids, or solids; however, the term most commonly refers to liquids. In this section, we will explore the role of solvents, their significance in chemical reactions, and the properties that make them effective. The primary function of a solvent is to facilitate the dissolution of solutes, leading to an even distribution of particles within a solution. Solvents are essential in various industries, including pharmaceuticals, paints, coatings, cosmetics, and food production. The ability of a solvent to dissolve substances depends on several factors, including its chemical nature, polarity, temperature, and pressure [53].

3.2.1.2 Classification of solvents

Solvents can be classified in several ways based on different criteria, including their chemical composition, polarity, and origin. The most common classification systems are based on the following categories:

(a) **Classification by polarity**

1. **Polar solvents**: These solvents, such as water and ethanol, exhibit significant dipole moments. They have electronegative atoms (such as oxygen or nitrogen) that create partial positive and negative changes. Polar solvents are effective at dissolving ionic and other polar substances due to their ability to stabilize charged species through solvation [77].

2. **Nonpolar solvents**: Examples include hexane and toluene. Nonpolar solvents have little or no dipole moment and are effective at solubilizing nonpolar compounds such as lipids and hydrocarbons [26].

(b) **Classification by origin**
1. **Natural solvents**: Derived from natural sources, these include solvents like ethanol from fermentation processes and water, which is naturally abundant.
2. **Synthetic solvents**: Many solvents are produced through chemical processes in laboratories or industrial settings. Examples include acetone, chloroform, and dichloromethane.

(c) **Classification by function**
1. **Aprotic solvents**: These do not donate hydrogen ions (protons) to solutes. Examples include dimethyl sulfoxide (DMSO) and acetonitrile. Aprotic solvents are used in chemical reactions where protic solvents could interfere with the product formation.
2. **Protic solvents**: Protic solvents can donate protons, such as water and alcohols. They are often used in reactions or processes needing proton transfer.

(d) **Classification by use**
1. **Industrial solvents**: Widely used in industries for a range of applications, such as paint thinners or adhesives.
2. **Laboratory solvents**: Solvents used in research settings for analytical techniques, chemical synthesis, and extraction processes, such as methanol and diethyl ether.

3.2.1.3 Properties influencing solvent behavior

Understanding the properties of solvents is vital for selecting appropriate solvents for specific applications. Key factors to consider include:

(a) **Polarity**: Solvents can be classified as polar or nonpolar based on their molecular structures. Polar solvents, such as water, have a significant dipole moment, allowing them to interact with ionic and polar compounds. Nonpolar solvents, like hexane, are more suitable for dissolving nonpolar substances.

(b) **Volatility**: This refers to how readily a solvent can vaporize at room temperature. High volatility solvents can evaporate quickly, which may be advantageous in certain applications like coatings but undesirable in others where prolonged exposure to the solvent is necessary.

(c) **Boiling and freezing points**: Solvents with low boiling points may be useful for processes requiring quick evaporation, whereas others with high boiling points might be suitable for applications involving elevated temperatures.

(d) **Viscosity**: The thickness or resistance to flow in a solvent can influence the rate of dissolution and mixing.

3.2.2 Traditional solvents and their environmental impact

3.2.2.1 Introduction

Solvents play a crucial role in various industrial processes, offering solutions for chemical reactions, extractions, and formulation of products. Traditional solvents, particularly organic solvents like benzene, toluene, and acetone, have been widely used across different industries, from pharmaceuticals to coatings [99]. However, the use of these solvents has raised significant environmental concerns, particularly concerning toxicity, air quality, and ecological balance. This chapter aims to explore the types of traditional solvents, their applications, and the associated environmental impacts, highlighting the need for sustainable alternatives.

3.2.2.2 Types of traditional solvents

Traditional solvents are generally classified into two categories: polar and nonpolar solvents.

(a) **Polar solvents**: Polar solvents are characterized by their high dielectric constant and their ability to dissolve ionic and polar substances. Common examples include:
   1. **Water**: Universally recognized as a solvent but has limitations in dissolving nonpolar substances
   2. **Ethanol**: Commonly used in alcoholic beverages and pharmaceuticals
   3. **Acetone**: Widely used as a nail polish remover and in cosmetic formulations

(b) **Nonpolar solvents:** Nonpolar solvents have low dielectric constants and are primarily used to dissolve nonpolar compounds. Notable examples include:
   1. **Benzene**: Used in the manufacturing of chemicals and as a solvent in the laboratory
   2. **Toluene**: Commonly found in paint thinners and adhesives
   3. **Hexane**: Frequently used in oil extraction and other chemical processes

3.2.2.3 Environmental impact of traditional solvents

(a) **Air pollution:** VOCs are a significant group of air pollutants released during the use of traditional solvents. VOCs can contribute to the formation of ground-level ozone and smog, which have adverse effects on human health and the environment. Studies have shown that emissions from solvents are a notable contributor to urban air quality issues [84, 68].

(b) **Water contamination:** Traditional solvents can enter water bodies through industrial discharges, spills, and improper disposal. Water pollution from solvents can lead to serious ecological consequences. For instance, benzene and toluene are known carcinogens that can harm aquatic life and reduce water quality for human consumption [25].

(c) **Soil degradation:** Improper disposal of solvent waste can lead to soil contamination, creating long-lasting ecological damage. The presence of harmful solvents in soil can inhibit plant growth and affect soil microbiota, leading to reduced agricultural productivity [40].

(d) **Human health risks:** Exposure to traditional solvents poses significant health hazards to workers and the general population. Chronic exposure to organic solvents can lead to neurological disorders, respiratory issues, and even cancer [45, 67, 70]. Regulatory frameworks, such as OSHA (Occupational Safety and Health Administration) standards, now place limits on permissible exposure levels to mitigate these risks.

3.2.2.4 Mitigation strategies

To address the environmental impact of traditional solvents, various strategies have been developed:

(a) **Green chemistry principles:** Green chemistry aims to reduce the environmental footprint of chemical processes. This includes the design of safer solvents, the use of renewable resources, and the implementation of more efficient processes **[3]**. For example, water and bio-based solvents are increasingly promoted as safer alternatives to traditional solvents.

(b) **Solvent recovery systems:** Industries are increasingly adopting solvent recovery systems that recycle solvents, significantly reducing waste and emissions. These systems not only promote sustainability but also offer economic benefits by minimizing the need for fresh solvent [15].

(c) **Regulatory frameworks:** Stricter regulations regarding solvent use and emissions have led industries to adopt cleaner technologies and practices. Regulatory bodies, such as the EPA in the United States, provide guidelines and standards that are critical in reducing environmental impacts [31].

3.2.3 Green solvents: characteristics and examples

In recent years, the pursuit of sustainable and environmentally friendly alternatives to traditional organic solvents has led to a growing interest in "green solvents." Green solvents are defined as solvents that minimize environmental impact and health-related risks, while still maintaining effectiveness in various applications including chemical synthesis, extraction, and formulation [65, 66]. The characteristics of green solvents often include biodegradability, nontoxicity, renewability, and minimal environmental persistence. This chapter highlights several classes of green solvents: water, ILs, supercritical fluids (SCFs), and bio-based solvents [39].

3.2.3.1 Water as a solvent

Water is often referred to as the "universal solvent" due to its ability to dissolve a wide variety of substances. It is nontoxic, inexpensive, and readily available, making it an attractive option as a green solvent. Water can facilitate many chemical reactions that typically require organic solvents, especially in the fields of organic synthesis and extraction processes.

One of the primary advantages of using water is its low environmental impact; it is biodegradable and poses no risk of persistent pollution [42]. Moreover, water-based reactions often generate fewer hazardous waste products compared to traditional organic solvent systems. For instance, water has been successfully employed in the synthesis of pharmaceutical compounds, allowing for reactions that are typically carried out using organic solvents [8, 37, 103].

However, water is not a universal substitute; its polarity limits its ability to dissolve nonpolar substances, and its high heat capacity restricts the temperature range for effective applications. Despite these limitations, innovative methodologies are continually being developed to expand the use of water as a solvent across various chemical reactions.

3.2.3.2 Ionic liquids

ILs are organic salts that exist in the liquid state at relatively low temperatures (often below 100 °C). They are composed entirely of ions, which bestow them with unique properties, including low vapor pressure and high thermal stability. ILs can be tailored for specific applications by modifying their cationic and anionic components, making them highly versatile solvents [103].

Due to their tunable nature, ILs have gained attention as solvents for a variety of green chemistry applications, including catalysis, extraction, and electrochemistry [4]. They can dissolve a wide range of organic and inorganic materials, including both polar and nonpolar compounds, thereby allowing for efficient reaction conditions [81]. Additionally, many ILs are nonvolatile and exhibit low toxicity, thereby reducing the risks associated with traditional solvents.

Despite their advantages, ILs can have considerable environmental impacts if not designed with sustainability in mind. The potential toxicity and biodegradability of specific ILs must be assessed to ensure they meet green chemistry principles [20, 51].

3.2.3.3 Supercritical fluids

SCFs are substances that exist at a temperature and pressure above their critical point, exhibiting properties of both liquids and gases. A commonly used SCF is $scCO_2$, which is widely regarded for its nontoxic, nonflammable, and low-cost characteristics. SCFs exhibit tunable solubility, high diffusivity, and low viscosity, attributes that enhance their effectiveness for extraction and reaction purposes [64].

$scCO_2$ is increasingly being utilized in extraction processes, especially in the food and pharmaceutical industries. For example, $scCO_2$ is employed for extracting essential oils and bioactive compounds from plant materials [10]. In addition to extraction, $scCO_2$ is also being explored as a reaction medium for various synthetic transformations, including polymerization and catalysis, promoting greener alternatives to conventional methods using organic solvents [78, 94].

While SCFs offer many advantages, challenges remain, including equipment costs and operational conditions, which can limit their applicability in certain contexts.

3.2.3.4 Bio-based solvents

Bio-based solvents are derived from renewable biological resources, including plant materials, and have emerged as an essential component of green chemistry. Common bio-based solvents include ethanol, acetone, and methyl acetate, which can often serve as suitable alternatives to petroleum-derived solvents [48]. These solvents are desirable not only for their reduced environmental impact but also for their potential sustainability, as they can be sourced from agricultural feedstocks and waste materials [80].

The use of bio-based solvents presents multiple advantages, such as biodegradability and reduced carbon footprint. Ethanol, derived from biomass, is widely employed in chemical reactions, particularly in pharmaceuticals [43]. Furthermore, increasing the use of bio-based solvents often aligns with government regulations aimed at reducing greenhouse gas emissions and promoting the circular economy [24, 33].

However, the production of bio-based solvents must be managed to avoid competition with food production and to ensure efficient utilization of agricultural resources. Additionally, thorough assessments of the environmental impact of producing and using bio-based solvents are critical to maintaining their status as green alternatives.

3.2.4 Evaluation of solvent selection: the Solvent Selection Guide

The selection of solvents is a crucial aspect in various fields, including organic synthesis, pharmaceuticals, and industrial processes. This chapter discusses the importance of solvent selection, how it influences the environmental and health impacts of chemical processes, and the development of the Solvent Selection Guide (SSG). The SSG is a tool designed to aid chemists in choosing the most appropriate solvents based on criteria such as toxicity, environmental impact, and performance in chemical processes. This chapter evaluates the SSG's influence on best practices in solvent selection.

Solvents play a significant role in numerous chemical processes, serving functions such as dissolving reactants, enhancing reaction rates, and facilitating extraction and purification [11]. However, their selection can significantly affect both the efficiency of chemical reactions, and the environmental and health impacts associated with their use. Recent trends, driven by stricter regulatory pressures and increased awareness of sustainability, underscore the necessity of adopting safer, more environmentally friendly solvents [34].

The SSG is an attempt to systematically evaluate and aid chemists in making informed decisions regarding solvent usage. This chapter evaluates solvent selection principles, explores the creation and effectiveness of the SSG, and considers its impact on contemporary chemistry practices.

3.3 The importance of solvent selection

3.3.1 Environmental and health considerations

The selection of solvents is influenced by a range of factors, including toxicity, flammability, and environmental impact. Common industrial solvents such as benzene, toluene, and chloroform have been scrutinized due to their harmful effects on human health, including carcinogenicity [21, 44]. The environmental ramifications of solvent use are also significant, particularly in terms of VOCs contributing to air pollution and ground-level ozone formation [30].

Choosing less hazardous solvents where feasible can help minimize health risks and environmental damage. Knowledge of solvent properties such as bioaccumulation potential, persistence, and ecological toxicity is crucial for sustainable solvent selection [29].

3.3.2 Performance in chemical processes

While environmental and safety considerations are paramount, the chosen solvent must also be effective in promoting desired chemical reactions. This includes considerations of solubility, reactivity, and the solvent's role in thermodynamic and kinetic aspects of the reaction [60]. A poorly chosen solvent may hinder the reaction efficiency, leading to increased costs and waste.

3.3.3 The Solvent Selection Guide (SSG)

The SSG was developed to facilitate informed decision-making regarding solvent use in chemical processes. It categorizes solvents based on several criteria:

1. **Toxicity**: Evaluates human health risks associated with exposure
2. **Environmental impact**: Assesses persistence, bioaccumulation, and ecological effects
3. **Performance**: Considers solubility, volatility, and other physical properties

The SSG utilizes a star rating system, where solvents are awarded stars based on their performance in these categories. An increase in stars indicates a more favorable solvent choice [23].

3.3.4 Application in practice

The SSG has been implemented in laboratories and industries aiming to enhance sustainability. For example, a study by de Lima et al. [86] demonstrated a significant reduction in hazardous solvent use in pharmaceutical formulations following the adoption of the SSG, leading to enhanced compliance with environmental regulations and improved safety protocols.

3.3.5 Challenges in solvent selection

Despite its benefits, the implementation of the SSG and other solvent selection tools faces several challenges. Many traditional organic synthesis processes rely on well-understood yet potentially hazardous solvents. Transitioning to safer alternatives may require extensive re-optimization of processes [72, 104].

Furthermore, the availability of data about solvent characteristics is not always comprehensive. The need for continued research into the properties and impacts of various solvents is critical to enhance the efficacy of the SSG and similar resources.

3.4 Catalysis in green chemistry

3.4.1 Definition and fundamentals of catalysis

3.4.1.1 Catalysis in green chemistry: definition and fundamentals of catalysis

Green chemistry aims to design chemical processes and products that minimize waste, reduce energy consumption, and use environmentally benign materials. Central to this discipline is catalysis, which provides an effective means to enhance reaction efficiency and selectivity. The importance of catalysis in green chemistry cannot be overstated, as its adoption can lead to significant reductions in environmental impact.

Catalysis is defined as the process of increasing the rate of a chemical reaction by adding a substance known as a catalyst. Catalysts facilitate reactions without being consumed in the process, allowing them to be used repeatedly.

There are two primary types of catalysis: heterogeneous and homogeneous.

(a) **Heterogeneous catalysis**: Involves catalysts that are in a different phase than the reactants, typically solid catalysts reacting with liquid or gas substrates. Examples include catalytic converters in automobiles that convert harmful gases into less toxic substances [105].
(b) **Homogeneous catalysis**: Involves catalysts that are in the same phase as the reactants, often in solution. This type is characterized by its ability to provide uniform reaction conditions, leading to high selectivity [106].

3.4.1.2 Fundamentals of catalysis

The fundamentals of catalysis can be understood through several key concepts:

(a) **Translation of energy landscapes:** Catalysts work by lowering the activation energy required for a reaction to proceed. This is typically represented by energy diagrams, where the free energy of the transition state is reduced in the presence of a catalyst, leading to an increased reaction rate [107].
(b) **Reaction mechanisms:** The mechanism of catalysis involves multiple steps, including adsorption, reaction, and desorption. In heterogeneous catalysis, reactants are adsorbed onto the surface of the catalyst, undergo a reaction, and then desorb as products.
(c) **Selectivity and activity:** Catalysts can be designed to enhance the selectivity of a reaction, directing it toward desirable pathways while inhibiting undesired ones. Factors such as catalyst composition, structure, and surface area play critical roles in determining both the activity and selectivity of the catalyst [108].

3.4.1.3 Role of catalysis in green chemistry

The application of catalysis in green chemistry significantly reduces the environmental impact of chemical processes. For instance, catalysts can minimize the formation of byproducts and waste through more direct synthetic pathways, reducing the need for extensive purification processes. Moreover, catalytic processes tend to operate under milder conditions – lower temperatures and pressures – leading to decreased energy consumption [3].

3.4.1.4 Case studies in green catalysis

Recent advancements in green chemistry underscore the role of catalysis in sustainable practices. For example:

(a) **Biocatalysis**: The use of enzymes as catalysts has gained popularity due to their specificity, efficiency, and favorable reaction conditions. Biocatalytic processes often involve fewer steps and lower energy inputs, making them ideal for green chemistry applications [109].

(b) **Photocatalysis**: The use of light to promote chemical reactions has opened new avenues for green chemistry, particularly in energy conversion and organic synthesis. Photocatalysts enable reactions under mild conditions, utilizing renewable energy [110].

3.4.2 Mechanism of action of homogeneous and heterogeneous catalysts

Catalysis plays a pivotal role in a myriad of chemical processes, significantly enhancing reaction rates without altering the final product. Catalysts are traditionally classified into two main categories: homogeneous and heterogeneous Table 3. This chapter discusses the fundamental differences between these types of catalysts, their mechanisms, applications, and advantages and disadvantages [5, 89, 90]

3.4.2.1 Homogeneous catalysts

Homogeneous catalysts are present in the same phase (typically liquid) as the reactants. They are often soluble within the reaction medium, allowing for intimate molecular interactions.

Mechanism of action: Homogeneous catalysis typically involves the formation of transient complexes between a catalyst and the reactants. These complexes lower the overall activation energy required for the reaction to occur. This process can be divided into several steps:

1. **Formation of the catalyst-reactant complex**: The reactants interact with the catalyst to form a temporary complex. This step is crucial as it often determines the selectivity and efficiency of the reaction.
2. **Reaction**: The complex then enters a transition state that leads to the formation of products.
3. **Reformation of the catalyst**: After the reaction, the catalyst is regenerated and can participate in further catalytic cycles.

Applications: Homogeneous catalysts are frequently employed in industries such as:

1. **Fine chemicals and pharmaceuticals**: Catalysts like Rh, Ru, and Pd complexes are used in reactions such as hydrogenation, cross-coupling, and various condensation reactions.
2. **Environmental chemistry**: Catalysts are utilized in processes like the oxidation of organic compounds or the treatment of wastewater.

3.4.2.1.1 Advantages and disadvantages

Advantages:

- High selectivity for specific reactions
- Easier to control reaction conditions (temperature, pressure) due to their liquid phase nature

Disadvantages:

- Difficulty in separating the catalyst from the product mixture, which may lead to challenges in recycling.
- Some homogeneous catalysts may be sensitive to reaction conditions, affecting stability.

3.4.2.2 Heterogeneous catalysts

Heterogeneous catalysts, by contrast, exist in a different phase (commonly solid) compared to the reactants (often gases or liquids). These catalysts interact with the reactants at the surface.

Mechanism of action: The mechanism by which heterogeneous catalysts operate can be described as follows:

1. **Adsorption**: Reactants adhere to the catalyst's surface, leading to a higher concentration at the active sites.
2. **Reaction**: The adsorbed reactants undergo a chemical reaction, facilitated by the surface properties of the catalyst.

3. **Desorption**: The products then detach from the surface and leave the catalyst free for another reaction cycle.

Applications: Heterogeneous catalysts are widely used in large-scale industrial processes, such as:
1. **Petrochemical industry**: Catalysts in processes like catalytic cracking and reforming
2. **Environmental applications**: Use of catalysts in exhaust systems to convert toxic gases (e.g., CO and NOx into less harmful substances).

3.4.2.2.1 Advantages and disadvantages

Advantages:
- Easier separation from reaction products, which aids in recycling and reduces waste.
- Generally, it is more stable and effective in high-temperature and high-pressure conditions [8].

Disadvantages:
- Lower selectivity compared to homogeneous systems.
- Reaction rates can be limited by the rate of mass transfer to and from the catalyst surface.

The following table 3.1 shows the difference between homogeneous and heterogeneous catalysis.

Table 3.1: Comparison of homogeneous and heterogeneous catalysis.

Characteristic	Homogeneous catalysis	Heterogeneous catalysis
Phase	Same phase as reactants (typically liquid)	Different phases from reactants (typically solid)
Mechanism	Complex formation and reaction	Adsorption and reaction at the surface
Separation	Difficult, often requires distillation or extraction	Easier, often via filtration or sedimentation
Selectivity	Higher	Generally lower
Stability	May be sensitive to reaction conditions	Generally, more robust in extreme conditions
Applications	Fine chemicals and pharmaceuticals	Petrochemical processing and environmental applications

3.4.3 The importance of catalysts in reducing waste and energy consumption

Catalysts play a fundamental role in enhancing chemical reactions by speeding them up without being consumed in the process. Their significance in mitigating environmental impacts, reducing energy consumption, and minimizing waste cannot be overstated. As global concerns about sustainability and climate change intensify, the development and use of catalysts have become pivotal in achieving greener industrial processes.

3.4.3.1 Catalysts in industrial applications

(a) **Types of catalysts:** Catalysts can be categorized mainly into homogeneous and heterogeneous catalysts.

1. **Homogeneous catalysts** operate in the same phase as the reactants, typically in a liquid solution.
2. **Heterogeneous catalysts**, on the other hand, involve reactants in one phase and the catalyst in a different phase, commonly seen in solid catalyst operations in gas- or liquid-phase reactions.

The choice of catalyst type influences not only reaction efficiency but also the scope for waste generation and energy usage [12].

3.4.3.2 Catalyst for energy reduction

Catalysts significantly lower the energy activation required for chemical reactions, enabling reactions to occur at lower temperatures. This reduction in temperature requirements directly correlates with lower energy consumption across industrial processes. For instance, in the Haber process for ammonia synthesis, the use of iron-based catalysts has allowed operational temperatures to drop, minimizing energy costs [46].

Recent advancements in catalyst design, including the development of nanocatalysts, have further facilitated reactions at even lower energy inputs. Research indicates that these materials can improve energy efficiency by up to 25% in certain processes.

3.4.3.3 Waste minimization

The application of catalysts helps reduce the production of unwanted byproducts, leading to less waste. Catalytic converters in automobiles are a prime example, converting toxic gases like carbon monoxide, hydrocarbons, and nitrogen oxides into less harmful emissions [17]. By enabling more complete reactions, catalysts minimize the formation of waste, ensuring more efficient use of raw materials.

Furthermore, in the chemical industry, processes such as asymmetric synthesis – where catalysts are used to produce specific enantiomers – can significantly reduce by-product formation. This specificity not only decreases waste but also maximizes the yield of desired products, aligning with the principles of green chemistry [111].

3.4.3.4 Enhancing green technologies

(a) **Renewable energy applications:** Catalysts are also instrumental in the development of renewable energy technologies. In hydrogen production through water splitting, catalysts can facilitate the breakdown of water into hydrogen and oxygen with minimal energy input, presenting a potential pathway for clean fuel production [16]. Solar fuel generation, involving the use of catalysts, is gaining traction as a method to convert solar energy into storable fuels, thus aiding in energy sustainability.

(b) **Carbon capture and utilization (CCU):** Another area where catalysts are proving invaluable is in CCU technologies. Catalysts can facilitate the conversion of captured carbon dioxide into useful chemicals and fuels, effectively closing the carbon loop. Research has shown that effective catalytic systems can mitigate emissions while generating commercially viable products, thus turning a potential waste stream into a resource [22].

3.4.4 Challenges and future directions

Despite the evident benefits, several challenges undermine the full potential of catalytic processes. For instance, catalyst deactivation over time can necessitate additional energy and material inputs for regeneration or replacement. Moreover, the environmental impact associated with the production of certain catalysts, particularly rare metals, raises concerns regarding sustainability.

The future of catalytic technology lies in the ongoing development of more sustainable materials and processes. Innovations such as biocatalysts derived from enzymes and the use of abundant and less harmful materials in catalyst design promise to further enhance the sustainability of chemical processes.

3.4.5 Green catalysts: characteristics and examples

Green catalysts have emerged as a vital component of sustainable chemistry, aiding in the development of environmentally friendly processes and providing routes for chemical reactions that minimize waste and reduce the use of hazardous materials. This chapter discusses three major types of green catalysts: biocatalysts and enzymatic reactions, solid-supported catalysts, and metal-free catalysts. Each category plays a significant role in advancing green chemistry principles by enhancing reaction efficiency and decreasing environmental impact.

3.4.5.1 Biocatalysts and enzymatic reactions

Biocatalysts, commonly enzymes, are biological molecules that facilitate biochemical reactions. They are highly specific, acting on particular substrates in mild conditions of temperature and pressure, which often results in higher selectivity and lower energy requirements compared to traditional chemical catalysts. This section discusses the characteristics of biocatalysts, their advantages, and examples of their usage:

(a) **Characteristics of biocatalysts**
   1. **Selectivity**: Enzymes exhibit exceptional substrate specificity, which minimizes side reactions and results in cleaner products.
   2. **Mild reaction conditions**: They often operate effectively at ambient temperatures and pressures, further reducing energy consumption.
   3. **Sustainability**: Derived from renewable resources, enzymes facilitate the transformation of biomass into valuable products.
   4. **Biodegradability**: Enzymes are naturally occurring and can be easily degraded in the environment, lessening pollution risks.

(b) **Examples**
   1. **Lipases in biodiesel production**: Lipases catalyze the transesterification of oils to produce biodiesel, a process that can take place at lower temperatures and pressures compared to traditional chemical methods. This application not only produces renewable fuel but also helps in utilizing waste oils, contributing to waste management.
   2. **Proteases in food processing**: Proteases are employed in the food industry to enhance flavors, modify textures, and improve nutritional value by breaking down proteins. Their use reduces the need for harsh chemicals, promoting a greener approach to food production.

3.4.5.2 Solid-supported catalysts

Solid-supported catalysts are another category of green catalysts where active catalytic materials are immobilized on solid supports. This strategy enhances the recyclability of the catalyst and simplifies separation processes, leading to less waste:

(a) **Characteristics of solid-supported catalysts**

1. **Recyclability**: The immobilization of catalysts allows for their reuse in multiple cycles, significantly reducing costs and waste.
2. **Easier separation**: Solid catalysts can be separated from reaction mixtures by simple filtration, eliminating the need for costly and environmentally damaging separation techniques.
3. **Controlled environment**: The support material can influence the reaction environment, enhancing catalytic activity and selectivity.

(b) **Examples**

1. **Supported metal catalysts in hydrogenation**: In hydrogenation reactions, metal catalysts like palladium or platinum are supported on materials such as carbon or alumina. These supported catalysts have been crucial in the synthesis of fine chemicals and pharmaceuticals while facilitating efficient separation and recycling.
2. **Zeolite catalysts in acid-catalyzed reactions**: Zeolites, as solid-supported catalysts, are used in processes like isomerization and cracking in the petrochemical industry. Their porous structure allows for selective reactions and easier product recovery, reducing the generation of by-products.

3.4.5.3 Metal-free catalysts

Metal-free catalysts have gained traction in green chemistry due to their ability to replace traditional metal-based catalysts, which can be toxic, expensive, or difficult to recycle. This section examines the characteristics and potential applications of metal-free catalysis:

(a) **Characteristics of metal-free catalysts**

1. **Nontoxicity**: Metal-free catalysts avoid the associated toxicity and environmental hazards of many metal catalysts.
2. **Simplicity and cost-effectiveness**: Typically, these catalysts are cheaper to produce and operate under milder conditions.
3. **Diversity of reaction mechanisms**: Metal-free catalysts can catalyze various types of reactions, including oxidation and reduction processes, often through unique reaction mechanisms.

(b) **Examples**
- **Carbon-based catalysts**: Graphene and carbon nanotubes have been utilized in various reactions, such as the oxidation of alcohols to carbonyls. Their high surface area and electronic properties enhance catalytic performance while remaining nontoxic.
- **Organic molecule catalysts**: Organocatalysts, such as proline and amines, have been employed in asymmetric synthesis. Their use in enantioselective reactions highlights the potential of organic catalysts in fine chemical synthesis without the drawbacks of metal-containing counterparts [61].

3.5 The role of solvents in catalytic reactions

Solvents are indispensable in the realm of catalytic reactions, playing a crucial role in determining reaction rates, mechanisms, and outcomes. The choice of solvent can influence various aspects, including the solubility of reactants, the stability of intermediates, and the overall efficiency of the catalytic process. Understanding how solvents affect reaction pathways is vital for the development of more efficient catalytic systems and for the optimization of various industrial processes.

3.5.1 Solvent properties impacting catalytic reactions

3.5.1.1 Polarity

Polarity is one of the most significant characteristics of a solvent that influences catalytic reactions. Polar solvents, such as water or acetonitrile, can stabilize charged transition states and intermediates, whereas nonpolar solvents, such as hexane, might not provide the same level of stabilization. Polar solvents are often preferred in reactions involving ionic species, as they can solve these charges, thereby lowering the energy barrier needed for reactions to proceed.

3.5.1.2 Dielectric constant

The dielectric constant of the solvent dictates its ability to moderate electrostatic interactions between charged species. A higher dielectric constant typically facilitates the dissociation of ionic species and stabilizes charge-separated states, promoting reaction pathways that involve ionic intermediates. For example, the transition state of many nucleophilic substitution reactions can be significantly stabilized by using solvents with a high dielectric constant.

3.5.1.3 Viscosity

The viscosity of the solvent also affects the diffusion rates of reactants and catalyzed species. Higher viscosity can impede the diffusion of reactants toward each other or toward the catalyst, slowing down the reaction. In homogeneous catalysis, lower viscosity solvents are usually preferred to enhance reactant mobility and increase reaction rates.

3.5.1.4 Hydrogen bonding

Solvents capable of hydrogen bonding can influence the stability of intermediates and the transition state. Hydrogen-bonding solvents can stabilize charged intermediates or transition states through solvation effects, thereby affecting the reaction mechanism. For example, in the S_N1 reaction mechanism, hydrogen bonding can stabilize carbocation intermediates, favoring the formation of these reactive intermediates.

3.5.2 Solvent effects on reaction mechanisms

Solvents can dramatically alter the mechanistic pathway of a reaction. The governing principle behind this change lies in the solvation effects that influence the formation and stability of reactive intermediates.

3.5.2.1 Homogeneous catalysis

In homogeneous catalysis, the solvent can dictate whether the reaction follows a concerted pathway or a stepwise mechanism. For instance, when using a polar aprotic solvent, such as DMSO, the reaction might favor a bimolecular pathway over a unimolecular pathway due to shifts in the activation energy associated with the solvation of transition states [8].

3.5.2.2 Bimolecular and unimolecular pathways

A classic example illustrating solvent effects is in the study of S_N1 versus S_N2 reactions. Polar protic solvents stabilize carbocations formed in S_N1 reactions, promoting this pathway. Conversely, polar aprotic solvents tend to favor S_N2 mechanisms by stabilizing the nucleophile without stabilizing the transition state extensively, thus facilitating a concerted mechanism [27, 63].

3.5.2.3 Lewis acid catalysis

Lewis acid-catalyzed reactions also exhibit solvent dependence. The reactivity of Lewis acids can be enhanced or suppressed by the solvent environment. For example, solvents that can donate lone pairs of electrons (like water) can interact with Lewis acids and shift equilibria toward deactivated species, thus impacting the overall catalytic activity [112].

3.5.3 Solvent effects during reaction progress

During the course of a reaction, solvent properties can influence not only the initial interactions but also the dynamics of the reaction as it progresses. For instance, changes in solvent polarity during a reaction could lead to variations in the reaction pathways followed by intermediate species.

3.5.3.1 Reaction reversibility

The ability of a solvent to influence the reversible nature of reactions, particularly in equilibrium processes, is notable. A solvent that stabilizes one side of the equilibrium more than the other can shift the reaction toward the products or reactants, effectively altering yields without major changes to the catalytic system.

3.5.3.2 Temperature and solvent compositions

The combined effects of temperature and varying solvent compositions can lead to substantial changes in reaction mechanisms. For instance, the addition of cosolvents or adjusting the temperature can enhance or inhibit solvation effects, shifting the balance between different mechanistic pathways [100].

3.5.4 Influence of solvent on catalyst performance

Catalysis is a fundamental area in chemistry used to increase the rate of chemical reactions by providing an alternative reaction pathway with lower activation energy. The performance of a catalyst is not solely dependent on its intrinsic properties but also on the reaction environment, particularly the solvent used in the reaction mixture. The choice of solvent can significantly affect the reaction kinetics, selectivity, and overall efficiency of catalytic processes.

3.5.4.1 Role of solvent in catalysis

Solvents play a critical role in the catalytic process by influencing factors such as solubility of reactants, stabilization of transition states, and reactivity of various species involved in the reaction. The interactions between the solvent and solute can alter the energy landscape of the reaction, which can lead to changes in reaction rates and product distributions.

3.5.4.2 Types of solvents and their effects

Solvents can be broadly categorized into two types: polar and nonpolar. Each category influences catalytic reactions differently:

(a) **Polar solvents**: These solvents, such as water, alcohol, and acetone, can stabilize charged transition states or intermediates, thus facilitating reactions that involve ionic species. For example, in the case of acid-catalyzed reactions, polar protic solvents tend to stabilize the carbocation intermediates, enhancing the reaction rate [52].

(b) **Nonpolar solvents**: Nonpolar solvents like hexane or toluene are less interactive compared to polar solvents. They typically solvate nonpolar reactants and do not stabilize charged intermediates. Their usage is often seen in reactions that favor nonpolar or hydrophobic species, which can sometimes lead to decreased polarity and lower reaction rates [19].

(c) **Solvent *effects on specific catalytic reactions:*** Catalytic hydrogenation reactions are a prime example of how solvent effects can significantly influence catalytic performance. The solvent can profoundly impact the adsorption of hydrogen onto the catalyst surface – essential for catalytic activity [41]. For instance, in the hydrogenation of olefins, the solvent can interact with both the substrate and catalyst, altering adsorption energies and consequently modifying the kinetics of the reaction.

    Enzymatic catalysis also demonstrates a clear solvent effect. Enzymes, which often use water as a solvent, show different catalytic efficiencies depending on the solvent properties. Studies have shown that various organic solvents can denature enzymes but may also lead to enhanced catalysis through increased reactant solubility [54].

(d) **Solvent hydrogen bonding and catalytic activity**: Hydrogen bonding interactions between solvent molecules and reactants can also significantly affect catalytic performance. Solvents that can form hydrogen bonds often enhance the rates of reactions involving polar transition states. For example, in acid–base catalysis, the presence of a hydrogen-bond-donating solvent can stabilize the transition state through additional interactions, thereby lowering the activation energy [96].

(e) **Miscibility and reaction medium:** The miscibility of the solvent with the reactants is a critical factor that can dictate the success of a catalytic reaction. In some cases, immiscible solvents can lead to phase separation, which may limit contact between reactants and catalysts, thus hindering the catalytic process. For instance, biphasic systems have been shown to improve selectivity in reactions such as the asymmetric hydrogenation of ketones by effectively separating product and reactant, thus pushing the equilibrium [36].
(f) **Modern trends in solvent use**: Recent trends in catalysis have emphasized greener alternatives to conventional solvents, such as ILs and SCFs. ILs, due to their tunable properties and ability to dissolve a wide range of substrates, have been shown to enhance reaction rates and selectivity in various catalytic processes [98].

Moreover, SCFs, particularly $scCO_2$, are being explored as sustainable solvents in catalytic reactions. Their unique properties of low viscosity and high diffusivity allow for enhanced mass transfer and reaction rates, which can significantly improve catalytic performance [73].

3.5.5 Case studies showcasing innovative solvent–catalyst interactions

The interaction between solvents and catalysts is a fundamental aspect of catalytic processes in industrial chemistry. Traditional solvent systems often pose environmental challenges, necessitating the development of eco-friendly alternatives. This chapter discusses various case studies that showcase innovative solvent–catalyst interactions, emphasizing eco-friendly solvent systems in catalytic processes.

3.5.5.1 Case study 1

Ionic liquids in asymmetric synthesis: ILs have emerged as promising solvents in catalysis due to their unique properties, including low vapor pressure, thermal stability, and the ability to dissolve a wide range of compounds. One significant study showcases the use of ILs in asymmetric synthesis reactions facilitated by chiral catalysts.

For instance, the use of 1-ethyl-3-methylimidazolium acetate in the asymmetric allylic alkylation of allyl acetates demonstrated higher enantioselectivity compared to traditional organic solvents. The IL's ability to stabilize transition states through specific ionic interactions with the chiral catalyst enhanced the reaction rate and selectivity, showcasing a synergistic effect between the solvent and the catalyst [38].

3.5.5.2 Case study 2

Supercritical fluids in green chemistry: SCFs, particularly $scCO_2$, provide a nontoxic, environmentally benign alternative for catalytic processes. One notable case is the use of $scCO_2$ as a solvent in the catalytic oxidation of hydrocarbons.

Jain et al. [50] demonstrated the catalytic oxidation of cyclohexane using a heteropolyacid catalyst in $scCO_2$. The supercritical state of CO_2 facilitated the solvation of both the reactant and catalyst, leading to increased reaction rates and selectivity. Additionally, the ability to easily separate the product from the solvent post-reaction makes SCFs an attractive option for reducing waste in chemical processes.

3.5.5.3 Case study 3

Aqueous phase reforming of biomass derivatives: The conversion of biomass derivatives to hydrogen using catalytic processes in aqueous phases presents an excellent example of an eco-friendly approach. One study explored the use of Ru-based catalysts in the aqueous phase reforming of glycerol.

The research conducted by Villa et al. [95] revealed that the use of water as a solvent significantly influenced catalyst performance. The study demonstrated that optimizing the solvent–catalyst interaction by altering the pH of the aqueous solution could enhance glycerol conversion rates and hydrogen yields. The eco-friendly nature of water as a solvent, combined with effective catalyst design, exemplifies the potential of utilizing renewable resources in catalytic processes.

3.5.5.4 Case study 4

Biocatalysis and green solvents: Biocatalysis has gained traction as an eco-friendly alternative to traditional chemical processes, particularly in organic synthesis. The use of green solvents, such as ethanol and water, in enzyme-catalyzed reactions reflects a significant advancement in solvent–catalyst interactions.

A notable case reported by Tran et al. [87] involved the use of lipases in the synthesis of biodiesel from vegetable oils using ethanol as the solvent. The study highlighted the solvent's role in enhancing enzyme activity and stability by facilitating better substrate solvation and increasing the reaction efficiency. The transition to green solvents not only demonstrated improved catalytic performance but also minimized the environmental impact of the process.

3.6 Developing sustainable processes: integration of solvents and catalysts in process design

The need for sustainability in chemical processes has never been more pressing. Faced with challenges such as climate change, resource depletion, and environmental degradation, the chemical industry must embrace practices that minimize negative impacts. One critical aspect of developing sustainable processes is the integration of solvents and catalysts in process design. This integration can significantly enhance the efficiency, selectivity, and environmental friendliness of chemical reactions.

3.6.1 Understanding solvents and catalysts

3.6.1.1 Solvents

Solvents are typically liquids that dissolve a solute, forming a solution. They play a critical role in facilitating chemical reactions, controlling reaction environments, and determining product yields. However, the choice of solvent can influence not only the efficiency of a reaction but also its environmental impact. Traditional solvents often derive from petrochemical sources and can be VOCs, contributing to air pollution and harmful emissions.

3.6.1.2 Catalysts

Catalysts are substances that increase the rate of a chemical reaction without being consumed in the process. They can lower activation energy and increase selectivity, leading to fewer by-products. The use of catalysts can significantly reduce the energy requirements of reactions, making them more efficient and cost-effective. However, the choice and design of catalysts are crucial for ensuring minimal environmental impact.

3.6.2 Reaction conditions optimization, continuous flow processes, and life cycle assessment of solvent–catalyst systems

The optimization of reaction conditions, the development of continuous flow processes, and the assessment of solvent–catalyst systems through life cycle evaluation are critical in advancing green chemistry and improving the efficiency of chemical processes. Embracing of these practices is vital for reducing waste, enhancing resource efficiency, and promoting sustainable development in chemical manufacturing. This chapter discusses the importance of optimizing reaction conditions, the advantages of continuous flow processes, and the role of life cycle assessment (LCA) in evaluating solvent–catalyst systems.

3.6.3 Integration of solvents and catalysts in process design

3.6.3.1 Reduction of hazardous solvent use

The integration of catalysts with solvent choice can lead to significant reductions in hazardous solvent applications. For example, exploring solvent-free reactions or utilizing solvents with a lower environmental footprint, such as bio-based solvents, can enhance sustainability. Reactions that utilize ILs or SCFs as solvents often demonstrate lower toxicity and reduced volatility, making them more environmentally friendly.

3.6.3.2 Process intensification

Combining solvents and catalysts can lead to process intensification, where the efficiency of a reaction is improved by optimizing reaction conditions (temperature, pressure, and concentration) while reducing energy and material inputs. For instance, the use of microwave-assisted catalytic reactions in organic synthesis can drastically reduce reaction times and solvent volumes, aligning with sustainable practices [85].

3.6.3.3 Enhanced selectivity and yields

The judicious choice of solvent in conjunction with catalysts can enhance the selectivity and yield of desired products. Understanding the interactions between solvents and catalysts can lead to optimized reaction pathways that maximize desired outcomes while minimizing waste. For example, the use of water as a solvent in conjunction with specific catalytic systems can often lead to higher selectivity for certain products compared to conventional organic solvents.

3.6.3.4 Energy efficiency

Integrating solvents and catalysts into the early stages of process design can lead to significant energy savings. Processes designed to operate at ambient conditions or that utilize waste heat from industrial processes can diminish energy consumption.

The choice of solvent can affect the thermodynamics of a reaction, influencing whether a reaction can proceed under milder conditions [83].

3.6.3.5 Green chemistry principles

The design of solvent–catalyst systems aligns well with the principles of green chemistry, which advocate for reducing waste, using safer alternatives, and improving efficiency. For instance, designing catalytic systems that operate effectively in aqueous media can support waste reduction by enabling simpler purification processes for the desired product.

3.6.3.6 Challenges and opportunities

Despite the clear benefits of integrating solvents and catalysts, several challenges remain. The development of novel solvents that are both effective and environmentally benign is still an area requiring research. Additionally, the industrial scalability of new catalytic systems is crucial; many laboratory-scale successes do not translate to industrial settings.

Research and innovation in this field can focus on the following areas:

- The development of sustainable, catalyst-supported, and solvent-free systems.
- Computational modeling to predict solvent–catalyst interactions and optimize process conditions.
- The exploration of biocatalysts and their applications in green solvents to promote eco-friendly chemical transformations.

3.6.4 Reaction conditions optimization

3.6.4.1 Importance of optimization

Optimizing reaction conditions is essential for improving yields, selectivity, and overall process efficiency. Factors such as temperature, pressure, catalyst type and concentration, solvent choice, and reaction time can significantly influence the outcomes of chemical reactions. Therefore, systematic optimization can lead to economic benefits, minimize environmental impact, and enhance product quality [35].

3.6.4.2 Techniques for optimization

Various techniques can be employed for optimizing reaction conditions. Traditional approaches include one-variable-at-a-time methods, which, while simple, can be inefficient and time-consuming, as they do not account for the interactions between variables. In contrast, statistical methods, such as design of experiments and response surface methodology, allow for more comprehensive evaluations of multi-parameter interactions [101]. Furthermore, model-assisted approaches and machine learning algorithms are increasingly being used to predict optimal conditions and accelerate the discovery of suitable reaction parameters [55].

3.6.5 Continuous flow processes and solvent recovery

Continuous flow processes represent a significant advancement in synthetic methodologies. Unlike traditional batch processes, which can lead to inefficiencies and excessive waste, continuous flow allows for precise control over reaction parameters, improved heat and mass transfer, and reduced processing times. These processes can enhance safety by minimizing risks associated with large-scale batch reactions, which are more susceptible to uncontrolled exothermic reactions [59].

3.6.5.1 Implementation and applications

The integration of continuous flow technology into industrial processes has facilitated the production of a wide range of chemicals, pharmaceuticals, and materials. Microreactors and plug flow reactors are examples of equipment that enable efficient mixing and precise control over reaction conditions in a continuous manner [28]. Additionally, continuous flow systems can be designed to incorporate real-time monitoring and feedback mechanisms, allowing for dynamic optimization of the reaction environment that can respond to deviations in operating conditions [75].

3.6.5.2 The role of solvent recovery

Reducing solvent waste is an essential component of sustainable chemical processes. Solvent recovery systems can significantly reduce the environmental impact of chemical production by minimizing waste disposal and recycling valuable solvents. Techniques such as distillation, membrane separation, and adsorption are commonly employed in solvent recovery [13].

3.6.5.3 Sustainability metrics

To assess the sustainability of solvent–catalyst systems, it is crucial to evaluate their performance in terms of efficiency, safety, and environmental impact. Metrics such as the environmental impact quotient, total annual cost, and other green chemistry principles provide valuable insight into the environmental credentials of solvent–catalyst systems [74].

3.6.6 Life cycle assessment of solvent–catalyst systems

LCA provides a comprehensive method for evaluating the environmental impacts of solvent–catalyst systems across their entire life cycle – from raw material extraction through to production, usage, and end-of-life [79]. This holistic approach allows for the identification of hotspots and areas for improvement in chemical processes.

3.6.6.1 LCA methodologies

Various methodologies exist for conducting LCAs, such as the ISO 14040 standards and the CML and ReCiPe methods, each offering different perspectives on environmental impacts [18]. Integrating LCA with other decision-making frameworks, such as technoeconomic analysis, can further enhance the understanding of the trade-offs involved in developing sustainable solvent–catalyst systems.

3.6.6.2 Case studies and applications

Recent case studies in the literature illustrate the value of LCA in optimizing solvent–catalyst systems. For example, studies have shown that integrating bio-based solvents and catalysts can significantly improve the sustainability profiles of chemical processes compared to conventional petroleum-derived systems [70].

3.7 Case studies in industry

Green chemistry represents a paradigm shift in chemical practice, aiming to design processes and products that minimize environmental impact while maximizing efficiency and safety. The principles of green chemistry are particularly relevant in the pharmaceutical, agrochemical, and materials industries, where traditional practices often result in significant waste and pollution. This chapter reviews case studies that illustrate the applications of green chemistry in these fields, focusing on pharmaceuticals, agrochemicals, and sustainable materials.

3.7.1 Green chemistry applications in pharmaceuticals

3.7.1.1 Case study: the use of biocatalysis in drug development

Biocatalysis has emerged as a cornerstone of green chemistry in the pharmaceutical industry. By utilizing enzymes as catalysts, companies can achieve high specificity and reduce the need for hazardous reagents. One notable example is the application of enzyme-catalyzed reactions in the synthesis of key pharmaceutical intermediates.

A case in point is the development of the anti-HIV drug, Saquinavir. Traditional synthetic routes involved harsh reaction conditions and toxic solvents. In contrast, the use of lipase-catalyzed acylation processes has allowed manufacturers to streamline production, reduce by-products, and enhance yields. According to a study by O'Reilly et al. [71], this biocatalytic approach not only reduced the environmental footprint of the synthesis but also improved economic efficiency.

3.7.2 Solvent and catalyst innovations in agrochemicals

3.7.2.1 Case study: switch to water as a solvent

In agrochemical formulations, the choice of solvents and catalysts significantly impacts toxicity and environmental sustainability. A major innovation has been the shift from organic solvents to water-based systems for pesticide formulation and application.

For instance, the development of a new class of herbicides based on water as a solvent demonstrates significant promise. In this case, the company Monsanto reworked the synthesis of glyphosate, a widely used herbicide. Traditionally synthesized using hazardous organic solvents that posed significant disposal issues, the new method leverages aqueous phase reactions and biocatalysts.

3.7.3 Catalyst innovations: use of ionic liquids

Another advancement in agrochemicals is the use of ILs as solvents and catalysts. ILs are salts that remain in liquid form at room temperature and have unique properties that make them excellent for facilitating reactions while being environmentally benign.

3.7.4 Sustainable materials and polymer chemistry

3.7.4.1 Case study: biopolymers from renewable resources

The development of sustainable materials has gained considerable attention, particularly in the plastics industry. Traditional polymers derived from petrochemicals are associated with significant environmental concerns, including pollution and resource depletion.

One compelling case study is the production of PLA, a biodegradable polymer derived from renewable resources, such as corn starch or sugarcane. Companies like NatureWorks have pioneered the commercial production of PLA, which offers a viable alternative to conventional plastics without sacrificing functionality. As reported by Avérous [7], PLA applications range from disposable items to packaging, demonstrating its versatility as a sustainable material. The life cycle analysis indicates that production and degradation processes of PLAs have a substantially lower carbon footprint compared to petroleum-based plastics.

3.7.5 Innovations in composite materials

Furthermore, advancements in composite materials have integrated renewable fibers into traditional polymer matrices. The incorporation of natural fibers, such as hemp or flax, into polymer composites has enhanced mechanical properties while reducing environmental impact. According to a study, these biocomposites exhibit similar performance to traditional materials but with less energy consumption during production

3.8 Challenges and future perspectives

3.8.1 Current limitations in solvent–catalyst research

The intersection of solvent and catalyst research is crucial in the advancement of green chemistry and sustainable reactions. However, several current limitations impede the progress in these areas.

3.8.1.1 Solvent toxicity and environmental impact

Despite ongoing efforts to identify "greener" solvents, many widely used solvents remain toxic or environmentally unfriendly. Traditional organic solvents, such as dichloromethane and acetone, are harmful to both human health and the environment. Research into alternatives like bio-based solvents is ongoing, but the efficacy and scalability of these environmentally benign options remain challenging.

3.8.1.2 Catalyst efficiency and selectivity

While various catalysts have been developed to improve reaction efficiencies, achieving high selectivity and turnover numbers continues to be an issue. Many heterogeneously catalyzed reactions suffer from leaching, requiring careful balance between catalyst activity and stability, as well as the potential for secondary reactions that can lead to undesired products [97].

3.8.1.3 Limited understanding of solvent–catalyst interactions

The role of solvents in catalytic processes is often undervalued or poorly understood. The interactions between solvents and catalysts can significantly influence reaction mechanisms and outcomes. Advanced computational methods and experimental techniques are needed to deepen our understanding of these interactions, forming a significant barrier in optimizing reactions.

3.8.1.4 Economic viability

One of the primary barriers in the transition to greener solvents and catalysts is economic viability. Many sustainable alternatives are more expensive, which can make them unattractive to industry stakeholders focused on cost-reduction and profitability. It is essential to find solutions that balance economic factors while promoting green chemistry practices.

3.8.2 The role of policy and regulation in promoting green chemistry

Policy and regulation play a pivotal role in advancing sustainable practices in solvent–catalyst research. Governments and regulatory bodies can implement frameworks that encourage the adoption of green chemistry principles, driving research and industry practices toward more sustainable options.

3.8.2.1 Incentives for green technology

Governments can provide financial incentives for companies investing in green solvents and catalysts. Tax deductions, grants, or subsidies can effectively prompt more industries to prioritize environmental considerations alongside their economic interests [92].

3.8.2.2 Setting standards and guidelines

The establishment of standards and guidelines for solvent and catalyst usage can help navigate the complex terrain of safety and environmental impact. Organizations like the American Chemical Society's Green Chemistry Institute play a vital role in creating metrics for assessing the greenness of chemical processes and products.

3.8.2.3 Educating stakeholders

Policy initiatives that focus on education and outreach can create a culture of sustainability within the chemical industry. Workshops, training programs, and collaboration with educational institutions can help disseminate knowledge about the importance of green chemistry and solvent–catalyst research [5].

3.8.3 The future landscape of solvents and catalysts in sustainable practices

The future of solvent–catalyst research looks promising, albeit challenging. Continuous innovation and interdisciplinary collaboration are essential for overcoming current limitations.

3.8.3.1 Emergence of biodegradable and renewable solvents

Research into bio-derived and biodegradable solvents is likely to increase. Such solvents can offer a more sustainable alternative to traditional solvents, potentially decreasing toxicity and environmental impact while utilizing renewable resources [32, 113].

3.8.3.2 Smart catalysts and adaptive processes

The field of catalysis is evolving with smart technologies – catalysts that can adapt to reaction conditions are gaining momentum. This evolution could lead to more efficient and selective processes, minimizing waste and maximizing resource use.

3.8.3.3 Integration of artificial intelligence (AI)

Artificial intelligence is increasingly utilized to optimize reaction conditions, virtual screening of solvents and catalysts, and acceleration of the discovery of new materials. This technology can analyze vast datasets to identify optimal solvent–catalyst pairs, leading to significant improvements in reaction efficiency and sustainability [69].

3.8.3.4 Global collaboration for sustainable practices

The advancement of solvent–catalyst research hinges on cross-sector collaboration and information sharing. Partnerships between academia, industry, and regulatory bodies can foster a shared vision of sustainability and promote the rapid development and adoption of green technologies.

3.9 Conclusions

Green chemistry focuses on designing chemical products and processes that minimize environmental harm by reducing waste and eliminating toxic substances. Solvents and catalysts play crucial roles in achieving sustainability, as traditional chemical processes often rely on hazardous solvents and stoichiometric reagents that contribute to pollution. Environmentally benign solvents, such as water, SCFs, and bio-based alternatives, offer safer and more sustainable options, reducing VOC emissions and enhancing process efficiency. Catalysts, particularly heterogeneous and biocatalysts, improve reaction rates, selectivity, and recyclability, minimizing waste and energy consumption. Integrating these green alternatives fosters a circular economy and promotes eco-friendly industrial practices. The urgency to implement green chemistry is greater than ever, requiring researchers to explore innovative solvent and catalytic systems while fostering interdisciplinary collaboration. Industry stakeholders must commit to sustainable practices by investing in renewable feedstock, waste reduction strategies, and solvent recovery technologies. Companies embracing green chemistry can enhance market competitiveness while addressing environmental concerns. Additionally, educational initiatives should equip future chemists with sustainability-focused skills. The combined efforts of researchers, industries, and academia are essential for advancing green chemistry, reducing toxic waste, and creating a more sustainable future for the planet.

References

[1] Adams, R. N., & Ingram, J. E. W. S. (2023). An introduction to catalysis through homogeneous and heterogeneous mechanisms. *The Journal of Chemical Physics*, 159(24), 244503.

[2] Amesho, K. T. T., Kadhila, T., Shangdiar, S., Shihomeka, S. P., Shopati, A. K., Edoun, E. I., & Mohan, C. (2024). Bioeconomy for sustainable environment management practices. In: Garg, V. K. & Kataria, N. (eds.), *Book: Bioeconomy for Sustainability*, Ed. 1. Springer: Singapore, pp. 41–63. https://doi.org/10.1007/978-981-97-1837-5_2

[3] Anastas, P. T., & Warner, J. C. (1998). *Green Chemistry: Theory and Practice*. England, UK: Oxford University Press.

[4] Anderson, J. A. (2012). Catalysis: From fundamentals to applications. *AnnualReview of Chemical and Biomolecular Engineering*, 3, 63–81.

[5] Aresta, M., & Angelini, A. (2019). Green chemistry education: Perspectives and challenges. *Chemistry*, 25(6), 862–879.

[6] Arya, H., Mohan, C., Pandey, S., Verama, M., & Kumar, V. (2021). Phytochemical screening of basella alba leaves extracts and evaluate its efficacy on sun burn (sun protection factor). *European Journal of Molecular & Clinical Medicine*, 8(1), 417–423.

[7] Avérous, L. (2017). Biodegradable polymers: Current trends and future prospects. *Polymer Reviews*, 57(1), 1–32.

[8] Cornils, B., & Herrmann, W. A. 2022. *Applied Homogeneous Catalysis with Organometallic Compounds*. Germany: Wiley-VCH.

[9] Bach, C., Pereira, D. S. R., Molin, M., & De Souza, C. A. (2013). Review of Green Solvents for Synthesis of Lung Tissues from the Literature of Green Chemistry. *Green Chemistry*, 15(7), 1377–1396.

[10] Barbosa, J. A. R. G., Lucena, R. F. P., & De Melo, L. T. (2019). Extraction of bioactive compounds with supercritical carbon dioxide: A review. *Molecules*, 24(10), 1989.

[11] Benson, F. (2018). *Essential Laboratory Experiments in Organic Chemistry*. USA: Academic Press.

[12] Bhan, A., & Davis, B. H. (2009). Heterogeneous catalysis and its applications: A review. *Chemical Engineering & Technology*, 32(11), 1737–1744.

[13] Bhanja, P., Shivhare, U. S., & Rao, R. P. (2021). Solvent recovery and sustainability: Approaches and applications. *Chemical Engineering Journal*, 405, 126766.

[14] Blaser, H (2000). Heterogeneous catalysis for fine chemicals production. *Catalysis Today*. 60(3–4). 161–165.

[15] Biosciences (2020). Solvent recovery systems: How to benefit your business. *Journal of Cleaner Production*, 254, 120–123.

[16] Bonin, J., & Artero, V. (2013). Water splitting: A highly efficient catalyst for hydrogen production. *Nature Chemistry*, 5, 71–76.

[17] Botre, C., Tosi, M., Mazzei, F., Bocca, B., Petrucci, F., & Alimonti, A. (2007). Automotive catalytic converters and environmental pollution: Role of the platinum group elements in the redox reactions and free radicals production. *International Journal of Environment and Health*, 1(1), 142. https://doi.org/10.1504/ijenvh.2007.012229

[18] Bryan, A., Fysikopoulos, A., & Vasile, A. (2022). Life cycle assessment: Methodologies and applications in sustainable chemistry. *Sustainable Production and Consumption*, 35, 177–190.

[19] Cavani, F., & Aresta, M.F. S. (1991). Recent advances in the use of catalytic processes in green chemistry. *Green Chemistry*, 1, 1–8.

[20] Cavani, F., Centi, G., & Perathoner, S. (2015). Ionic liquids in the chemical industry. *Chemical Society Reviews*, 44(20), 7698–7718.
[21] Chakrapani, I. S., Rajasekhar, M., Bhaskar, L. V. K. S., & Mohan, C. (2023). Smart investigation on nanocomposites composed of carbon dioxide derives, repeatable biological polymers- A review. *Journal of Polymer & Composites*, 10, 66–77.
[22] Cho, W., Yu, H., & Mo, Y. (2017). CO2 conversion to chemicals and fuel for carbon utilization. *Recent Advances in Carbon Capture and Storage*, https://doi.org/10.5772/67316
[23] Cosgrove, D. J. (2019). Improving solvent selection processes with the solvent selection guide. *Green Chemistry*, 21, 4323–4332.
[24] D'Amore, M., Dall'Osto, G., & Vicalvi, R. A. (2016). The role of biofuels in the bioeconomy: A sustainability assessment. *Biofuels, Bioproducts and Biorefineries*, 10(2), 220–228.
[25] Davis, P. S., Harris, E. E., & Thomas, L. (2018). Water quality impacts of solvent contamination. *Environmental Science and Technology*, 52(22), 12954–12963.
[26] Dean, J. A. (1990). *Lange's Handbook of Chemistry* , (ed.). McGraw-Hill Education.
[27] De Oliveira, O. R. M. F., De Castro, M. L. L. D., & Adami, M. J. M. (2020). Homogeneous and heterogeneous catalysis: General concepts and applications. *Chemical Reviews*, 5, 2342–2373.
[28] Duan, J., Wang, Y., & Wang, J. (2019). Continuous flow synthesis of pharmaceuticals: Recent advances and future perspectives. *Chemical Engineering Research and Design*, 144, 205–221.
[29] Eldin, M. R., Alsharif, M. H., & Alabed, B. B. (2018). Selection of green solvents: The chemicals that do not harm the environment. *Journal of Environmental Science and Health*, 53(12), 1232–1240.
[30] EPA (2019). Volatile organic compounds' impact on air quality. *Environmental Protection Agency*, Retrieved from www.epa.gov
[31] EPA. (2021). Volatile organic compounds' impact on air quality. Retrieved from https://www.epa.gov/indoor-air-quality-iaq/volatile-organic-compounds-impact-air-quality
[32] Faber, K. (2011). Biotransformations in organic chemistry. *Springer Science & Business Media*. UK.
[33] Fernandes, J., Kalluri, S., Alsuweidi, M., Mayakrishnan, V., Mohan, C., & Madhavan, A. A. (2024). Band gap engineering of g-C3N4/CuS and its application in solar still. *Chemical Physics Impact*, 100684,100684, https://doi.org/10.1016/j.chphi.2024.100684
[34] Fletcher, G., Garzón-Roca, J., & Zubizarreta, M. J. (2008). Towards the development of a comprehensive effective solvent selection guide. *Green Chemistry*, 10, 1485–1490.
[35] Fogassy, E., Hossain, S., & Zhang, Q. (2016). Reaction condition optimization in chemical synthesis. *Journal of Organic Chemistry*, 81(15), 6465–6470.
[36] Fujita, S., & Matsumoto, K.H. K. (2001). Asymmetric hydrogenation of ketones using layered silicate-supported catalysts. *Journal of Catalysis*, 89(1), 116–123.
[37] Gani, R. (2011). Using water as a solvent: How green is it?. *Green Chemistry*, 13(12), 3185–3191.
[38] García, J. J., De la Hoz, A., & Vázquez, J. L. (2014). ionic liquids as solvents in asymmetric synthesis. *Green Chemistry*, 16(4), 2065–2073, 10.1039/c3gc42547g
[39] Gautam, R., Jindal, R., Mohan, C., & Mishra, A. K. (2024). Sustainable innovations as a solution to energy crisis and industrial pollution. In: *Book: Promoting Multi-Sector Sustainability with Policy and Innovation*. USA: IGI Global. https://doi.org/10.4018/979-8-3693-2113-3.ch007
[40] Ghosh, S., Kumar, J., & Sharma, A. (2019). Soil contamination: Effects of organic solvents on soil microbiota. *International Journal of Environmental Science*, 14(1), 34–48.
[41] Graham, P. H., & Henson, E. C. (1998). Preferential hydrogenation of olefins before aromatic desorption: A solvent effect. *Journal of Catalysis*, 62(2), 503–509.
[42] Greener, J., & Scott, A. J. (2002). Water as a green solvent. *Chemical Society Reviews*, 31(3), 239–249.
[43] Gros, F., Maurin, S., & Coudray, C. (2019). The use of ethanol as a green solvent. *Current Opinion in Green and Sustainable Chemistry*, 19, 8–14.
[44] Gupta, K. (2012). *Solvent Effects in Chemistry: A Comprehensive Review*. Nova Science Publishers.

[45] Gupta, N., Sharma, S., & Rani, P. (2021). Health risks associated with traditional solvents: A review. *Occupational Medicine*, 71(4), 276–283.

[46] Haber, F., & Bosch, C. (1909). The Synthesis of Ammonia from Its Elements. *Zeitschrift Für Physikalische Chemie*, 65, 203–213.

[47] Hart, S. L., & Milstein, M. (1999). Creating sustainable value. *Academy of Management Executive*, 13(2), 56–67.

[48] Huang, H., Usman, A. M., & Wang, Y. (2017). Recent advances in bio-based solvents. *Renewable and Sustainable Energy Reviews*, 79, 370–398.

[49] International Labour Organization (ILO) (2021). *World Employment and Social Outlook 2021: Trends 2021*. Genève, Switzerland.

[50] Jain, A., Sharma, R., & Bansal, S. (2017). Supercritical carbon dioxide as a green solvent for catalytic oxidation: A review. *Chemical Engineering Journal*, 313, 1482–1498. doi:10.1016/j.cej.2016.12.061.

[51] Joshi, P. B., Durve, N. C., & Mohan, C. (2023). Full blown green metrics. In: Garg, V. K. Yadav, A. Mohan, C. Yadav, S. & Kumar, N. (eds.), *Book: Green Chemistry Approaches to Environmental Sustainability*. Elsevier, pp. 109–129. https://doi.org/10.1016/B978-0-443-18959-3.00013-6

[52] Kamlet, M. J., Abboud, J. L. M., & Taft, R. W. (1983). The solvatochromic comparison method. 1. A new approach to the selection of solvents for chemical reactions. *Journal of the American Chemical Society*, 105(24), 7037–7044.

[53] Kirk-Othmer (2000). *Encyclopedia of Chemical Technology (4th*, ed.), Hoboken, New Jersey: Wiley-Interscience.

[54] Klibanov, A. M. (2001). Stability of enzymes in organic solvents. *Trends in Biotechnology*, 19(10), 428–435.

[55] Kotthaus, J. R., Scherf, K. G., & Gläser, R. (2020). Machine learning applications for optimizing heterogeneous catalysis. *Industrial &Engineering Chemistry Research*, 59(44), 19546–19557.

[56] Kumar, N., Kumar, V., & Chowdhary, Y. (2022). A review on synthesis methods of tricyclic 1, 2, 3, 4-tetrahydrocarbazoles. *World Journal of Advanced Research and Reviews*, 13(1), 160–171.

[57] Kumar, V., Pal, M., & Dhiman, N. (2019). Determination of sun protection factor in different extract of Eulaliopsis binata. *Plant Archives*, 19(2), 185–187.

[58] Kumar, V., Sharma, A. K., Rajput, S. K., Pal, M., & Dhiman, N. (2018). Pharmacognostic and pharmacological evaluation of Eulaliopsis binata plant extracts by measuring in vitro/in vivo safety profile and anti-microbial potential. *Toxicology Research*, 7(3), 454–464.

[59] Little, T. A., Lutz, J. A., & Roberts, R. W. (2015). Continuous flow processes in the chemical industry: A review. *Chemical Engineering Journal*, 271, 7–29.

[60] López-Muñoz, F., García-García, P., & Valencia, J. (2010). The role of solvents in the syntheses of pharmaceutical compounds. *Pharmaceutical Research*, 27(3), 525–535.

[61] MacMillan, D. W. C. (2011). The advent and development of organocatalysis. *Nature*, 469(7331), 304–312.

[62] Mathew, D., Thomas, B., & Devaky, K. S. (2020). Green catalysis, green chemistry, and organic syntheses for sustainable development. In: Esteso, M. A. Ribeiro, A. C. F. & Haghi, A. K. (eds.), *Chemistry and Chemical Engineering for Sustainable Development*, 139–165.

[63] Marcus, Y. (2016). Effect of solvent on ionization constants and solvation energy. *Journal of Chemical Education*, 93(3), 482–487.

[64] Meyer, A., Dusselier, M., & Sels, B. (2013). Supercritical fluids: The next generation of green solvents. *Chemical Society Reviews*, 42(7), 2479–2490.

[65] Mohan, C., & Kumar, V. (2021). A comparative study of SARS, MERS with COVID-19. *Coronaviruses*, 2(3), 364–368.

[66] Mohan, C., Chauhan, S., Shekhar, S., Kaushik, P., Madhu, M., Sharma, M., & Kumar, V. (2020). A brief study on rate of morphological and genomic similarities between SARS and COVID-19. *International Journal of Pharma Research and Health Sciences*, 8(3), 3167–3171.

[67] Mohan, C., Kumar, V., & Kumari, S. (2018). synthesis, characterization, and antibacterial activity of schiff bases derived from thiosemicarbazide, 2-acetylthiophene and thiophene-2-aldehyde. *International Research Journal of Pharmacy*, 9, 153–158.

[68] Mohan, C., Kumar, V., Kumari, N., Kumari, S., Yadav, J., Gandass, T., & Yadav, S. (2020). Synthesis, characterization and antibacterial activity of semicarbazide based Schiff bases and their Pb (II), Zr (IV) and U (VI) complexes. *Advanced Journal of Chemistry, Section B*, 2, 187–196.

[69] Ochoa, L., & Martínez, J. (2023). Integrating artificial intelligence in catalysis: A promising future. *Catalysis Today*, 382, 165–176.

[70] Ochoa, M., Halasz, A., & Bley, K. (2021). Evaluating the sustainability of biobased solvents in chemical processes. *Green Chemistry*, 23(19), 7457–7470.

[71] O'Reilly, E., Gunter, H., & McKenzie, K. (2015). Water in organic synthesis: The importance of keeping it simple. *Chemical Society Reviews*, 44(18), 6689–6703.

[72] Penny, C. A., Wong, H. M., & Ding, J. (2020). Optimizing reaction conditions in chemistry: A comparison between traditional and sustainable solvents. *Journal of Cleaner Production*, 252, 119763.

[73] Pérez, R., & Ruiz, F. J.C. D. (2015). Supercritical fluids in organic synthesis. *Ricerca Scientifica*, 10(2), 116–125.

[74] Pérez-Santos, M. C., Remón, S., & Pérez, A. L. (2020). Sustainability metrics in solvent-catalyst systems. *Environmental Science & Technology*, 54(4), 2060–2070.

[75] Prabhu, A. S., Stringer, S. J., & Yip, H. Y. (2017). Continuous flow involvement of real-time monitoring for dynamic optimization in chemical processes. *Chemical Engineering Science*, 169, 159–170.

[76] Rani, G., Chitkara, Y., Kumar, S., & Mohan, C. (2024). Nanomaterials in healthcare for waste management and environmental sustainability. In: Mohan, C., Jeet, S., Dixit, S., & Carabineiro, S. A. (eds.), *Book: Exploring Waste Management in Sustainable Development Contexts*. USA: IGI Global, pp. 143–153. 10.4018/979-8-3693-4264-0.ch010

[77] Reichardt, C. (2010). *Solvent Effects in Organic Chemistry*. Wiley-VCH.

[78] Rocca, G., Contaldi, F., & Storta, I. (2013). Supercritical fluids for industrial applications: Green approaches for extraction and reactions. *Chemical Engineering Transactions*, 32, 1285–1290.

[79] Römer, J., Möller, J., & Dürre, P. (2019). Life cycle assessment for process optimization: Approaches and applications. *Resources, Conservation and Recycling*, 145, 189–197.

[80] Sandhir, V., Kumar, V., & Kumar, V. (2021). Prognosticating the spread of COVID-19 pandemic based on optimal arima estimators. *Endocrine, Metabolic & Immune Disorders-Drug Targets (Formerly Current Drug Targets-ImmuneEndocrine & Metabolic Disorders)*, 21(4), 586–591.

[81] Seddigi, Z., Wu, S., & Malaghini, G. (2012). Solutions of ionic liquids: A new approach for green chemistry. *Green Chemistry*, 14(4), 922–939.

[82] Sheldon, R. A. (2016). The E Factor 25 years on: 25 reasons why E Factor matters. *Green Chemistry*, 18(20), 5344–5355.

[83] Sheldon, R. A. (2016). The E factor: Fifteen years on. *Green Chemistry*, 18(13), 3670–3680.

[84] Soleimani, B., Endo, T., & Tajima, M. (2020). Emission factors for conventional organic solvent use. *Atmospheric Environment*, 243, 117–124.

[85] Stadler, A., & Kappe, C. O. (2005). Microwave-assisted solid-phase synthesis. *Microwave Assisted Organic Synthesis*, 177–221.

[86] de Lima, T. B. M. H., Iron, J. S., & Pessoa, C. M. (2021). Adoption of the solvent selection guide in drug development: Case study in a pharmaceutical company. *Journalof Pharmacological Sciences*, 145(2), 111–121.

[87] Tran, C., Surena, A., & Leclercq, G. (2018). Lipase-catalyzed biodiesel production using ethanol as solvent: Enhanced activity and kinetics. *Renewable Energy*, 119, 149–160. doi:10.1016/j.renene.2017.11.063.

[88] Turner, N. J. (2011). Asymmetric biocatalysis: An update. *Nature Chemistry*, 3(8), 607–614.

[89] Witcomb, D. J. M., & Whiting, D. W. W. W. (2021). A review of heterogeneous catalysis and intrinsic kinetics. *Catalysis Reviews: Science and Engineering*, 53, 526–585.

[90] U.S. Environmental Protection Agency (EPA). (2023). *Green Chemistry*. Retrieved from https://www.epa.gov/greenchemistry

[91] United Nations (UN) (2015). *Transforming Our World: The 2030 Agenda for Sustainable Development.*

[92] United Nations Environment Programme (2022). Policy Recommendations for the Promotion of Green Chemistry.

[93] United Nations Framework Convention on Climate Change (UNFCCC) (2015). *The Paris Agreement*.

[94] Vafaeva, K. M., Chhetri, A., Sudan, P., Mishra, M., Pakkiraiah, B., & Mohan, C., 2024, 'polymer matrix nanocomposites for sustainable packaging: A green approach', *E3S Web of Conferences*, vol. 511, pp. 01008. https://doi.org/10.1051/e3sconf/202451101008

[95] Villa, A., Ochoa-Gómez, J., & Oliva, A. (2016). Aqueous phase reforming of glycerol for hydrogen production: Lessons learned and future directions. *Bioresource Technology*, 211, 59–65. doi:10.1016/j.biortech.2016.03.135.

[96] Wang, C., & Zhang, Y.S. B. (2012). Stabilization of charged transition states by solvent hydrogen bonding. *Journal of Physical Chemistry B*, 116(21), 6091–6099.

[97] Wang, H., & Zhang, Z. (2023). Recent advances in heterogeneous catalysis: Challenges and opportunities. *ACS Catalysis*, 13(2), 778–792.

[98] Wasserscheid, P., & Keim, W. (2000). Ionic liquids – New "green" solvents for organic synthesis. *Angewandte Chemie International Edition*, 39(21), 3772–3789.

[99] Yadav, R., Sanduja, M., Kumar, V., Sharma, K., Khan, S., & Kumar, K. (2024). Microwave-assisted synthetic pathways of pyrrole: A comprehensive review. *Asian Journal of Organic Chemistry*, 13(12), e202400401.

[100] Yamazaki, S., et al. (2017). Effects of temperature and mixed solvent systems on reaction mechanisms in organic reactions. *Organic & Biomolecular Chemistry*, 15, 2827–2835.

[101] Zhang, Y., Frohn, J., & Schmidt, T. (2019). Statistical methods for optimizing reaction conditions in organic synthesis. *Organic & Biomolecular Chemistry*, 17(7), 1307–1320.

[102] Mohan, B., et al. (2023). Covalent organic frameworks (COFs) and metal-organic frameworks (MOFs): Sustainable synthesis using water and green solvents. *Journal of Cleaner Production*, 2023.

[103] Zhao, D., Liao, Y., & Zhang, Z. (2018). Ionic liquids: Applications in catalysis. *Chemical Reviews*, 117(10), 6590–6620. https://doi.org/10.1021/acs.chemrev.7b00581

[104] Byrne, F. P., Jin, S., Paggiola, G., Petchey, T. H. M., Clark, J. H., Farmer, T. J., Hunt, A. J., McElroy, C. R., & Sherwood, J. (2016). Tools and techniques for solvent selection: green solvent selection guides. *Sustainable Chemical Processes*, 4, 7. https://doi.org/10.1186/s40508-016-0051-z

[105] Gambarotta, A., Papetti, V., and Dimopoulos Eggenschwiler, P. (2019). Analysis of the effects of catalytic converter on automotive engines performance through real-time simulation models. *Frontiers in Mechanical Engineering*, 5 (Article 48), 1–17. https://doi.org/10.3389/fmech.2019.00048

[106] Halpern, J. (1974). Homogeneous Catalysis by Coordination Compounds. In: *Advances in Chemistry*, American Chemical Society: Washington, DC.

[107] Berg, J. M., Tymoczko, J. L., Gatto, G. J., & Stryer, L. (2015). *Biochemistry*, 8th ed. W. H. Freeman and Company.

[108] Dumesic, J. A., Rudd, D. F., Aparicio, L. M., Rekoske, J. E., & Treviño, A. A. (2012). The Microkinetics of Heterogeneous Catalysis. *Chemical Engineering Science*, 66(5), 1175–1182. https://doi.org/10.1016/j.ces.2011.06.048

[109] Bornscheuer, U. T., Huisman, G. W., Kazlauskas, R. J., Lutz, S., Moore, J. C., & Robins, K. (2012). Engineering the third wave of biocatalysis. *Nature*, 485(7397), 185–194. https://doi.org/10.1038/nature11117

[110] Deng, H., Li, X., & Wang, Y. (2020). Photocatalysts for energy conversion and organic synthesis: A green chemistry approach. *Chemical Engineering Journal*, 397, 125431. https://doi.org/10.1016/j.cej.2020.125431

[111] Walsh, P. J. (2007). A Green Chemistry Approach to Asymmetric Catalysis. *Chemical Reviews*, 107(6), 201–216. doi: 10.1021/cr078380v

[112] Mohammadpoor-Baltork, I., Moghadam, M., Tangestaninejad, S., & Kia, R. (2001). Solvent effects in Lewis acid-catalyzed reactions. *Journal of Molecular Catalysis A: Chemical*, 170(1–2), 49–58. https://doi.org/10.1016/S1381-1169(01)00264

[113] Sharma, V., et al. (2022). Review: Deep eutectic solvents as promising pretreatment agents for lignocellulosic biomass—Greener alternatives to ionic liquids and conventional solvents. *Bioresource Technology*, 363, 127966. https://doi.org/10.1016/j.biortech.2022.127966

Namita Das*, Manjusha Panthanal, and Geetha Kannan

Chapter 4
Overview of biopolymers for sustainable environment

Abstract: Biopolymers, which are derived from renewable resources, have demonstrated potential in mitigating the environmental problems associated with conventional plastics. Plastics are a global problem since traditional garbage is not managed effectively. Proteins, polysaccharides, and bio-based polymers like polylactic acid and polyhydroxyalkanoates are examples of biopolymers, both natural and synthetic. The chapter looks at how biopolymers are made using renewable resources, highlighting the role that biomass and sustainable agriculture have played in this transition. It also considers the properties of biopolymers, such as mechanical strength and composability, when comparing their effects on the entire life cycle to those of conventional plastics. Applications of biopolymers are explored, with a focus on how they could help encourage more environmentally friendly procedures in the building, medical, packaging, and agricultural sectors. The infrastructure required for biopolymer recycling and composting, together with the waste management challenges they raise, are discussed, emphasizing the importance of proper degradation and disposal routes. The economic and policy implications of the adoption of biopolymers, including cost considerations, market demand, and governmental regulations, are also covered in this chapter. The limitations and challenges of increasing production volume, in addition to technological developments that enhance the functionality of biopolymers – like bio-based nanocomposites and advanced synthesis methods are discussed. The future of biopolymers is discussed in the next section of the chapter, along with potential solutions to existing problems and emerging trends that make biopolymers well-positioned to assist in achieving the objectives of global sustainability.

Keywords: Synthetic, bio-based biopolymers, biopolymer composites, waste management, government regulation

***Corresponding author: Namita Das**, General Education Department, City University Ajman, Ajman, UAE, e-mail: namitaprabhan@gmail.com
Manjusha Panthanal, Geetha Kannan, Chemistry Department, Bits Pilani, Dubai, UAE

https://doi.org/10.1515/9783111618906-004

4.1 Introduction to biopolymers

The term "biopolymer" comes from the Greek words "bio" and "polymer," which denote nature or living things. Biopolymers are large macromolecules composed of several repeating units. The long chain of polymeric biomolecules is usually produced by living organisms. The monomer units of these polymers are connected by covalent bonds to create substantial structures. DNA, RNA, fatty acids, gelatin, keratin, cellulose, starch, and so on are a few types of biopolymers [49]. Genetic material can be passed down from one generation to the next via a variety of biopolymers, including DNA and RNA. The two primary components that make up plants are cellulose and starch. These polymers have been on the planet for billions of years, in contrast to synthetic plastics. Whereas synthetic polymers have similar and arbitrary arrangements, biopolymers are organized in a well-defined structure. Thus, one characteristic of biopolymers is the biological breakdown of large chains into smaller chains.

Due to their biodegradable, biocompatible, and exceptionally resilient nature, biopolymers have taken up a significant amount of space in every known area of research, manufacturing, and production, including the pharmaceutical or therapeutic sector, edible industry, textile department, agronomics, cosmetics, and automotive industries. These biopolymers – which include chitosan, cellulose, collagen, keratin, pectin, carrageenan, alginate, dextran, curdlan, and starch – are frequently used by researchers across a range of disciplines. Biopolymers can come from natural sources and can be completely synthesized from microorganisms or chemically derived from biological ingredients. Based on their type of origin, polymer backbone, and repeating monomer units, biopolymers can be categorized into three classes. Biopolymers have undoubtedly earned attention and respect in the biomedical industry due to their remarkable applications in wound healing, tissue engineering, medicine transfer, medical implants, and gene therapy. Chitosan, cellulose, dextran, alginate, pectin, curdlan, and starch are examples of biopolymers that are frequently used.

Because of their chemical flexibility, which allows for the creation of materials with a wide range of mechanical and physical properties, biopolymers have shown enormous promise. Since their inception, biopolymers have proven to be useful in a wide range of biomedical and other technical applications, such as long-term implants, wearable electronics, regenerative medicine, controlled medication delivery, food packaging, and the construction sector. Rapid developments in materials science and technology have created new avenues for exploring the innovative uses of biopolymers in the biological and engineering domains [39].

4.2 Classification and types of biopolymers

A vast variety of polymers are produced by living things as an important component of their dry matter, cellular makeup, and morphology. These biopolymers are fundamental to an organism's life cycle because they support its vital cellular and metabolic process. Biopolymers can be classified into three categories as shown in the Figure 4.1. More comprehensive details about the various types of biopolymers, their sources, and their uses is provided in Tables 4.1 and 4.2.

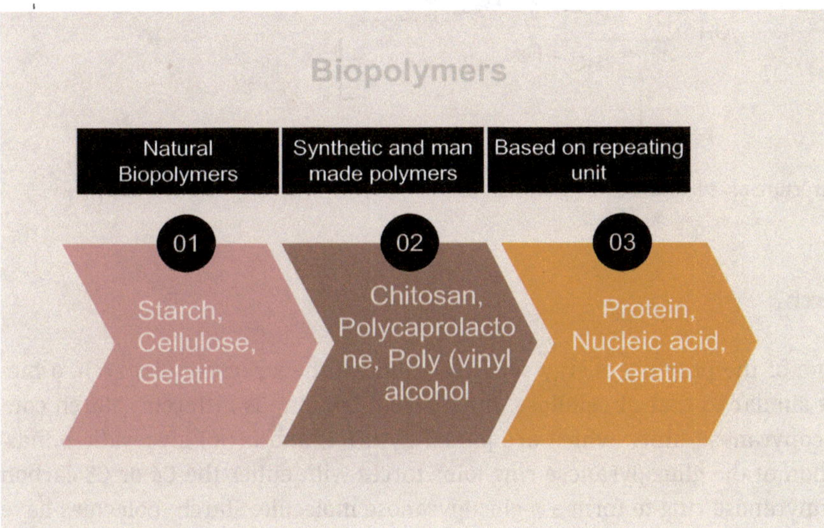

Figure 4.1: Classification of biopolymers.

4.2.1 Natural biopolymers

4.2.1.1 Cellulose

Cellulose is the first thermoplastic polymer found in plants. Cellulose's chemical formula is $(C_6H_{10}O_5)_n$, where n is the number of repeating units. The IUPAC name for cellulose is (6,5)-trihydroxy-2-(hydroxymethyl)-6-(-2-(hydroxymethyl))-3*S*-4,5,6-trihydroxy-oxan-3-yl, or oxyoxan-3,4,5-triol. Its linear structure is composed of 100–1,000 repeating units of (1–4) *b*-linked glucose units. D-Glucose is the glucose unit that has the (1e5) OH and CH_2OH groups in the same plane (Figure 4.2) [51, 52]. When cellulose reacts with either organic or inorganic acid, cellulose fibers are produced [46]. These are called "derivatives of cellulose," and their properties are determined by the hydroxyl group of cellulose [29]. In order to prepare ester and ether derivatives, it typically reacts with a reagent. Celluloid is produced by the reaction of nitrocellulose with camphor. Cellulose has outstanding chemical qualities but mediocre thermal ones. It is a crystalline powder that has no taste or odor. Because of its 100% crystalline structure, it exhibits no glass transition temperature. As a homopolymer, cellulose is harmless and decomposable. Its cross-linked structure gives it exceptional tensile and compressive strength [8, 9, 21].

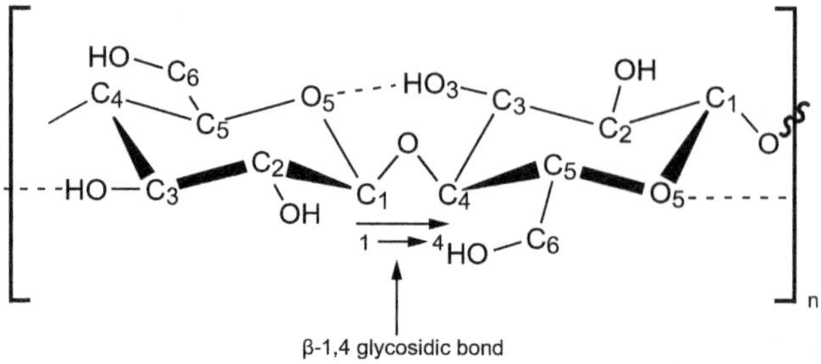

Figure 4.2: Structure of cellulose with 1,4-glycosidic bond.

4.2.1.2 Starch

Starch is one of the primary energy sources. Although the structure of starch, a biopolymer, is similar to that of cellulose, the internal bonding is different. Starch contains D-glucopyranose units, which are joined by α(1/4) and α(1/6) glycosidic bonds. The C1 carbon of the glucopyranose ring joins forces with either the C4 or C6 carbon of the glucopyranose ring to form a D-glucopyranose molecule. Starch molecules have a single free reducing end due to the aldehyde group (Figure 4.3) [59]. Glucose contains two types of starch polymerization: amylose and amylopectin. Compared to amylose, a significant linear carbohydrate, amylopectin is a larger molecule with a more branching shape. The essential carbohydrate amylose is defined as having a linear structure made up of α(1/4) glycosidic bond. Despite having the *H* atom in its helical shape, it is hydrophobic. Amylose has a stronger propensity to gel when starch granules are heated. Compared to amylose, amylopectin has a bigger and more branching structure. α(1/4) and α(1/6) glycosidic bonds make up its molecules. Only 4–6% of α(1–4) links exist, yet all molecules have α(1–6) linkage [8, 9, 18, 32].

4.2.1.3 Gelatin

Gelatin comes from the meat industry and is a naturally occurring biopolymer made from raw collagen. Gelatin is also known as hydrolysate, denatured collagen, and collagen. There are 18 kinds of necessary amino acids in it. Gelatin comes in two varieties based on content and method: When raw collagen is hydrolyzed in an acidic environment, type-A gelatin is created. It is composed of 18.5% nitrogen. Collagen that in-

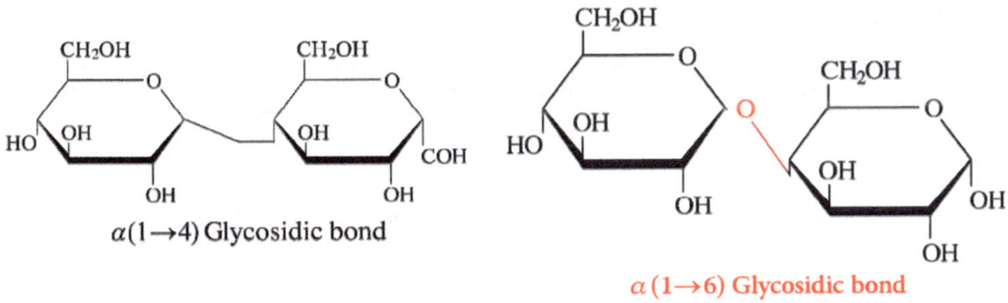

Figure 4.3: Structure of cellulose with 1,4-glycosidic bond.

Table 4.1: Different categories of biopolymers.

Category	Source	Biopolymers	References
Polysaccharides	Animals	Hyaluronic acid and chitosan	[61]
	Plants	Starch, agar, and cellulose	[60]
	Lipids	Fatty acids and oil	[37]
Proteins	Animals	Fish, collagen, silk, and gelatin	[19]
	Plants	Soy protein, zein, and gluten	[13]
Microorganisms	Polyester	PHA, PCL, and PHB	[38]
Synthetic		PVA and PVAc	[26]

Table 4.2: Different biopolymers, sources, and their applications.

Biopolymers	Sources	Applications	References
Chitosan	Animal	Food industry, agricultural, and pharmaceutical	[17]
Collagen	Animal	Tissue engineering, cosmetic and food industry	[45]
Starch	Plant	Textile and paper industry, and drug delivery system	[55]
Cellulose	Plant fibers	Biomedical and drug delivery	[53]
PTT	Petroleum-based monomers	Textile industry and automotive parts	[54]
PLA	Cornstarch, wheat, and sugar	Food packaging and drug delivery	[27]
PHB	Produced and stored by bacteria	Tissue scaffolding and nerve regeneration	[41]

Table 4.2 (continued)

Biopolymers	Sources	Applications	References
PGA	Aliphatic polyester	Drug delivery and dental/orthopedic application	[42]
PCL	Petroleum-based products	Surgical implants and drug delivery devices	[6]

cludes 18% nitrogen atoms without the amide group is hydrolyzed alkalinely to create type-B gelatin (eCONHe).

4.2.2 Synthetic biopolymers

4.2.2.1 Chitosan

Chitosan is made up of repeated modified glucose subunits and is the largest structural polymer based on nitrogen. Overall, it can be found in the cells of several fish, insects, and vertebrates. Chitin and cellulose are structurally similar due to the presence of hydroxyl and amine substituents on monomer units. The hydroxyl group is replaced by a methyl amide group. Since it is produced by organisms, living things and vertebrates may easily digest it [1, 14]. Vertebrates can absorb chitosan in addition to regularly consuming insects [28, 35]. This is due to the fact that symbiotic bacteria and protozoa may break down fibrous chitin into smaller constitutional glucose units. *N*-Acetylglucosamine monomeric units, which are connected by (1e4) glycosidic bonds, are acetylated to produce it [59].

4.2.2.2 Polycaprolactone

A synthetic polyester biopolymer called polycaprolactone (PCL) is made by polymerizing caprolactone through ring opening when a catalyst, like stannous octate, is present. PCL is utilized in biomedical applications and is known by its IUPAC nomenclature, (1,7)-polyoxepan-2-one. Additionally, it is added to starch to reduce its cost, boost its biodegradability, and strengthen its impact [62].

4.2.2.3 Poly(vinyl alcohol)

Polyvinyl acetate, the precursor of poly(vinyl alcohol) (PVA), is hydrolyzed to produce the thermoplastic biopolymer. It is broken down by biological microorganisms. It is very soluble in water due to its increased crystallinity. PVA is used to make a wide range of polymer end products, such as liquors, surgical threads, and food packaging. It is a strong, ductile, and extremely flexible polymer. Figure 4.4 show the structure of the Chitosan, polycarprolactone and poly vinyl alcohol.

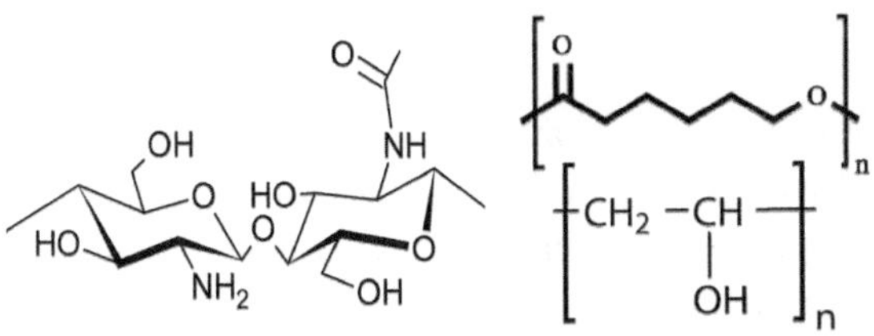

Figure 4.4: Structure of (a) chitosan, (b) polycaprolactone, and (c) poly(vinyl alcohol).

4.2.3 Repeating biopolymers

4.2.3.1 Proteins

Proteins are natural repeating biopolymers composed of amino acid monomers linked through peptide bonds. They play vital structural and functional roles in living organisms, forming enzymes, hormones, and cellular components. Each protein's sequence and structure determine its specific biological function.

In contrast, polyvinyl alcohol (PVA) is a synthetic biopolymer that exhibits biodegradability similar to natural protein-based polymers. It is produced by the hydrolysis of polyvinyl acetate (PVAc), which removes acetate groups to yield hydroxyl groups along the polymer chain. This modification enhances hydrophilicity and water solubility, making PVA suitable for various biomedical and environmentally friendly applications such as drug delivery films, biodegradable packaging, and water-soluble adhesives [15, 44].

4.2.3.2 Keratin

Keratin is a fibrous structural protein and a naturally occurring biopolymer found in hair, wool, feathers, nails, and horns. It is composed of repeating amino acid units, primarily cysteine, which forms strong disulfide linkages responsible for its high tensile strength, durability, and resistance to degradation. Keratin plays a crucial role in providing mechanical protection and structural stability in biological systems.

In comparison, polyvinyl alcohol (PVA), though a synthetic biopolymer, shares certain desirable properties such as biodegradability, flexibility, and film-forming ability. PVA is produced through the hydrolysis of polyvinyl acetate (PVAc) and is widely used in biomedical and industrial applications like biodegradable films, coatings, and adhesives. While keratin is naturally derived and protein-based, PVA serves as a synthetic analog for developing biocompatible and environmentally sustainable materials inspired by natural polymers like keratin. [11, 50].

4.2.3.3 Nucleic acid

The earliest known biomolecule is nucleic acid. It transports and stores all cellular data as well as energy. Hereditary information is stored in DNA within all polymeric chain segments, including genes. Every nucleic acid component is crucial to the delivery of genetic information. It is a polymer made up of nucleotides, which are tiny monomeric units [24, 64].

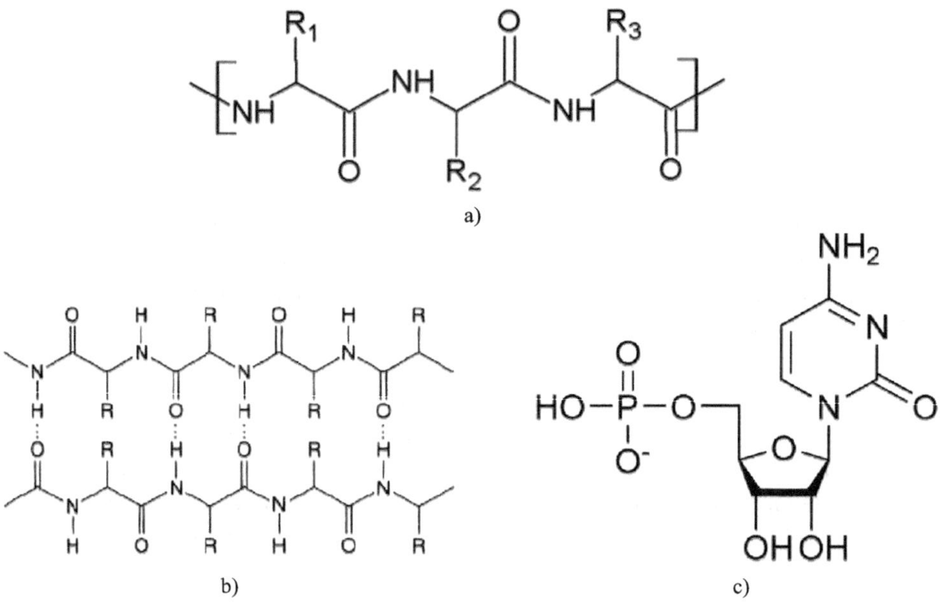

Figure 4.5: Chemical structure of (a) protein,(b) keratin, and (c) nucleic acid.

4.3 Renewable sources for biopolymer production

The primary benefit of biopolymers, which are natural polymers made by bacteria, plants, and animals, is their biodegradability. Lignocellulosic feedstock has become a huge, renewable, and promising source of chemicals [25]. Carbohydrate polymers (cellulose and hemicelluloses) and aromatic polymers (lignin and tannins) make up the majority of this biomass. These polymers can be separated and utilized to create hydrogels without competing with other biopolymers that are used in food, such as starch. Lignin is the second most abundant and intriguing terrestrial biopolymer, after cellulose [16, 40]. In actuality, 150 billion tons of lignin are produced annually by plants on the Earth. The bulk of this complex aromatic macromolecule is composed of three different phenyl propane monomers, *p*-coumaryl, coniferyl, and sinapyl alcohols, which are cross-linked employing many stable bondage types [44]. Numerous methods and raw resources, including wood, specialist crops, and agricultural waste, can be used to extract lignin [15]. PLA is one of the best polymeric substitutes for other petropolymers because of its exceptional thermomechanical properties, biodegradability, biocompatibility, renewable nature, and affordability (Figure 4.6). At first, most of its uses were in the biomedical industry and for temporary purposes like packaging, especially because PLA has 191 biodegradable qualities. Remarkably, PLA is increasingly seen as a valuable bio-sourced polymer substitute in long-term applications like electronics and automobiles because of the depletion of petroleum supplies [29].

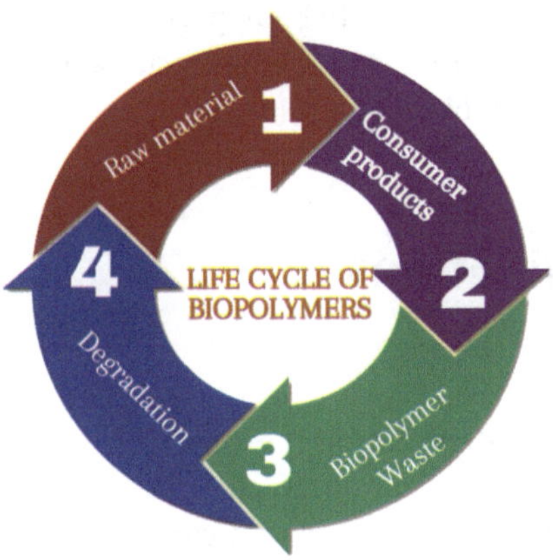

Figure 4.6: Life cycle of a biopolymer.

4.4 Properties of biopolymers and its environmental impact

Since biopolymers have also been linked to serious health and environmental issues such as greenhouse gas emissions and unfavorable land use change, it is necessary to evaluate the actual health and environmental impact of their use [7]. The cultivation of the crop is one aspect of life cycle assessments (LCAs) on biopolymers made from agricultural products [33]. This covers the fuel needed for agricultural tasks like harvesting, pesticide application, and ploughing. It should also cover the production and delivery of necessary supplies including pesticides, fertilizers, and herbicides [36]. Along with nitrogen-based emissions from fertilizer use, land use, and water use may also be significant concerns. For common biopolymers, additional procedures to incorporate are manufacture and grinding [58, 63].

4.5 Applications of biopolymers in sustainability

By lowering plastic contamination and the detrimental ecotoxicological effects on aquatic and terrestrial creatures, bio-based polymers complement the approach of creative agricultural breakthroughs and result in more productive farming. However, sophisticated polymer membranes serve as agrochemical capturing agents, preventing environmental intoxication. The variety of polymer uses in agriculture around the world will prevent the near-term eradication of current polymer-based technology. However, more "green" polymers ought to be used in this industry in accordance with ecological tendencies in contemporary agriculture [34]. A significant step toward sustainable and ecologically friendly farming is the application of biodegradable polymers in seed coating. Coatings that are biodegradable reduce the possibility of plastic buildup in soil and water, promoting thriving ecosystems and reducing the harmful consequences of plastic pollution. Notwithstanding the many benefits, issues with scalability, production costs, and customizing material qualities for particular uses still exist. The goal of research and development is to remove these obstacles and increase the availability and effectiveness of biodegradable coatings [47, 56].

4.6 Impact of government regulations and incentives

4.6.1 Market drivers for sustainable biopolymers

The market for biodegradable polymers is anticipated to expand at a compound annual growth rate (CAGR) of 24.8% from $4.3 billion in 2021 to $14.4 billion in 2027. The market for bio-based polymers is anticipated to expand at a CAGR of 24.1% from $6.7 billion in 2021 to $22.0 billion in 2027. The following are the main drivers of the market expansion for sustainable biopolymers:

a. Growing customer inclination for bioplastic
b. Green flexible materials increasingly being used for main packaging

The market for sustainable biopolymers is divided into segments based on geography, application, and chemistry. By the end of 2027, the polylactic acid (PLA) segment will control the market for bio-based and biodegradable polymers. The largest market share is held by Europe. Amcor PLC, BASF SE, Evonik, ADM, Berry Global, Braskem, Cargill, FKuR, and others are important players in the market (BCC Research: Global Markets Reports and Industry Analysis, n.d.).

4.6.2 Government regulations and incentives

The market will grow more rapidly if strict restrictions are put in place to monitor the usage of conventional polymers and make sure they stay within the allowed limit. The market expansion is also being aided by the simple access to feedstock, which includes natural fibers, sugar, and starch, and is necessary for the production of biopolymers. A transparency market research (TMR) analysis projects that the global biopolymers market will grow at a 14% CAGR from 2017 to 2025, from US$2,422.1 million in 2016 to US$7,774.7 million by the end of 2025.

The global market for biopolymers will expand as a result of the tight laws governing the use of plastic and the increased awareness of the need to safeguard the environment. The use of biopolymer-based goods in a variety of industries, including electronics and automotive, will be fueled by the strict laws governing carbon emissions and plastic consumption in many countries worldwide. The market will also be driven by the growing demand for bio-based packaging due to shifting lifestyles and the increased use of sustainable and eco-friendly products. The market for biopolymers will expand as a result of all these causes.

Furthermore, the industry participants are making significant investments in newer biopolymers as a consequence of which the worldwide biopolymers market will expand.

4.6.3 Cost-effectiveness and market adoption of biopolymer technologies

Although the majority of research that identifies or thoroughly examines the uses of biopolymers does not show the need for recycling to make the materials sustainable or connect them to environmental damage, a sizable portion of the literature is devoted to the negative effects of biopolymers on the environment [4].

Since biopolymers are made from natural materials, their primary disadvantage at the moment is their expensive cost, which ranges from US$4,000/metric ton to US$15,000/metric ton, while synthetic polymers cost US$1,500 [4]. However, as more and more manufacturers from China, Japan, Thailand, India, the USA, and Europe adopt biopolymers with new manufacturing methods, it is anticipated that the cost of biopolymers would decrease. In comparison to conventional plastic, biopolymers require less fossil fuel, produce fewer greenhouse gas emissions, and are less harmful to the environment [48, 57]. Additionally, there are less disposal issues because biopolymers are completely biodegradable, breaking down through microbial processes to produce biomass, CO_2, CH_4, and H_2O [3, 22].

4.7 Biopolymers and the circular economy

4.7.1 Integration of biopolymers into circular economy models

According to the circular economy model, all goods, resources, and raw materials ought to be utilized for as long as feasible. The concept is that used goods should be repurposed as raw materials for new manufacturing rather than ending up as waste [12].

4.7.2 Closing the loop: biopolymers in reuse, recycling, and composting

A modern, environmentally conscious economy has challenges in reducing waste output, increasing the use of biodegradable and/or biobased feedstock, and developing recycling techniques. Sorting, mechanical recycling, chemical recycling and the more recent enzymatic depolymerization of biopolymers are the primary recycling processes that are studied. As of now, the only practical way to recycle biodegradable PLA seems to be mechanically processing post-industrial trash. Nonbiodegradable bio-based polymers, including bio-PA, bio-PET, and bio-PE can be recycled in existing facilities by mixing them with their counterparts derived from fossil fuels. It is also possible to recycle poly(ethylene furanoate) by mixing up to 2% of it with industrial PET. Bio-based polymers have drawn interest from governments, businesses, and consumers in recent years as a means of lowering municipal solid waste. Their creation and application have developed interest. Nevertheless, these materials have potential environmental advantages that only become apparent when their energy is recovered by incineration, recycling, or composting. The closure of the carbon cycle loop is made possible by these end-of-life circumstances.

4.7.3 Biopolymers as part of sustainable supply chains

The development of sustainability in contemporary industry depends on biopolymer composites. The creation of novel bio-based materials is greatly impacted by the current world situation because of its crucial benefits, which help replace hazardous and synthetic materials. By lowering waste, landfills, and toxic emissions, the use of renewable bio-based materials in engineering and construction will increase sustainability and create a cleaner, greener world [20].

4.8 Technological innovations in biopolymers

4.8.1 Advances in biopolymer synthesis and processing

The biodegradability, environmental friendliness, and sustainability of biopolymer materials have drawn interest [46]. A lot of research has been done on recent advancements in the synthesis, properties, and industrial applications of biopolymers. Biopolymers – made from renewable resources – have garnered a lot of attention due to their potential to replace conventional petrochemical-based plastics and their environmental benefits. The synthesis of biopolymer explores cutting-edge methods, such as chemical alterations, microbial fermentation, and new biotechnological strategies, for producing biopolymers [29].

4.8.2 Next-generation biopolymers: improved properties and applications

The worldwide polymer industry is undergoing a major transformation. Rapid advancements are being made in the creation of a new class of polymers derived from renewable resources, or bio-based polymers. These novel bioplastics include a rising variety of commercially successful biopolymers, such as polylactides and polyhydroxyalkanoates (PHAs), which complement traditional biopolymers like cellulosic and natural rubber. Due to the quick advancements in biotechnology, many more bioplastics will soon be available [5]. Biochemical transformation and molecular specificity make them perfect for generating high-purity monomers, which are required to create high-molecular-weight polymer molecules. Polyethylene, polybutlylene, and poly (ethylene terephthalate) are well-known polymers that are produced using innovative biochemical processes, which start with renewable resources rather than fossil ones, are one of the most recent advancements [20]. By optimizing certain stimuli to increase therapeutic efficacy, smart materials can now offer amazing capabilities like regulated treatments [31]. Dual-stimuli or multistimuli-responsive smart materials, which may activate several functions on a single smart nanomaterial, have recently attracted increased attention in an attempt to enhance disease diagnostics [28, 35]. Biopolymer-based nanocomposites incorporating adsorbents such graphene oxide, carbon nanotubes, and nanoclays could be very advantageous for water treatment applications [30].

4.8.3 Bio-based nanocomposites: strengthening biopolymer performance

Research on new eco-friendly and biodegradable composites is highly appealing due to the current environmental policy and the strong desire to create new materials with improved performance [1, 14]. Traditional nonbiodegradable polymers may be replaced by biopolymer-based nanocomposites due to their high biodegradability, advantageous mechanical properties, and simplicity of application [52]. Advanced physical, structural, and functional properties, as well as affordability, biodegradability, and biocompatibility, are some of the wonderful qualities of polysaccharide-based nanocomposites [6]. Among other industries, food, water treatment, and medicinal sectors have made use of these bio-based nanocomposites [10].

4.9 Challenges and limitations of biopolymers

4.9.1 Scalability and production costs

Biopolymers have seen a resurgence in recent decades [4]. The market for bio-based plastics has emerged as a result of the public desire for environmentally friendly and sustainable procedures and the growing depletion of fossil fuels [5]. Nonbiodegradable bioplastics in particular are highly intriguing materials because they combine the advantages of enhanced resource efficiency and lower carbon footprint during manufacture with the ability to withstand microbial destruction [22, 48].

4.9.2 Next-generation biopolymers: improved properties and applications

Biopolymers have recently shown previously unheard-of advantages over traditional plastic polymers, so long as they are applied in ways that improve functionality and produce additional benefits for human life. As a result, biopolymers have garnered a lot of attention because of their important role in daily life and their unique, adjustable properties, which make them appealing for a variety of uses. When creating materials for use in biomedical applications, biopolymers can yield materials with adjustable qualities including biodegradability, biocompatibility, renewability, affordability, and availability [3]. Furthermore, by altering the polymer components, smart biopolymers can be made, leading to more target-oriented applications [12].

4.9.3 Compatibility with existing industrial and waste infrastructure

Because of their biocompatibility and biodegradability, biopolymers are aesthetically pleasant and merit more research consideration. It is vital to assess the true environmental impact of bioplastic use in the first place, though because its use has been connected to a number of environmental issues, including greenhouse gas emissions and unfavorable land use change [20].

4.10 Future prospects for biopolymers

4.10.1 Emerging trends in biopolymer research and development

Cost analysis of biopolymers is challenging since production costs are significantly influenced by the variety of their properties (source, purity, molecular weight, degree of esterification, degree of deacetylation, and ratio of guluronic to mannuronic acid) [30, 52]. Although the price of the basic resource may have an impact on the ultimate cost, other crucial factors – like energy-related products or sustainability concerns – are becoming increasingly significant (due to the growing restrictions) [31].

4.10.2 Potential role of biopolymers in global sustainability goals

Bio-based polymers are materials derived from renewable resources [10]. Their biodegradable properties have recently drawn the attention of researchers and manufacturers worldwide due to the need and desire for substitutes for polymers made from fossil fuels. The use of biodegradable polymers creates a sustainable industry. However, the majority of petrochemical-based synthetic polymers are not biodegradable, and their raw materials will eventually run out. Despite these advantages, bio-based polymers still account for a small percentage of the global plastic market at the moment [51].

4.10.3 The path forward: overcoming challenges to widespread adoption

Conventional petrochemical polymers may eventually be replaced by biodegradable bio-based polymers derived from biomass, including forestry, marine, plant, and animal materials [56]. Decades of research and development have focused on potential biodegradable biobased polymers, such as PLA, succinate polymers, and PHAs [31, 51].

However, problems like the limitations imposed by the environment on the rates of biodegradation of biodegradable biobased polymers need to be addressed. These resources have been evaluated, citing cost, production capabilities, and practicality as barriers to commercialization [23].

4.11 Conclusion

The expanding intersection of molecular biology and polymer chemistry is one of the most promising avenues. This interface opens up new and practical uses for these kinds of biopolymers by fusing the functional properties of polymers with strong control over polymer structure. This pattern is very apparent in polymers made on proteins [23]. The first uses of these bioengineered materials will be for pharmaceutical and medical purposes because of the production expenses. However, as molecular biology instruments advance, as production costs are reduced, high-volume applications will also be realized [43]. For instance, goods made from plants, such as polyesters and studies on proteins, are already underway.

References

[1] Aaliya, B., Valiyapeediyekkal Sunooj, K., & Lackner, M. (2021). Biopolymer composites: A review. *International Journal of Biobased Plastics*, 3(1), 40–84. https://doi.org/10.1080/24759651.2021.1881214

[2] Katiyar, V., Kumar, A., & Mulchandani, N. (2020). *Advances in Sustainable Polymers*. Springer, n.d. doi: 10-1007978-981-15-1251-3.

[3] Ahari, H., Golestan, L., Anvar, S. A. A., Cacciotti, I., Garavand, F., Rezaei, A., Sani, M. A., & Jafari, S. M. (2022). Bio-nanocomposites as food packaging materials; the main production techniques and analytical parameters. *Advances in Colloid and Interface Science*, 310, 102806. https://doi.org/10.1016/J.CIS.2022.102806

[4] Akinsemolu, A. A., Idowu, A. M., & Onyeaka, H. N. (2024a). Recycling technologies for biopolymers: current challenges and future directions. *Polymers*, 16(19), 2770. https://doi.org/10.3390/POLYM16192770

[5] Akinsemolu, A. A., Idowu, A. M., & Onyeaka, H. N. (2024b). Recycling technologies for biopolymers: Current challenges and future directions. *Polymers*, 16(19), 2770. https://doi.org/10.3390/POLYM16192770

[6] Aliotta, L., Cinelli, P., Coltelli, M. B., Righetti, M. C., Gazzano, M., & Lazzeri, A. (2017). Effect of nucleating agents on crystallinity and properties of poly (lactic acid) (PLA). *European Polymer Journal*, 93, 822–832. https://doi.org/10.1016/J.EURPOLYMJ.2017.04.041

[7] Anjana, R. G., Shree, S., Sharma, A., Panesar, P. S., & Goswami, S. (2021). Recent approaches for enhanced production of microbial polyhydroxybutyrate: Preparation of biocomposites and applications. *International Journal of Biological Macromolecules*, 182, 1650–1669. https://doi.org/10.1016/J.IJBIOMAC.2021.05.037

[8] Arif, Z. U., Khalid, M. Y., Ahmed, W., Arshad, H., & Ullah, S. (2022a). Recycling of the glass/carbon fibre reinforced polymer composites: A step towards the circular economy. *Polymer-Plastics Technology and Materials*, 61(7), 761–788. https://doi.org/10.1080/25740881.2021.2015781

[9] Arif, Z. U., Khalid, M. Y., Sheikh, M. F., Zolfagharian, A., & Bodaghi, M. (2022b). Biopolymeric sustainable materials and their emerging applications. *Journal of Environmental Chemical Engineering*, 10(4), 108159. https://doi.org/10.1016/J.JECE.2022.108159

[10] Azadi, E., Dinari, M., Derakhshani, M., Reid, K. R., & Karimi, B. (2024). Sources and extraction of biopolymers and manufacturing of bio-based nanocomposites for different applications. *Molecules*, 29(18), 4406. https://doi.org/10.3390/MOLECULES29184406

[11] Balart, R., Garcia-Garcia, D., Fombuena, V., Quiles-Carrillo, L., & Arrieta, M. P. (2021). Biopolymers from natural resources. *Polymers*, 13(15), 2532. https://doi.org/10.3390/POLYM13152532

[12] Bansal, P., & Purwar, R. (2023). Nanofibers' utility for rejuvenation of heavily contaminated environments. *Nanobiotechnology for Bioremediation: Fundamentals and Mechanisms*, 23–50. https://doi.org/10.1016/B978-0-323-91767-4.00004-5

[13] Bilal, M., & Iqbal, H. M. N. (2019). Naturally-derived biopolymers: Potential platforms for enzyme immobilization. *International Journal of Biological Macromolecules*, 130, 462–482. https://doi.org/10.1016/J.IJBIOMAC.2019.02.152

[14] Biswal, T. (2019). Biopolymers for tissue engineering applications: A review. *Materials Today: Proceedings*, 41, 397–402. https://doi.org/10.1016/J.MATPR.2020.09.628

[15] Cazón, P., Velazquez, G., & Vázquez, M. (2020). Characterization of mechanical and barrier properties of bacterial cellulose, glycerol and polyvinyl alcohol (PVOH) composite films with eco-friendly UV-protective properties. *Food Hydrocolloids*, 99, 105323. https://doi.org/10.1016/J.FOODHYD.2019.105323

[16] Chakrapani, I. S., Rajasekhar, M., Bhaskar, L. V. K. S., & Mohan, C. (2023). Smart investigation on nanocomposites composed of carbon dioxide derives, repeatable biological polymers- A review. *Journal of Polymer & Composites*, 10, 66–77.

[17] Chiellini, E., Corti, A., D'Antone, S., & Solaro, R. (2003). Biodegradation of poly (vinyl alcohol) based materials. *Progress in Polymer Science*, 28(6), 963–1014. https://doi.org/10.1016/S0079-6700(02)00149-1

[18] Devadas, V. V., Khoo, K. S., Chia, W. Y., Chew, K. W., Munawaroh, H. S. H., Lam, M. K., Lim, J. W., Ho, Y. C., Lee, K. T., & Show, P. L. (2021). Algae biopolymer towards sustainable circular economy. *Bioresource Technology*, 325, https://doi.org/10.1016/J.BIORTECH.2021.124702

[19] Díez-Pascual, A. M. (2022). Biopolymer composites: Synthesis, properties, and applications. *International Journal of Molecular Sciences*, 23(4), 2257. https://doi.org/10.3390/IJMS23042257

[20] Getahun, M. J., Kassie, B. B., & Alemu, T. S. (2024). Recent advances in biopolymer synthesis, properties, & commercial applications: A review. *Process Biochemistry*, 145, 261–287. https://doi.org/10.1016/J.PROCBIO.2024.06.034

[21] Gowthaman, N. S. K., Lim, H. N., Sreeraj, T. R., Amalraj, A., & Gopi, S. (2021). Advantages of biopolymers over synthetic polymers: Social, economic, and environmental aspects. *Biopolymers and Their Industrial Applications*, 351–372. https://doi.org/10.1016/B978-0-12-819240-5.00015-8

[22] Jayakumar, A., Radoor, S., Kim, J. T., Rhim, J. W., Parameswaranpillai, J., Nandi, D., Srisuk, R., & Siengchin, S. (2022). Titanium dioxide nanoparticles and elderberry extract incorporated starch based polyvinyl alcohol films as active and intelligent food packaging wraps. *Food Packaging and Shelf Life*, 34, https://doi.org/10.1016/J.FPSL.2022.100967

[23] Jha, S., Akula, B., Enyioma, H., Novak, M., Amin, V., & Liang, H. (2024). Biodegradable biobased polymers: A review of the state of the art, challenges, and future directions. *Polymers*, 16(16), 2262. https://doi.org/10.3390/POLYM16162262

[24] Kajal, S., & Thakur, S. (2024). Coexistence of microplastics and heavy metals in soil: Occurrence, transport, key interactions and effect on plants. *Environmental Research*, 262, 119960. https://doi.org/10.1016/J.ENVRES.2024.119960

[25] Karanastasis, A. A., Safin, V., & Pitet, L. M. (2022). Bio-based upcycling of poly(ethylene terephthalate) waste for the preparation of high-performance thermoplastic copolyesters.

Macromolecules, 55(3), 1042–1049. https://doi.org/10.1021/ACS.MACROMOL.1C02338/SUPPL_FILE/MA1C02338_SI_001.PDF

[26] Kashirina, A., Yao, Y., Liu, Y., & Leng, J. (2019). Biopolymers as bone substitutes: A review. *Biomaterials Science*, 7(10), 3961–3983. https://doi.org/10.1039/C9BM00664H

[27] Khalid, M. Y., & Arif, Z. U. (2022). Novel biopolymer-based sustainable composites for food packaging applications: A narrative review. *Food Packaging and Shelf Life*, 33, https://doi.org/10.1016/J.FPSL.2022.100892

[28] Khalid, M. Y., Arif, Z. U., Ahmed, W., & Arshad, H. (2022). Recent trends in recycling and reusing techniques of different plastic polymers and their composite materials. *Sustainable Materials and Technologies*, 31, https://doi.org/10.1016/J.SUSMAT.2021.E00382

[29] Kim, D. H., Yu, A., & Goh, M. (2021). Oxidative chemical depolymerization of thermoset epoxy resin for green recycling. *Journal of Industrial and Engineering Chemistry*, 96, 76–81. https://doi.org/10.1016/J.JIEC.2021.01.047

[30] Kolya, H., & Kang, C. W. (2023). Next-generation water treatment: Exploring the potential of biopolymer-based nanocomposites in adsorption and membrane filtration. *Polymers*, 15(16), 3421. https://doi.org/10.3390/POLYM15163421

[31] Kumar, A., Mishra, R. K., Verma, K., Aldosari, S. M., Maity, C. K., Verma, S., Patel, R., & Thakur, V. K. (2023). A comprehensive review of various biopolymer composites and their applications: From biocompatibility to self-healing. *Materials Today Sustainability*, 23, 100431. https://doi.org/10.1016/J.MTSUST.2023.100431

[32] Kumari, P., Kumari, N., Mohan, C., Chinglenthoiba, C., & Amesho, K. T. T. (2023). Environmentally benign approach to formulate nanoclay/starch hydrogel for controlled release of zinc and its application in seed coating of Oryza Sativa plant. *International Journal of Biological Macromolecules*, 257, 128278.https://doi.org/10.1016/j.ijbiomac.2023.128278

[33] Kumar, V., Pal, M., & Dhiman, N. (2019). Determination of sun protection factor in different extract of Eulaliopsis Binata. *Plant Archives*, 19(2), 185–187.

[34] Lewicka, K., Szymanek, I., Rogacz, D., Wrzalik, M., Łagiewka, J., Nowik-Zając, A., Zawierucha, I., Coseri, S., Puiu, I., Falfushynska, H., & Rychter, P. (2024). Current trends of polymer materials' application in agriculture. *Sustainability*, 16(19), 8439. https://doi.org/10.3390/SU16198439

[35] Lim, B. K. H., & Thian, E. S. (2022). Biodegradation of polymers in managing plastic waste – A review. *The Science of the Total Environment*, 813, https://doi.org/10.1016/J.SCITOTENV.2021.151880

[36] Liu, J., He, J., Xue, R., Xu, B., Qian, X., Xin, F., Blank, L. M., Zhou, J., Wei, R., Dong, W., & Jiang, M. (2021). Biodegradation and up-cycling of polyurethanes: Progress, challenges, and prospects. *Biotechnology Advances*, 48, https://doi.org/10.1016/J.BIOTECHADV.2021.107730

[37] Liu, J., Sun, L., Xu, W., Wang, Q., Yu, S., & Sun, J. (2019). Current advances and future perspectives of 3D printing natural-derived biopolymers. *Carbohydrate Polymers*, 207, 297–316. https://doi.org/10.1016/J.CARBPOL.2018.11.077

[38] Mitura, S., Sionkowska, A., & Jaiswal, A. (2020). Biopolymers for hydrogels in cosmetics: Review. *Journal of Materials Science. Materials in Medicine*, 31(6), 50. https://doi.org/10.1007/S10856-020-06390-W

[39] Mohan, C., Kumar, V., Kumari, N., Kumari, S., Yadav, J., Gandass, T., & Yadav, S. (2020). Synthesis, characterization and antibacterial activity of semicarbazide based Schiff bases and their Pb (II), Zr (IV) and U (VI) complexes. *Advanced Journal of Chemistry, Section B*, 2, 187–196.

[40] Morales, A., Labidi, J., Gullón, P., & Astray, G. (2021). Synthesis of advanced biobased green materials from renewable biopolymers. *Current Opinion in Green and Sustainable Chemistry*, 29, 100436. https://doi.org/10.1016/J.COGSC.2020.100436

[41] Murariu, M., Dechief, A. L., Ramy-Ratiarison, R., Paint, Y., Raquez, J. M., & Dubois, P. (2015a). Recent advances in production of poly(lactic acid) (PLA) nanocomposites: A versatile method to tune

crystallization properties of PLA. *Nanocomposites*, 1(2), 71–82. https://doi.org/10.1179/2055033214Y.0000000008

[42] Murariu, M., Dechief, A. L., Ramy-Ratiarison, R., Paint, Y., Raquez, J. M., & Dubois, P. (2015b). Recent advances in production of poly(lactic acid) (PLA) nanocomposites: A versatile method to tune crystallization properties of PLA. *Nanocomposites*, 1(2), 71–82. https://doi.org/10.1179/2055033214Y.0000000008

[43] Musa, Y., & Bwatanglang, I. B. (2020). Current role and future developments of biopolymers in green and sustainable chemistry and catalysis. *Sustainable Nanocellulose and Nanohydrogels from Natural Sources*, 131–154. https://doi.org/10.1016/B978-0-12-816789-2.00006-7

[44] Numata, K., Abe, H., & Iwata, T. (2009). Biodegradability of poly(hydroxyalkanoate) materials. *Materials*, 2(3), 1104–1126. https://doi.org/10.3390/MA2031104

[45] Polman, E. M. N., Gruter, G. J. M., Parsons, J. R., & Tietema, A. (2021). Comparison of the aerobic biodegradation of biopolymers and the corresponding bioplastics: A review. *Science of the Total Environment*, 753, 141953. https://doi.org/10.1016/J.SCITOTENV.2020.141953

[46] Ragitha, V. M., & Edison, L. K. (2022). Safety issues, environmental impacts, and health effects of biopolymers. *Handbook of Biopolymers*, 1–27. https://doi.org/10.1007/978-981-16-6603-2_54-1

[47] Rani, G., Chitkara, Y., Kumar, S., & Mohan, C. (2024). Nanomaterials in healthcare for waste management and environmental sustainability. In: *Book: Exploring Waste Management in Sustainable Development Contexts. IGI Global, USA*, pp. 143–153. doi: 10.4018/979-8-3693-4264-0.ch010

[48] Silvestre, C., Duraccio, D., & Cimmino, S. (2011). Food packaging based on polymer nanomaterials. *Progress in Polymer Science (Oxford)*, 36(12), 1766–1782. https://doi.org/10.1016/J.PROGPOLYMSCI.2011.02.003

[49] Singh, R., Gautam, S., Sharma, B., Jain, P., & Chauhan, K. D. (2021a). Biopolymers and their classifications. *Biopolymers and Their Industrial Applications*, 21–44. https://doi.org/10.1016/B978-0-12-819240-5.00002-X

[50] Singh, R., Gautam, S., Sharma, B., Jain, P., & Chauhan, K. D. (2021b). Biopolymers and their classifications. *Biopolymers and Their Industrial Applications*, 21–44. https://doi.org/10.1016/B978-0-12-819240-5.00002-X

[51] Singh, S., & Chunglok, W. (2022). Biopolymers towards green and sustainable development. *Biopolymers Towards Green and Sustainable Development*, https://doi.org/10.2174/97898150793021220101

[52] Sumrith, N., Rangappa, S. M., Dangtungee, R., Siengchin, S., Jawaid, M., & Pruncu, C. I. (2019). Biopolymers-based nanocomposites: Properties and applications. *Bio-Based Polymers and Nanocomposites: Preparation, Processing, Properties & Performance*, 255–272. https://doi.org/10.1007/978-3-030-05825-8_12

[53] Taniguchi, I., Yoshida, S., Hiraga, K., Miyamoto, K., Kimura, Y., & Oda, K. (2019a). Biodegradation of PET: Current status and application aspects. *ACS Catalysis*, 9(5), 4089–4105. https://doi.org/10.1021/ACSCATAL.8B05171/ASSET/IMAGES/MEDIUM/CS-2018-05171U_0012.GIF

[54] Taniguchi, I., Yoshida, S., Hiraga, K., Miyamoto, K., Kimura, Y., & Oda, K. (2019b). Biodegradation of PET: Current status and application aspects. *ACS Catalysis*, 9(5), 4089–4105. https://doi.org/10.1021/ACSCATAL.8B05171

[55] Tarique, J., Sapuan, S. M., Khalina, A., Sherwani, S. F. K., Yusuf, J., & Ilyas, R. A. (2021). Recent developments in sustainable arrowroot (Maranta arundinacea Linn) starch biopolymers, fibres, biopolymer composites and their potential industrial applications: A review. *Journal of Materials Research and Technology*, 13, 1191–1219. https://doi.org/10.1016/J.JMRT.2021.05.047

[56] Thiagamani, S. M., Krishnasamy, S., & Siengchin, S. (2019). Challenges of biodegradable polymers: An environmental perspective. *Applied Science and Engineering Progress*, 12(3), 149. https://doi.org/10.14416/J.ASEP.2019.03.002

[57] Tiwari, A., Niveita, S. V., Mohan, C., & Bendi, A. (2024). Waste management from agricultural field: Sustainable Practices and framework. In: *Book: Exploring Waste Management in Sustainable Development Contexts. IGI Global, USA*, pp. 154–169. doi: 10.4018/979-8-3693-4264-0.ch011

[58] Vafaeva, K. M., Chhetri, A., Sudan, P., Mishra, M., Pakkiraiah, B., & Mohan, C. (2024). Polymer matrix nanocomposites for sustainable packaging: A green approach. *E3S Web of Conferences*, 511, 01008. https://doi.org/10.1051/e3sconf/202451101008

[59] Wang, H., Wei, D., Ziaee, Z., Xiao, H., Zheng, A., & Zhao, Y. (2015). Preparation and properties of nonleaching antimicrobial linear low-density polyethylene films. *Industrial and Engineering Chemistry Research*, 54(6), 1824–1831. https://doi.org/10.1021/IE504393T

[60] Wen, K., Wei, L., Ren, Z., Wang, B., & Lu, J. (2021). Enhanced photocatalytic degradation of cationic and anionic dyes by Ag-modified g-C_3N_4 composite: Insights on different mechanisms under visible light. *Journal of Materials Research*, 36(7), 1549–1560. https://doi.org/10.1557/S43578-021-00203-8

[61] Yaashikaa, P. R., Senthil Kumar, P., & Karishma, S. (2022). Review on biopolymers and composites – Evolving material as adsorbents in removal of environmental pollutants. *Environmental Research*, 212, 113114. https://doi.org/10.1016/J.ENVRES.2022.113114

[62] Yang, K., Du, J., Zhang, Z., Liu, D., & Ren, T. (2020). Facile and eco-friendly preparation of super-amphiphilic porous polycaprolactone. *Journal of Colloid and Interface Science*, 560, 795–801. https://doi.org/10.1016/J.JCIS.2019.11.012

[63] Yates, M. R., & Barlow, C. Y. (2013). Life cycle assessments of biodegradable, commercial biopolymers – A critical review. *Resources, Conservation and Recycling*, 78, 54–66. https://doi.org/10.1016/J.RESCONREC.2013.06.010

[64] Yu, H., Hou, J., Dang, Q., Cui, D., Xi, B., & Tan, W. (2020). Decrease in bioavailability of soil heavy metals caused by the presence of microplastics varies across aggregate levels. *Journal of Hazardous Materials*, 395, 122690. https://doi.org/10.1016/J.JHAZMAT.2020.122690

Basant Kumar Singh, Panmei Gaijon, Sudipta Ghosh,
M. Ramananda Singh, Ranjana Dixit, Neeraj Kumari, Jitendra Kumar,
and Arun Kant*

Chapter 5
Integrated technologies for environmental remediation by using green materials

Abstract: This chapter aims to remediate toxic inorganic molecules (i.e., inorganic pigment, metal ion, anions, etc.), organic molecules/compounds (i.e., hazardous organic compounds, dyes, etc.), and other contaminants from aqueous wastewater as well as from soil and air using green materials. Designing chemical processes and a variety of materials that minimize the use of hazardous substances and do not generate hazardous substances is green chemistry. Plant-based, biodegradable, and recycled materials are known as green materials, and are used as barrier membranes to function as adsorbent materials for different types of contaminants.

Various analytical techniques and technologies based on photocatalysis, adsorption, absorption, chemical reaction, filtration, and centrifugation are used to remediate the contaminant from different natural sources of the environment.

In our environment, natural sources such as water, soil, and air have a significant role in human life as well as for the existence of life of others. The human population and its activities toward industrialization and unsustainable agricultural practices lead to air, water, and soil pollution, which is a major global concern.

In the case of water, out of 10 hazardous metals, Hg (mercury), Pb (lead), As (arsenic), and Cd (cadmium) are major threats to mankind. Herbicides, pesticides, fertilizers, and oil spills are also contaminants of the water body. These contaminants and metals are nonbiodegradable in nature and due to this, they can persist for a long period in a natural aquatic body.

Main source of toxic waste material is from municipalities, industrial effluents and sewage from village. In the case of air, filters, ionizers, activated carbon, some oxidant materials, ultraviolet germicidal irradiation, and so on are used to remove dust particulates and other airborne contaminants. In the case of soil, soil washing by

***Corresponding author: Arun Kant**, Kirori Mal College, University of Delhi, Delhi, India, e-mail: arun107941@kmc.du.ac.in
Basant Kumar Singh, Ram Swarath College, Munger University, Bihar, India
Panmei Gaijon, Sudipta Ghosh, M. Ramananda Singh, Kirori Mal College, University of Delhi, Delhi, India
Ranjana Dixit, Ramjas College, University of Delhi, Delhi, India
Neeraj Kumari, School of Basic and Applied Sciences, K. R. Mangalam University, Gurugram, India
Jitendra Kumar, Sardar Vallabhbhai Patel College, V.K.S. University, Bhabua, India

https://doi.org/10.1515/9783111618906-005

using environmentally friendly surfactants and other green solvents is used to remediate its contaminants.

The green method and green material are likely effective for treating wastewater, soil, air and other commercial wastes. Overall, green method intends to project a path of rapid development that provides a broad overview of recent applications of green material as future opportunities that are involved in the field of wastewater treatment.

Keywords: Remediation, green material, industrial wastewater and gas, economic evaluation, pollution, nanotechnology

5.1 Introduction

Environmental degradation is one of the most critical challenges facing humanity today, largely driven by rapid industrialization, urbanization, and unsustainable exploitation of natural resources [75]. Pollution of air, water, and soil, along with biodiversity loss, has far-reaching consequences for ecosystems and human health [31]. These challenges demand innovative and sustainable solutions, particularly in environmental remediation, the process of removing contaminants from the environment and restoring ecological balance [86]. Traditional remediation methods often rely on chemical and mechanical processes, which, while effective, can be resource-intensive and lead to secondary pollution [46, 64]. In contrast, green materials offer a sustainable alternative, aligning with the principles of environmental conservation and circular economy [4]. These materials are typically derived from renewable sources, are biodegradable, and possess minimal ecological footprint, making them ideal for addressing pollution in a sustainable manner [70]. Examples of green materials include bio-based polymers, algae-derived sorbents, plant-based adsorbents, and nanomaterials developed using eco-friendly synthesis techniques [66, Zhang et al., 2019]. These materials not only reduce reliance on fossil fuels but also minimize the risk of introducing harmful byproducts during their life cycle [47].

The role of integrated technologies in enhancing the potential of green materials cannot be overstated. By combining diverse scientific disciplines, such as nanotechnology, bioengineering, and green chemistry, integrated technologies facilitate the development of solutions that are both innovative and environmentally friendly [61]. For instance, integrating biochar, a carbon-rich material derived from organic waste, with nanomaterials enhances its adsorption capacity for pollutants, making it highly effective in water purification [13, 89]. Similarly, microbial bioremediation, when combined with green catalysts, improves the efficiency of soil detoxification, enabling the removal of complex contaminants in an eco-friendly manner [52].

This chapter explores the interplay between green materials and integrated technologies in addressing environmental challenges. By focusing on the synergy between sustainability and innovation, it provides a comprehensive overview of how these approaches are transforming environmental remediation. The chapter also highlights the versatility of green materials in tackling various types of pollution, from heavy metals in soil to persistent organic pollutants in water [48, 73]. It also discusses how integrated technologies enhance the functionality and efficiency of [55]. The need for a multidisciplinary approach to environmental challenges has never been greater. Integrated technologies provide the tools to tailor green materials to specific environmental contexts, ensuring that remediation strategies are both targeted and efficient [88]. By minimizing resource consumption and maximizing impact, these solutions align with global sustainability goals, such as the United Nations Sustainable Development Goals (SDGs) [78].

In conclusion, the integration of green materials and advanced technologies represents a paradigm shift in environmental remediation, offering innovative solutions to pressing ecological challenges. By leveraging this synergy, we can move closer to achieving a sustainable future, where environmental restoration is achieved without compromising the health of ecosystems or the planet [25]. This chapter serves as a foundation for understanding the transformative potential of these approaches, setting the stage for a deeper exploration of their applications and implications in the subsequent sections.

5.1.1 Green materials and their importance

Green materials are substances derived from renewable, sustainable, and environmentally friendly sources that aim to minimize ecological impacts throughout their life cycle. These materials are designed to reduce reliance on finite resources, decrease energy consumption during production, and ensure biodegradability or recyclability at the end of their use. Green materials have hues of importance such as environmental preservation, human health benefits, economic viability, and alignment with policy.

Examples of green materials include:

(a) **Bio-based materials:** Derived from natural resources like cellulose, starch, or lignin.

(b) **Eco-friendly polymers:** Examples are polylactic acid (PLA) and other biodegradable plastics.

(c) **Recycled materials:** Produced from waste streams, they reduce landfill dependency and conserve virgin resources.

5.2 Overview of environmental remediation and its challenges

Environmental remediation refers to the processes aimed at removing pollutants, contaminants, or toxins from environmental media such as soil, water, and air, and restoring them to a state that supports health and biodiversity [2].

Environmental remediation employs various techniques to remove or neutralize pollutants from soil, water, and air. The choice of method depends on factors such as pollutant type, contamination extent, and environmental conditions. The primary categories of remediation techniques are shown in Figure 5.1, and are described further.

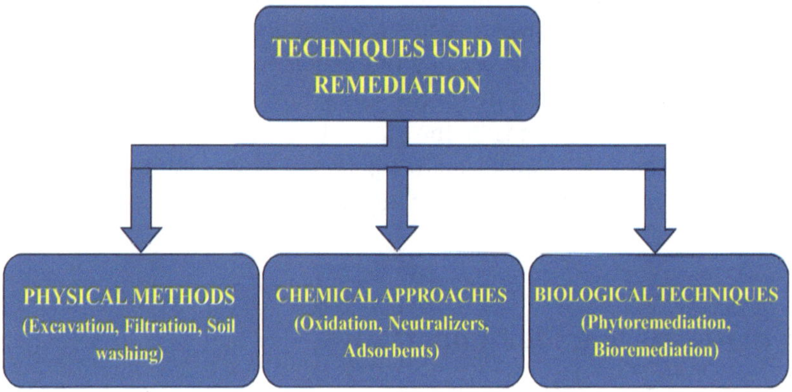

Figure 5.1: Representation of the flowcharts for the primary categories of remediation techniques.

5.2.1 Physical methods

Physical methods are foundational techniques in environmental remediation, focusing on the physical separation or removal of contaminants. Common approaches include excavation, which physically removes contaminated soil for treatment elsewhere, and filtration, effective in purifying water by removing suspended solids or pollutants. Advanced techniques like soil washing use aqueous solutions to extract contaminants, ideal for heavy metals and hydrophobic compounds. These methods are often used in combination with other techniques for enhanced efficiency and are widely regarded for their immediate effectiveness in reducing contaminant loads [80]. Some of the physical methods are as follows.

5.2.1.1 Excavation

This method involves physically removing contaminated soil or sediment and transporting it to a treatment facility. It is often used for heavy metal contamination in localized areas [80].

5.2.1.2 Filtration

Applied for water remediation, filtration techniques use physical barriers to remove suspended solids or contaminants. Advanced methods, such as membrane filtration, can separate nanoparticles and pathogens from water sources.

5.2.1.3 Soil washing

Soil washing involves using water or chemical solutions to extract contaminants from soil. It is particularly effective for heavy metals and hydrocarbons [92].

5.2.2 Chemical approaches

Chemical approaches target contaminants by altering their chemical composition or reactivity. Oxidation techniques, such as Fenton's reaction, break down organic pollutants using oxidants like hydrogen peroxide. Neutralization balances pH in acidic or alkaline environments using materials like lime or soda ash. Adsorption employs substances such as activated carbon or zeolites to bind and isolate contaminants. These methods excel in treating persistent organic pollutants and heavy metals but often require careful handling to prevent secondary impacts [5]. Some of the chemical approaches are described further.

5.2.2.1 Oxidants

Chemical oxidation, such as Fenton's reaction, uses agents like hydrogen peroxide to degrade organic pollutants. This technique is widely employed for groundwater remediation [5].

5.2.2.2 Neutralizers

Alkaline or acidic substances are used to neutralize pH imbalances caused by contaminants like mine tailings. Lime, for instance, can treat acidic mine drainage [30].

5.2.2.3 Adsorbents

Materials such as activated carbon, biochar, and zeolites capture pollutants through surface interactions. These are effective in removing heavy metals, dyes, and organic contaminants [31].

5.2.3 Biological techniques

Biological techniques leverage natural organisms and processes to remediate contaminated environments. Phytoremediation uses plants like *Brassica juncea* or *Pteris vittata* to absorb heavy metals or organic pollutants from soil and water. Similarly, bioremediation employs microorganisms, such as *Pseudomonas* species, to degrade hydrocarbons, pesticides, or industrial solvents. Engineered microbes can enhance pollutant degradation efficiency. These methods are eco-friendly, cost-effective, and suitable for large-scale applications. However, they require longer timescales and are sensitive to environmental conditions, necessitating careful optimization [23, 57]. Some of the biological techniques are as follows.

5.2.3.1 Phytoremediation

Plants such as *Brassica juncea* (Indian mustard) or *Pteris vittata* (Chinese brake fern) are used to absorb heavy metals and contaminants from soil and water. This eco-friendly method is suitable for large-scale remediation [57, 85].

5.2.3.2 Bioremediation

Microorganisms, such as *Pseudomonas* species or *Mycobacterium*, degrade organic pollutants, including oil spills and chlorinated compounds. Engineered microbes can enhance this process [23].

5.2.4 Role of integrated technologies in advancing sustainability

Integrated technologies involve the convergence of diverse scientific and technological disciplines to enhance the functionality and sustainability of solutions. In environmental remediation, this integration amplifies the potential of green materials through synergistic applications. Some important roles of integrated technologies are as follows.

5.2.4.1 Enhancing efficiency

Combining nanotechnology with green adsorbents improves pollutant capture and degradation rates.

5.2.4.2 Scalability and adaptability

Technologies like 3D printing enable precise customization of green materials for diverse applications.

5.2.4.3 Reducing environmental impact

Green chemistry innovations minimize hazardous by-products during material synthesis.

5.2.4.4 Facilitating circular economy

Integrated recycling and upcycling systems ensure that materials can be reused multiple times, reducing waste.

By leveraging such integrations, the transition toward sustainability becomes not only feasible but also economically and ecologically advantageous. This multidisciplinary approach is vital for addressing complex environmental challenges comprehensively. Some significance examples of integration in action are as follows.

Nanotechnology in water purification:
Green-derived nanomaterials act as highly efficient adsorbents for heavy metals.
Microbial technology for soil remediation [83, 87]:
Coupling biochar with engineered microbes accelerates pollutant breakdown.
Hybrid photocatalysts:
Combining green polymers with photocatalytic nanoparticles enhances the breakdown of organic pollutants in wastewater.

5.2.5 Challenges in environmental remediation

Green materials offer a promising alternative, addressing many challenges by providing cost-effective, efficient, and environmentally compatible solutions. The challenges in environmental remediation are shown in Figure 5.2.

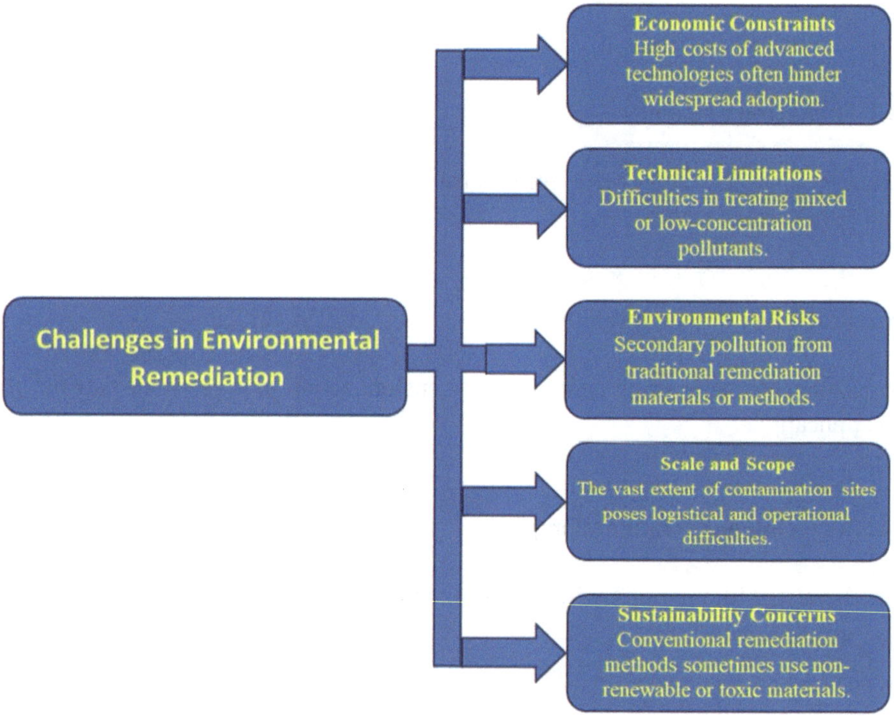

Figure 5.2: Representation of the flowcharts for challenges in environmental remediation.

5.2.6 Overview of green materials

Green materials, often referred to as sustainable or eco-friendly materials, are those designed to minimize environmental impact throughout their life cycle. These materials are typically derived from renewable resources, exhibit biodegradability, and align with the principles of sustainability and circular economy [50]. Unlike conventional materials, green materials are engineered to reduce pollution and resource depletion, making them crucial in mitigating the adverse effects of industrial activities on the environment [22]. One of the defining features of green materials is their renewable origin. For instance, bio-based polymers such as PLA are derived from agricultural sources like corn and sugarcane, offering a sustainable alternative to petroleum-based plastics [47].

Similarly, algae-derived sorbents have been recognized for their ability to adsorb heavy metals from water, showcasing the versatility of natural resources in addressing pollution [73]. These materials not only reduce the dependency on finite resources but also contribute to reducing carbon emissions during production [31].

Another important category of green materials is nanomaterials synthesized through green chemistry approaches. Green nanomaterials, such as silver nanoparticles synthesized using plant extracts, are gaining traction for their environmental and biomedical applications due to their minimal ecological footprint [61]. These materials are designed with both functionality and sustainability in mind, enabling applications in fields ranging from water treatment to energy storage. In addition to their eco-friendly attributes, green materials play a pivotal role in fostering the circular economy. Materials such as biochar, a carbon-rich byproduct of biomass pyrolysis, exemplify this principle by converting agricultural waste into a valuable resource for soil remediation and carbon sequestration [13]. Similarly, recycled materials like reclaimed metals and composites are integral to reducing waste and conserving raw materials in construction and manufacturing [42]. The adoption of green materials is also driven by global initiatives and regulations promoting sustainable development. For example, the United Nations SDGs emphasize the need for sustainable production and consumption patterns, further highlighting the importance of green materials in achieving environmental and economic goals [78]. These materials are instrumental in transitioning industries toward low-carbon and resource-efficient systems [25]. Despite their numerous advantages, green materials face challenges in terms of scalability and cost-effectiveness. However, advances in integrated technologies, such as combining bio-based materials with nanotechnology, have shown promise in overcoming these limitations [89]. By enhancing material properties and performance, these technologies are paving the way for broader adoption of green materials in various applications [55].

In conclusion, green materials represent a cornerstone of sustainable development, offering innovative solutions to environmental challenges, while fostering a balance between economic growth and ecological preservation. Their integration into industrial practices, supported by technological advancements and policy frameworks, is essential for building a resilient and sustainable future [90].

5.2.7 Characteristics of green materials

The characteristics of green materials such as sustainability, biodegradability, and low carbon footprint are discussed below. Sustainability is about deriving from renewable resources such as plants, agricultural by-products, or recycled materials. Green materials reduce dependency on finite resources and align with the circular economy [22, 50]. Biodegradability is about green materials decomposing naturally without leaving harmful residues, ensuring minimal environmental impact [13, 70].

Low carbon footprint talks of using energy-efficient processes, resulting in fewer greenhouse gas emissions [18, 31, 78]. Some important green materials with examples are represented in Figure 5.3.

5.5.2.8 Applications of green materials

Green materials serve diverse applications across various industries. In construction, renewable resources like bamboo and straw enable the creation of eco-friendly, sustainable building designs. In packaging, biodegradable plastics reduce environmental plastic pollution without compromising the functionality [16, 61]. For environmental remediation, materials like biochar and nanocellulose are highly effective in adsorbing pollutants and restoring contaminated soil and water, offering innovative solutions for sustainable cleanup efforts [13, 73]. These applications highlight the potential of green materials to address pressing environmental challenges (Figure 5.3).

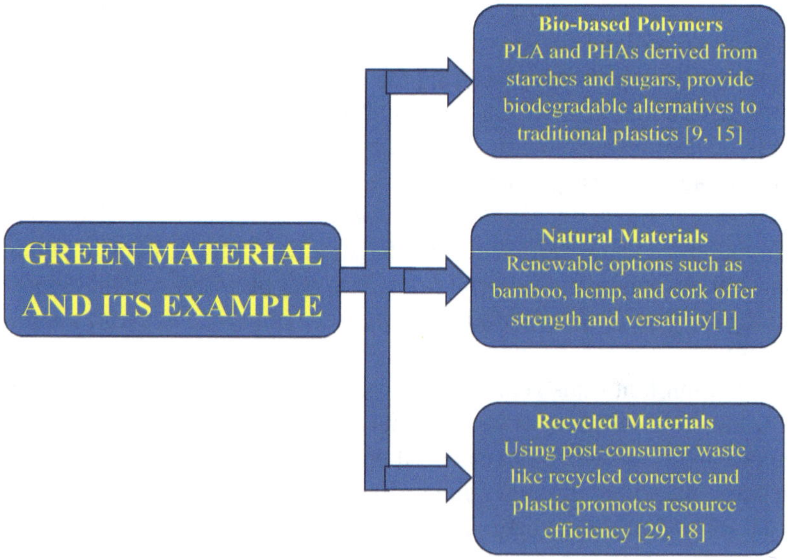

Figure 5.3: Important green materials and their examples represented in the flowchart.

5.2.9 Benefits of green materials

Green materials offer numerous environmental, economic, and regulatory benefits. They promote environmental preservation by reducing pollution and conserving resources through the use of renewable sources [64, 78]. These materials provide economic advantages by lowering costs via energy efficiency and recycling processes. Additionally, they ensure regulatory compliance, meeting rigorous global sustainability standards and policies. These benefits highlight their role in advancing a more sustainable future by addressing environmental challenges, while supporting economic and policy goals.

5.2.10 Challenges in adopting green materials

Adopting green materials faces several challenges. Economic feasibility is a major barrier, as high initial costs for green technologies and materials deter widespread adoption [89]. Performance metrics are often underdeveloped compared to conventional materials, limiting confidence in their effectiveness. Additionally, awareness and availability issues arise due to insufficient consumer knowledge and limited market access, hindering broader utilization of green alternatives. Addressing these challenges requires investment in innovation, improved data on material performance, and targeted outreach to increase awareness and accessibility.

5.3 Key environmental remediation techniques

Environmental remediation involves methods to remove or mitigate pollution from soil, water, and air, aiming to restore natural ecosystems. Key techniques integrate physical, chemical, biological, and emerging technologies to address diverse environmental challenges.

5.3.1 Mechanisms of adsorption

The adsorption mechanism involves the interaction between the adsorbate molecules and the adsorbent surface. Adsorption may be monolayer or multilayer, depending upon the interaction. Green adsorbents remove dyes and HOCs through several mechanisms, as follows:

I. **Physical adsorption**
II. **Chemical adsorption**
III. **Ion exchange**
IV. **Hydrophobic interactions**

5.3.1.1 Physical adsorption

Physical adsorption is a surface process where adsorbate molecules adhere to an adsorbent through weak van der Waals forces [67]. It is reversible, does not involve chemical bonding, and is effective at low temperatures [17]. This method is extensively used in environmental remediation due to its simplicity and cost-effectiveness [26]. Adsorbents like activated carbon, biochar, and other green materials are particularly efficient in removing contaminants, such as dyes, HOCs and heavy metals, from aqueous solutions [6]. Its eco-friendly nature ensures minimal environmental impact [42]. A schematic representation of physical adsorption is shown in Figure 5.4.

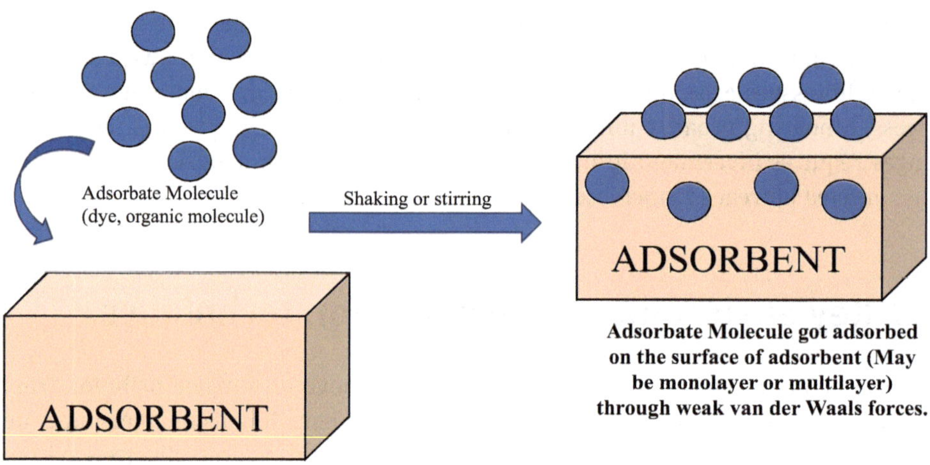

Figure 5.4: Schematic representation of physical adsorption between the adsorbate and an adsorbent.

5.3.1.2 Chemical adsorption

Chemical adsorption (chemisorption) involves the formation of strong chemical bonds between the adsorbate molecules and the adsorbent surface, often resulting in higher binding energies compared to physical adsorption. This process is typically specific to the adsorbate and adsorbent, occurring at elevated temperatures and resulting in monolayer adsorption [17]. Chemisorption is widely employed for removing heavy metals, dyes, and organic pollutants due to its high selectivity and capacity [36]. Common green adsorbents for chemisorption include functionalized biochar, modified clays, and biosorbents, with tailored surface chemistry [6]. A schematic representation of chemisorption is shown in Figure 5.5.

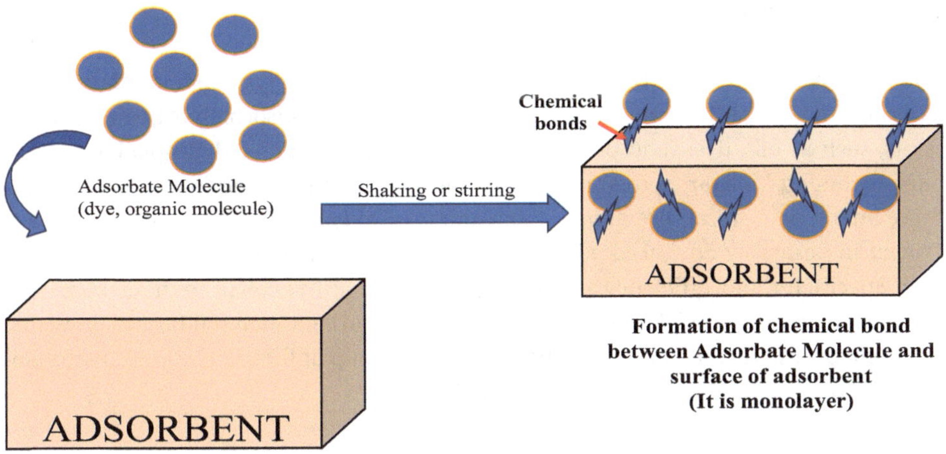

Figure 5.5: Schematic representation of chemisorption, that is, chemical bond formation between the adsorbate and an adsorbent.

5.3.1.3 Ion exchange

Ion exchange is a reversible chemical process where unwanted ions in a solution are replaced with desirable ions from an adsorbent, such as a resin or mineral surface. It is widely employed in water treatment for the removal of heavy metals, nitrates, and other contaminants [36]. Ion exchangers, including zeolites, biochar, and modified clays, exhibit high selectivity and regeneration potential, making them suitable for environmental applications [59]. This method is particularly effective in removing cations like lead, cadmium, and arsenic from aqueous systems [6]. A schematic diagram representing the ion exchange of some metal ions is shown in Figure 5.6.

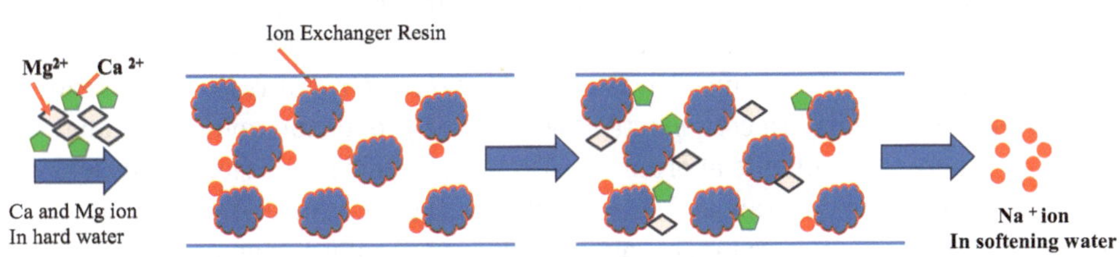

Figure 5.6: Schematic diagram representing the ion exchange of calcium and magnesium ion with sodium ion in ion exchanger resin.

5.3.1.4 Hydrophobic interactions

Hydrophobic interactions occur when nonpolar substances aggregate to minimize their exposure to water. These forces play a significant role in adsorbing organic contaminants, such as oils, dyes, and pesticides, from aqueous solutions. Hydrophobic adsorbents, including biochar, activated carbon, and modified polymeric materials, exploit these interactions to capture nonpolar pollutants effectively [27]. These interactions are crucial in designing advanced adsorbents for wastewater treatment, as they provide high specificity and efficiency for organic contaminant removal without requiring chemical modification [34]. Chitosan is used as an adsorbent material for MB, in which hydrophobic interaction takes place between $-CH_3$ group of the MB dye with hydrophobic $-CH_3$ region of the chitosan [3], shown in Figure 5.7.

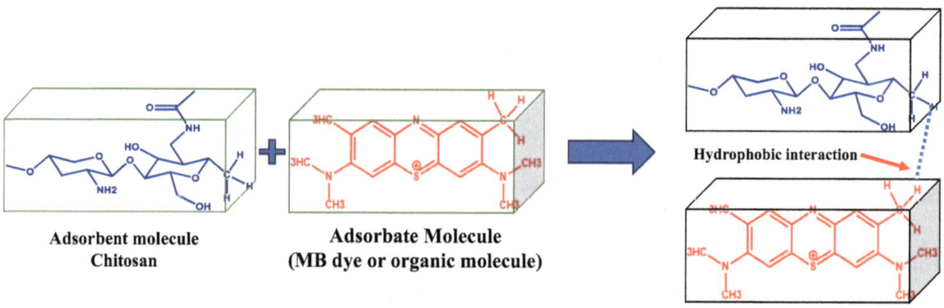

Figure 5.7: Schematic diagram representing the hydrophobic interactions between the adsorbent molecule chitosan and the adsorbate molecule MB dye or organic molecule.

5.3.2 Advantages of green adsorbents

Green adsorbents offer numerous advantages in water remediation. They are sustainable, since they are derived from renewable resources, often utilizing agricultural or industrial waste, which minimizes waste and conserves natural resources. Economically, green adsorbents are cost-effective, with lower production and operational expenses compared to synthetic alternatives. Their high efficiency ensures excellent removal rates for heavy metals and dyes, even at trace concentrations, making them highly effective in water treatment. Additionally, their environmental compatibility, characterized by biodegradability and nontoxicity, ensures minimal secondary pollution, supporting eco-friendly remediation practices.

5.3.3 Green materials used as adsorbents

Water contamination by dyes and hazardous organic compounds (HOCs) represents a critical environmental challenge. These pollutants, often originating from industrial activities such as textile dyeing, pharmaceutical manufacturing, and chemical manufacturing, are persistent, toxic, and resistant to degradation. Traditional remediation techniques, while effective, often involve significant environmental and economic costs.

5.4 Integrated technologies for green material synthesis

Integrated technologies leverage interdisciplinary methods to synthesize green materials efficiently, merging advances in nanotechnology, biotechnology, material science, and computational tools. These methods prioritize sustainability, minimize environmental footprints, and improve material properties for diverse applications.

5.4.1 Green chemistry approaches

Green material synthesis emphasizes eco-friendly practices, utilizing green chemistry principles to minimize environmental impacts. Two key approaches are described further.

5.4.1.1 Solvent-free or aqueous synthesis

Solvent-free synthesis represents a significant advancement in green chemistry by eliminating the need for hazardous organic solvents, thereby reducing toxic emissions, waste generation, and environmental risks. One notable example is mechanochemical synthesis, which employs grinding or milling techniques to drive chemical reactions without using a liquid medium, making it a sustainable alternative for material production [21, 58, 69]. This method has been particularly effective in synthesizing catalysts, composites, and metal-organic frameworks.

Aqueous synthesis, in contrast, utilizes water as a safe, nontoxic, and abundant solvent. This approach is highly versatile, supporting the synthesis of nanoparticles, polymers, and other functional materials under mild conditions. Aqueous-based reactions often minimize energy requirements and avoid the introduction of harmful byproducts, aligning with the principles of atom economy and environmental safety [43,

44, 49]. For example, aqueous synthesis has been instrumental in developing nanomaterials for applications in water purification and biomedical fields.

Both solvent-free and aqueous methods demonstrate the potential to revolutionize material synthesis by addressing sustainability challenges, reducing ecological footprints, and advancing cleaner industrial practices. These approaches also emphasize the importance of designing scalable and cost-effective solutions in alignment with green chemistry principles.

5.4.1.2 Use of renewable raw materials

The shift toward renewable raw materials in material synthesis significantly advances sustainability goals by replacing fossil-based precursors with plant-derived oils, starches, or lignocellulosic biomass. This approach not only conserves finite resources but also promotes carbon neutrality by utilizing carbon-sequestering feedstocks. For example, lignocellulosic biomass derived from agricultural residues serves as a versatile raw material for bio-based chemicals and fuels, reducing reliance on petroleum-based products.

Bio-based polymers like PLA and polyhydroxyalkanoates (PHAs) exemplify this transition. These polymers are synthesized from renewable sources such as corn starch, sugarcane, and agricultural waste, offering biodegradable alternatives to conventional plastics while significantly lowering their carbon footprints [41, 68, 71]. The production of PLA, for instance, involves fermenting starches into lactic acid, which is then polymerized into PLA, a material with diverse applications in packaging and textiles.

Moreover, using renewable feedstocks fosters economic and environmental resilience by valorizing agricultural by-products and reducing greenhouse gas emissions. The integration of these resources into green material synthesis ensures a sustainable supply chain and aligns with circular economy principles. Overall, renewable raw materials offer an innovative pathway toward achieving eco-friendly material production while mitigating environmental challenges.

5.4.1.3 Nanotechnology integration

Nanotechnology enables the creation of eco-friendly nanostructures with high surface areas, tunable properties, and enhanced reactivity. These nanostructures are synthesized using green methods such as aqueous-phase reactions, biotemplates, and plant extracts, which avoid toxic chemicals and reduce energy use. For example, green-synthesized silver nanoparticles have been used for antimicrobial applications [51]. Similarly, bio-based nanocellulose derived from agricultural residues offers a sustainable platform for various applications [38]. Nanostructures like titanium dioxide

(TiO_2) and zinc oxide (ZnO) serve as effective photocatalysts, degrading organic pollutants in water via solar energy-driven reactions [28] Similarly, nanomaterials such as graphene oxide and carbon nanotubes are used as adsorbents for removing heavy metals and dyes due to their high adsorption capacities and selectivity [82]. The integration of nanotechnology with green materials enhances their performance and aligns with sustainability goals [62].

5.4.1.4 Bioengineering innovations

Bioengineering leverages microorganisms and enzymes to develop eco-friendly materials. Microbial synthesis involves using bacteria, fungi, or algae to produce biopolymers like PHAs and PLA from renewable feedstocks such as agricultural waste [14, 65, 56, 91]. Enzymatic synthesis, on the other hand, utilizes biocatalysts to polymerize or modify materials under mild, energy-efficient conditions, reducing the reliance on toxic chemicals [72]. These routes are instrumental in creating biodegradable materials for applications like packaging, water treatment, and environmental remediation, promoting sustainability through innovative biological processes [45].

5.5 Applications in environmental remediation

Green materials, synthesized through integrated technologies, offer innovative solutions for addressing pollution across various ecosystems. Their applications leverage biodegradability, sustainability, and efficiency in mitigating environmental contaminants.

5.5.1 Water pollution and its effect

Water pollution is the contamination of water bodies, including rivers, lakes, oceans, and groundwater, by harmful substances. These pollutants compromise water quality, making it unsuitable for drinking, agriculture, and recreation, while causing severe environmental and health consequences. Different types of water pollution are discussed below.

Chemical pollution: This is caused by toxic chemicals from industrial discharges, agricultural runoff (pesticides and fertilizers), and household waste (detergents and oils). Mercury contamination from industrial effluents in rivers harms aquatic ecosystems and bioaccumulates in the food chain [89].

Biological pollution: This is introduced by harmful microorganisms, such as bacteria, viruses, and parasites, through untreated sewage or agricultural runoff. Diseases like cholera and dysentery are caused from sewage-contaminated water [78]
Nutrient pollution: Excess nitrogen and phosphorus from fertilizers and waste leads to eutrophication, causing algae blooms and oxygen depletion. Dead zones in the Gulf of Mexico are caused due to agricultural runoff [60].
Plastic pollution: Plastics and microplastics accumulate in aquatic systems, harming marine life and ecosystems. Marine organisms ingest microplastics in the Pacific Ocean gyre [77].
Thermal pollution: The discharge of heated water by industries alters aquatic ecosystems by reducing the dissolved oxygen levels. Thermal releases from power plants impact riverine biodiversity [39]
Oil pollution: Oil spills from tankers and drilling operations suffocate marine life and damage ecosystems. The 2010 BP oil spill caused extensive ecological and economic damage in the Gulf of Mexico [63].

Water pollution mitigation requires stringent regulations, advanced treatment technologies, and public awareness to ensure ecosystem health and sustainable water use.

5.5.2 Water purification

Green materials improve cost-effectiveness, reduce secondary pollution risks, and align with global sustainability goals. Their multifaceted applications highlight their potential in achieving large-scale environmental restoration. Wastewater treatment: This involves implementing proper treatment systems to clean industrial, agricultural, and domestic wastewater before releasing it into water bodies.

Nanomaterial-based membranes, such as those utilizing graphene oxide and nanocellulose, are highly effective in water purification. These membranes exhibit exceptional mechanical strength, high surface area, and tunable porosity, making them ideal for removing heavy metals, pathogens, and organic pollutants. For instance, graphene oxide membranes efficiently filter nanoscale contaminants, including lead and arsenic, while offering high water permeability [19]. Nanocellulose, derived from plant fibers, not only filters heavy metals but also removes bacterial contaminants sustainably [37].

Biodegradable filters made from plant-based materials, such as coconut coir and natural fibers, are cost-effective and eco-friendly alternatives. These filters are employed in sustainable water treatment systems to adsorb pollutants like dyes, pesticides, and oil residues. Coconut coir, with its high lignin content and porous structure, has demonstrated excellent performance in removing organic compounds and metals [11].

Hybrid photocatalysts combine green polymers and nanoparticles like titanium dioxide (TiO_2) for wastewater treatment. They efficiently degrade persistent organic pollutants, including dyes and pharmaceutical residues, under sunlight or UV light. For example, biochar-supported TiO_2 nanoparticles exhibit enhanced photocatalytic activity due to their synergistic effects, driving pollutant breakdown with minimal energy input [92–93].

5.5.3 Remediation of dye

Synthetic dyes, including azo, reactive, and anthraquinone dyes, are widely used in industries but are harmful to aquatic ecosystems and human health. The complex aromatic structures of these dyes make them stable and difficult to break down. Green adsorbents such as biochar, algae-based materials, and biopolymers have shown great promise for dye removal.

Biochar are produced through the pyrolysis of agricultural or forestry waste. Biochar possesses a high surface area and functional groups that enhance adsorption capacity for dye molecules [1]. Activated biochar, modified with acids or metal oxides, further improves dye removal efficiency. Algae-derived materials are effective due to their polysaccharide-rich cell walls, which provide active binding sites for dye molecules [81]. Algae biomass is abundant, cost-effective, and can be regenerated for multiple adsorption cycles. Chitosan and other biopolymers are widely used for removing anionic dyes due to their amino groups that interact with dye molecules. Other biodegradable polymers, like alginate and starch derivatives, offer additional environmentally friendly options [46, 53, 74].

5.5.4 Remediation of hazardous organic compounds

In the case of HOCs such as pesticides, phenols, and pharmaceuticals, they often accumulate in water bodies, posing risks to ecosystems and human health. These contaminants are characterized by their persistence and bioaccumulation potential. Green materials like activated carbons, cellulose-based adsorbents, and clay-organic composites are widely studied for HOC removal.

Activated carbons from biomass are agricultural residues such as coconut shells, sawdust, and sugarcane bagasse, which are converted into activated carbon with high microporosity, enabling efficient adsorption of organic pollutants [35]. Surface modifications further enhance selectivity for specific contaminants.

Functionalized cellulose, derived from plant fibers, has hydroxyl groups that allow the adsorption of phenolic and other hydrophobic compounds. The incorporation of nanoparticles enhances its capacity for removing a wide range of HOCs [40].

Montmorillonite and other natural clays, when combined with bio-based polymers, form hybrid materials that efficiently adsorb organic contaminants from aqueous solutions [9].

5.5.5 Remediation of heavy metals

Heavy metal contamination in water is a severe environmental and public health issue, stemming from industrial effluents, mining, and agricultural activities. Metals such as lead (Pb), cadmium (Cd), chromium (Cr), and mercury (Hg) are toxic, even at trace levels, and their nonbiodegradable nature makes them persistent in the environment. To address this, green materials have emerged as effective, sustainable, and eco-friendly adsorbents for heavy metal removal from aqueous solutions. Pyrolysis of agricultural waste to produce biochar, offers a porous structure and abundant functional groups like hydroxyl and carboxyl, enabling efficient heavy metal adsorption. Activated biochar further enhances adsorption through chemical activation [1]. Biochar's affordability and scalability make it a widely studied adsorbent for metals like Pb, Cd, and Cu. Chitosan, a biopolymer derived from crustacean shells, is known for its amino and hydroxyl groups, which form complexes with metal ions. Chitosan's adsorption efficiency for metals such as Cr(VI) and Ni(II) increases when chemically modified or combined with nanomaterials [54, 62].

Algae biomass contains polysaccharides and proteins with active binding sites that sequester heavy metals from water. Algal biosorbents like *Spirulina* and *Sargassum* are abundant, low-cost, and effective for metals such as Zn and Hg [81–82]. Nanocellulose is functionalized with chemical groups like thiol and amine for enhanced metal adsorption. Its high surface area and tunable surface chemistry allow efficient removal of Cu(II), Cd(II), and Pb(II) ions [40]. Bentonite and montmorillonite are combined with bio-based polymers to enhance adsorption performance. These hybrid materials effectively capture metal ions through ion exchange and surface complexation mechanisms [9].

5.5.6 Soil pollution

Soil pollution refers to the contamination of soil by harmful substances such as heavy metals, pesticides, industrial chemicals, and waste. It depletes soil fertility, disrupts ecosystems, and poses risks to human and animal health through food chain contamination. Major causes include agricultural runoff, industrial waste disposal, and urbanization.

Remediation strategies involve bioremediation, phytoremediation, and the use of biochar to detoxify pollutants and restore soil quality [12, 91]. Sustainable practices are essential to minimize further degradation and ensure soil health for agriculture and biodiversity conservation.

5.5.6.1 Soil remediation using green materials

Soil remediation is essential for restoring soil functionality, enhancing biodiversity, and mitigating pollutant migration into water sources. Green materials are gaining prominence in soil remediation due to their sustainability, cost-effectiveness, and ability to minimize secondary pollution [79]. Green materials, including biochar, clay minerals, and plant-based adsorbents, are effective in immobilizing contaminants like heavy metals, pesticides, and hydrocarbons. For instance, biochar a carbon-rich material derived from agricultural waste adsorbs heavy metals such as lead, cadmium, and zinc, reducing their mobility and bioavailability [1]. Its porous structure and high surface area enhance soil retention capacities and nutrient availability, promoting soil fertility.

5.5.7 Phytoremediation and microbial applications

Green materials complement biological methods such as phytoremediation, which uses plants to extract or stabilize soil contaminants. Plants like *Brassica juncea* and *Helianthus annuus* can accumulate heavy metals in their biomass, reducing soil toxicity [84]. The integration of microbial agents with biochar improves the breakdown of organic pollutants, enhancing soil recovery. For instance, bacteria such as *Pseudomonas putida* and fungi such as *Trichoderma* degrade hydrocarbons, aiding in bioremediation processes [32].

5.5.8 Air pollution

Air pollution occurs when harmful substances, such as particulate matter (PM), nitrogen oxides (NO_x), sulfur dioxide (SO_2), carbon monoxide (CO), and volatile organic compounds (VOCs), are released into the atmosphere, primarily from industrial emissions, vehicle exhaust, and fossil fuel combustion. These pollutants contribute to respiratory illnesses, smog formation, and climate change [80]. Green technologies offer sustainable solutions, including phytoremediation, where plants absorb pollutants [32], algae-based biofilters for CO_2 capture, and urban green spaces to improve air quality [7, 15, 76].

5.5.8.1 Air quality improvement using green coatings and paints

Green coatings and paints have emerged as effective solutions for improving air quality by adsorbing VOCs and PM. These coatings are typically made from natural, non-toxic materials, such as plant-based resins, and are designed to bind with harmful chemicals present in the air, thereby reducing indoor and outdoor pollution. Studies have shown that these coatings can capture VOCs such as formaldehyde, toluene, and xylene, which are common in indoor environments, and reduce their harmful effects on human health [8]. Additionally, green paints are often free from toxic solvents, contributing to a healthier living environment.

5.5.8.2 Air quality improvement using bio-based filters

Bio-based filters, made from natural fibers like hemp, coconut coir, or cotton, are becoming popular for use in air purifiers to capture fine dust, allergens, and PM. These filters are not only environmentally friendly but also highly effective in improving indoor air quality. The porous structure of these fibers helps in trapping airborne pollutants, including dust, pollen, and other fine PM, which can have adverse effects on respiratory health [83]. Moreover, they are biodegradable and can be easily disposed of without posing an environmental risk, making them a sustainable alternative to synthetic air filters.

5.5.9 Waste and its management

Recyclable composites are materials designed to be reused, repurposed, and recycled after their initial use, reducing waste and conserving resources. These composites, often used in packaging and construction, are made from materials like plant fibers, biopolymers, and other renewable resources that can be easily disassembled or repurposed. Their recyclability not only minimizes landfill waste but also supports a circular economy.

For example, biodegradable composites in packaging reduce reliance on single-use plastics, providing both environmental and economic benefits by extending the life cycle of the materials used [20].

Upcycling is the process of converting waste materials into higher-value products, contributing to waste reduction and resource conservation. One prominent application is the transformation of post-consumer plastics into biocomposites or energy-efficient materials. This process helps to reduce the environmental impact of plastic waste while creating useful materials for construction, automotive, and other industries. Through innovative techniques such as chemical modification and blending with natural fibers, waste plastics are upcycled into materials with enhanced properties, making them suitable for high-performance applications [10, 24, 33].

5.6 Future prospects

The integration of green materials and advanced technologies in environmental remediation holds immense promise for driving sustainability. Future developments will likely focus on enhancing efficiency, scalability, and environmental compatibility, while addressing current challenges. We have considered some important points as follows.

5.6.1 Advanced nanomaterials

Continued innovation in eco-friendly nanotechnology will lead to materials with enhanced properties for pollution control, such as higher adsorption capacities and targeted pollutant removal.

5.6.2 Biotechnological innovations

Advances in synthetic biology and bioengineering will enable the production of customized microorganisms and enzymes for more efficient bioremediation processes.

5.6.3 Hybrid systems

The combination of biological, chemical, and physical remediation methods, supported by green materials, will create integrated solutions for complex contamination scenarios.

5.6.4 Carbon capture and storage

Materials such as biochar and algae-based composites will become critical for scalable carbon capture and storage applications, addressing climate change.

5.6.5 Zero-waste approaches

Green materials will play a central role in the circular economy, transforming waste into resources and minimizing ecological footprints.

5.7 Conclusion

In conclusion, green materials and integrated technologies hold the potential to redefine environmental remediation, making them more efficient, scalable, and sustainable. Embracing these innovations is crucial for creating a healthier and more resilient planet for future generations. With continued research, investment, and collaboration, the vision of sustainable environmental management can become a global reality.

The integration of green materials and advanced technologies represents a paradigm shift in environmental remediation, combining innovation with sustainability. These materials, derived from renewable resources and enhanced by interdisciplinary technologies such as nanotechnology and biotechnology, offer solutions that are not only effective but also environmentally benign. Their applications span soil, water, and air remediation, addressing diverse challenges such as heavy metal contamination, wastewater treatment, and carbon capture. However, the widespread adoption of these solutions is hindered by economic, scalability, and regulatory challenges. High production costs and limited performance data often make stakeholders hesitant, while regulatory frameworks lag behind technological advancements. Despite these barriers, the global demand for sustainable solutions continues to grow, driven by public awareness, environmental policies, and international agreements.

References

[1] Ahmad, M., et al. (2014). Biochar as a sorbent for contaminant management in soil and water. *Environmental Science & Pollution Research*. 21(11), 8091–8107.

[2] Ali, H., et al. (2013). Phytoremediation of heavy metals – Concepts and applications. *Chemosphere*.

[3] Amor, I. B., Hemmami, H., Laouini, S. E., Zeghoud, S., Benzina, M., Achour, S., Naseef, A., Alsalme, A., & Barhoum, A. (2023). Use of insect-derived chitosan for the removal of methylene blue dye from wastewater: Process optimization using a central composite design. *Materials*, 16, 5049. https://doi.org/10.3390/ma16145049

[4] Anastas, P. T., & Warner, J. C. (1998). *Green Chemistry: Theory and Practice*. Oxford University Press.

[5] Andreozzi, R., et al. (1999). Advanced oxidation processes (AOP) for water purification and soil remediation. *Catalysis Today*. 53(1), 51–59.

[6] Babel, S., & Kurniawan, T. A. (2003). Low-cost adsorbents for heavy metals uptake from contaminated water. *Journal of Hazardous Materials*, 97(1–3), 2019–2243. https://doi.org/10.1016/S0304-3894(02)00263-7

[7] Bai, H., Zhu, H., & Li, W. (2021). Algae-Based Biofilters for CO2 and Pollutant Capture. *Journal of Cleaner Production*, 8, 13253–13280. https://doi.org/10.1016/j.egyr.2022.09.143
[8] Bashir, S., et al. (2018). Role of plants and biochar in heavy metal remediation. *Ecotoxicology and Environmental Safety*.
[9] Bergaya, F., et al. (2006). *Handbook of Clay Science*. Elsevier.
[10] Bhat, P., et al. (2020). Circular economy and upcycling plastics. *Environmental Progress & Sustainable Energy*, 39(1), e13350. https://doi.org/10.1002/ep.13350
[11] Bhatnagar, A., et al. (2010). Coconut coir as a low-cost adsorbent. *Advances in Colloid and Interface Science*.
[12] Bolan, N., Hoang, S. A., Beiyuan, J., et al. (2021). Environmental applications of biochar: Current status and perspectives. *Journal of Hazardous Materials*, 414, 125439. https://doi.org/10.1016/j.jhazmat.2021.125439
[13] Bolan, N., et al. (2020). Biochar: Its role in soil remediation. *Science of the Total Environment*, 717, 137293.
[14] Brandt, E., & Gruber, P. R. (2015). Biopolymers from microbes: An overview. *Biotechnology Advances*, 33(8), 1284–1297. https://doi.org/10.1016/j.biotechadv.2015.02.008
[15] Chaurasia, R., Pandey, R., & Tiwari, M. (2020). Role of phytoremediation in air pollution control. *Environmental Science and Pollution Research*, 27(15), 17312–17324. https://doi.org/10.1007/s11356-020-08281-2
[16] Chen, G. Q., & Patel, M. K. (2012). Plastics derived from biological sources: Present and future. *Chemical Reviews*, 112(4), 2082–2099. https://doi.org/10.1021/cr200162d
[17] Chen, X., Chen, G., Chen, L., Chen, Y., Lehmann, J., McBride, M. B., & Hay, A. G. (2015). Adsorption studies of dyes and heavy metals on biochar. *Bioresource Technology*, 179, 57–62. https://doi.org/10.1016/j.biortech.2014.12.074
[18] Chinomso Iroegbu, A. O., & Ray, S. S. (2021). Bamboos: From Bioresource to Sustainable Materials and Chemicals. *Sustainability*, 13(21), 12200. https://doi.org/10.3390/su132112200)
[19] Choi, W., Choi, J., Bang, J., & Lee, J. (2013). Graphene-based membranes for water purification. *Science*, 342(6159), 1337–1340. https://doi.org/10.1126/science.1236098
[20] Chowdhury, I. H., Mohamed, A. A., Manjusri, M., & Amar, K. M. (2021). Sustainable biocomposites from recycled bale wrap plastic and agave fiber: Processing and property evaluation. *ACS Omega*, 6, 2856–2864. https://doi.org/10.1021/acsomega.0c05186
[21] Clark, J. H., Luque, R., & Matharu, A. S. (2016). Mechanochemistry and solvent-free synthesis: Green chemistry in action. *Green Chemistry*, 18(2), 391–406. https://doi.org/10.1039/C5GC02445A
[22] Clark, J. H., & Luque, R. (2014). Sustainable material synthesis. *Nature Chemistry*, 6, 133–139.
[23] Das, N., & Chandran, P. (2011). Microbial degradation of petroleum hydrocarbon contaminants. *Biotechnology Research International*, 2011, 1–13.
[24] Davis, M., Hillmyer, M. A., & Lodge, T. P. (2018). Upcycling plastic waste into bio-composites. *Waste Management*, 76, 147–160. https://doi.org/10.1016/j.wasman.2018.03.035
[25] European Environment Agency (2022). The future of sustainable innovation. Available at EEA.
[26] Foo, K. Y., & Hameed, B. H. (2010). Insights into the modeling of adsorption isotherm systems. *Chemical Engineering Journal*, 156(1), 2–10. https://doi.org/10.1016/j.cej.2009.09.013
[27] Fu, F., Wang, Q., & Tang, B. (2014). Application of advanced adsorption technologies in water treatment. *Journal of Environmental Management*, 145, 1–10. https://doi.org/10.1016/j.jenvman.2014.06.016
[28] Fujishima, A., Rao, T. N., Tryk, D. A., (2000). TiO_2 photocatalysis. *Journal of Photochemistry and Photobiology C: Photochemistry Reviews*, 1(1), 1–21. https://doi.org/10.1016/S1389-5567(00)00002-2
[29] Gautam, R., Jindal, R., Mohan, C., & Mishra, A. K. (2024). Sustainable innovations as a solution to energy crisis and industrial pollution. *Book: Promoting Multi-Sector Sustainability with Policy and Innovation*, IGIGlobal USA. https://doi.org/10.4018/979-8-3693-2113-3.ch007

[30] Genty, T., et al. (2012). Neutralization of acidic mine drainage using lime. *Journal of Elimelech*, M., & Phillip, W. A. (2011). The future of membrane desalination: Energy, technology, and the environment. *Science*, 333(6043), 712–717. Environmental Management 101, 143–150.

[31] Ghisellini, P., et al. (2016). Circular economy implementation: Key challenges and opportunities. *Journal of Cleaner Production*, 114, 11–32.

[32] Gupta, P., Singh, R., & Kumar, A. (2018). Air pollution: Causes, effects, and control. *Journal of Environmental Health Research*, 18(2), 123–135.

[33] Gupta, N., Sharma, R., & Kumar, V. (2019). Green composites for sustainable construction. *Materials Science and Engineering*, 561, 012123. https://doi.org/10.1088/1757-899X/561/1/012123

[34] Hameed, B. H., Ahmad, A. L., & Latiff, K. N. A. (2007). Adsorption of basic dye (methylene blue) onto activated carbon prepared from rattan sawdust. *Dyes and Pigments*, 75(1), 143–149. https://doi.org/10.1016/j.dyepig.2006.05.039

[35] Hameed, B. H., et al. (2007). Agricultural waste-derived activated carbon for dye removal. *Journal of Hazardous Materials*. 154(1–3), 37–46.

[36] Helfferich, F. (1995). *Ion Exchange*. Dover Publications.

[37] Isogai, A. (2020). Nanocellulose: Sustainable water purification. *Journal of Materials Chemistry A*, 8(3), 1066–1073. https://doi.org/10.1039/C9TA12520A

[38] Joseph, J. B., & Gene, R. P. (2010). Technology development for the production of biobased products from biorefinery carbohydrates—the US Department of Energy's "Top 10" revisited. *Critical Review, The Royal Society of Chemistry, Green Chemistry*, 12, 539–554. doi: 10.1039/b922014c

[39] Kaushal, S. S., et al. (2010). Effects of thermal pollution on river systems. *River Research and Applications*. 26(8), 932–947.

[40] Klemm, D., et al. (2011). Nanocelluloses: A new family of nature-based materials. *Angewandte Chemie International Edition*. 50(24), 5438–5466.

[41] Koller, M., & Braunegg, G. (2015). Advances in PHA production using renewable feedstocks. *Bioengineering*, 2(2), 94–108. https://doi.org/10.3390/bioengineering2020094

[42] Kumar, A., & Sharma, S. (2021). Scaling challenges in green material technologies. *Sustainable Materials and Technologies*, 28, e00294.

[43] Kumar, M., Bolan, S., Padhye, L. P., Konarova, M., Foong, S. Y., Lam, S. S., Wagland, S., Cao, R., Li, Y., Batalha, N., & Ahmed, M. (2023). Retrieving back plastic wastes for conversion to value added petrochemicals: Opportunities, challenges and outlooks. *Applied Energy*, 345, 121307. https://doi.org/10.1016/j.apenergy.2023.121307

[44] Kumar, N., Kumar, V., & Chowdhary, Y. (2022). A review on synthesis methods of tricyclic 1, 2, 3, 4-tetrahydrocarbazoles. *World Journal of Advanced Research and Reviews*, 13(1), 160–171.

[45] Kumar, A., & Sharma, S. (2021). Scaling challenges in green material technologies. *Sustainable Materials and Technologies*, 28, e00294.

[46] Kumar, S., Kalra, G., Bhardwaj, H. K., Rajoria, Y. K., Kumar, D., & Boadh, R. (2022) Internet of medical thing and FIS evaluation for selecting and delivering the best health insurance coverage. *Journal of Pharmaceutical Negative Results*, 3438–3446. https://doi.org/10.47750/pnr.2022.13.S08.422

[47] Leung, D. Y. C., et al. (2010). Biodegradable materials for pollution control. *Progress in Polymer Science*. 35(10), 1232–1246.

[48] Mohan, C., Kumar, V., & Kumari, S. (2018). Synthesis, characterization, and antibacterial activity of schiff bases derived from thiosemicarbazide, 2-acetylthiophene and thiophene-2-aldehyde. *International Research Journal of Pharmacy*, 9, 153–158.

[49] Mohan, D., & Pittman, C. U. (2016). Green materials in sustainable chemistry. *Environmental Science and Technology*, 50(7), 4287–4301. https://doi.org/10.1021/acs.est.5b05657

[50] Mohanty, A. K., Misra, M., & Drzal, L. T. (2002). Sustainable bio-composites from renewable resources: Opportunities and challenges in the green materials world. *Journal of Polymers and the Environment*, 10(1), 19–26.

[51] Mie, R., Samsudin, M. W., Din, L. B., Ahmad, A., Ibrahim, N., & Adnan, S. N. (2014). Green synthesis of silver nanoparticles and their antimicrobial properties. *Journal of Nanoscience and Nanotechnology*, 14(3), 2627–2637. https://doi.org/10.1166/jnn.2014.9141
[52] Mulligan, C. N., et al. (2001). Bioremediation: Environmental clean-up using biological systems. *Environmental Microbiology*. 3(6), 321–329.
[53] Ngah, W. S. W., & Fatinathan, S. (2010). Adsorption of dyes using chitosan-based adsorbents. *Chemical Engineering Journal*, 158(2), 337–346.
[54] Ngah, W. S. W., et al. (2011). Chitosan adsorption for water treatment. *Carbohydrate Polymers*. 83(4), 1446–1456.
[55] Pandey, A., et al. (2019). Integrating green materials in environmental remediation. *Environmental Progress & Sustainable Energy*. 38(3), 1–12.
[56] Patel, S. K. S., et al. (2018). Microbial factories for biodegradable materials. *Trends in Biotechnology*, 36(12), 1237–1250. https://doi.org/10.1016/j.tibtech.2018.08.010
[57] Pilon-Smits, E. (2005). Phytoremediation. *Annual Review of Plant Biology*, 56, 15–39. https://doi.org/10.1146/annurev.arplant.56.032604.144214
[58] Poliakoff, M., & Licence, P. (2007). Sustainable synthesis through green chemistry. *Nature*, 450(7171), 810–812. https://doi.org/10.1038/450810a
[59] Qiu, H., et al. (2009). A review of ion exchange processes for water treatment. *Colloids and Surfaces A: Physicochemical and Engineering Aspects*, 336(1–3), 2–14. doi: 10.1016/j.colsurfa.2008.10.003
[60] Rabalais, N. N., et al. (2010). Hypoxia in the Gulf of Mexico: Natural and human influences. *Bioscience*. 60(2), 123–135.
[61] Rajput, V., et al. (2021). Artificial intelligence in green material synthesis. *Computational Materials Science*, 191, 110280.
[62] Rajput, N., Rajoria, Y. K., Aarya, D. D., Singh, A. P., & Chauhan, A. (2025, April). Impact on FMCG production in the period of pandemic with the q-fuzzy demand rate and channel financing policy. *AIP Conference Proceedings*, 3283(1), 040044. AIP Publishing LLC.
[63] Ramseur, J. L. (2010). The Deepwater Horizon oil spill: Background and legislative proposals. *Congressional Research Service Report*, R41262.
[64] Reinhard, M., et al. (2017). Innovative chemical techniques for groundwater remediation. *Hydrogeology Journal*. 25(4), 1007–1021.
[65] Richard, A. G., & Bhanu, K. (2002). Biodegradable polymers for the environment. *Pub Med*, 297(5582), 803–807. doi: 10.1126/science.297.5582.803
[66] Royal Society of Chemistry. (2020). Advances in green material chemistry. Available at RSC.
[67] Ruthven, D. M. (1984). *Principles of Adsorption and Adsorption Processes*. Wiley-Interscience.
[68] Satlewal, A., Agrawal, R., Bhagia, S., Das, P., & Ragauskas, A. J. (2018). Agricultural waste for renewable materials. *Renewable & Sustainable Energy Reviews*, 93, 583–599. doi: 10.1016/j.rser.2018.05.036
[69] Sharma, S., & Tyagi, V. (2016). Advances in renewable feedstocks for sustainable material synthesis. *ACS Sustainable Chemistry & Engineering*, 4(9), 4663–4680. https://doi.org/10.1021/acssuschemeng.6b01010
[70] Sharma, S., et al. (2020). Biodegradable materials in environmental applications. *Materials Today Sustainability*, 7, 100032.
[71] Sheldon, R. A. (2017). Biocatalysis in green and sustainable chemistry. *Catalysis Today*, 280(Part 1), 2–13. https://doi.org/10.1016/j.cattod.2016.09.015
[72] Sheldon, R. A. (2016). Biocatalysis and green chemistry. *Catalysis Today*, 259(Part B), 2–13. https://doi.org/10.1016/j.cattod.2015.06.009
[73] Singh, R. P., et al. (2020). Applications of nanomaterials in remediation technologies. *Journal of Hazardous Materials*, 397, 122796.

[74] Singh, A. P., Chauhan, A., Aarya, D. D., & Rajoria, Y. K. (2025). Fuzzy economic order quantity models with carbon emission considerations and planned discounts. In: *Recent Trends in Engineering and Science for Resource Optimization and Sustainable Development*, CRC Press, pp. 255–258.

[75] Tang, X., et al. (2018). Recent developments in green remediation technologies. *Environmental Technology & Innovation*, 10, 149–159.

[76] Tanner, C. J., Zwick, S., & Ross, L. (2019). Urban Green spaces and air quality improvement: A comprehensive review. *Urban Ecosystem Research*. doi: 10.1007/s13280-019-01164-3

[77] Thompson, R. C., et al. (2004). Lost at sea: Where is all the plastic?. *Science*. 304(5672), 838.

[78] UNEP (2021). State of water pollution and global remediation strategies. United Nations Environment Programme Report.

[79] United Nations Environment Programme (2021). Sustainability in action. Available at UNEP.

[80] USEPA. (2013). *Environmental Remediation: Excavation Methods*.

[81] Wang, J., & Chen, C. (2009). Biosorbents for heavy metal removal. *Biotechnology Advances*, 27(2), 195–226.

[82] Wang, S., et al. (2019). Graphene-based materials in adsorbent applications. *Environmental Science & Technology*, 53(14), 7939–7957. https://doi.org/10.1021/acs.est.9b01022

[83] Wang, Y., et al. (2019). Green synthesis of nanoparticles for soil remediation. *Nanomaterials*, 9(9), 1380. https://doi.org/10.3390/nano9091380

[84] Xu, R., et al. (2021). Green nanotechnology for environmental applications. *Environmental Nanotechnology*, 17, 100536. https://doi.org/10.1016/j.ennano.2021.100536

[85] Yadav, R., Sanduja, M., Kumar, V., Sharma, K., Khan, S., & Kumar, K. (2024). Microwave-assisted synthetic pathways of pyrrole: A comprehensive review. *Asian Journal of Organic Chemistry*, 13(12), e202400401.

[86] Zhang, L., et al. (2016). Innovative technologies for sustainable remediation. *Environmental Science & Technology*. 50(4), 1868–1880.

[87] Zhang, Y., et al. (2019). Biochar's role in soil remediation. *Soil Science Society of America Journal*. 83(4), 1211–1225.

[88] Zhang, X., et al. (2018). Nanotechnology in environmental remediation. *Environmental Science: Nano*. 5(1), 27–43.

[89] Zhang, X., et al. (2019a). Innovative soil washing methods for heavy metal remediation. *Environmental Research*, 172, 403–412.

[90] Zhang, Y., et al. (2018). Biochar's role in soil remediation. *Soil Science Society of America Journal*. 83(4), 1211–1225.

[91] Zhang, Y., et al. (2020). Enzyme-based approaches for green manufacturing. *ACS Sustainable Chemistry & Engineering*, 8(5), 16124–16135. https://doi.org/10.1021/acssuschemeng.0c04148

[92] Zhang, W., et al. (2019). Environmental impact of industrial chemical discharge. *Journal of Environmental Studies*. 23(3), 345–360.

[93] Lisowski, P., Colmenares, J. C., Mašek, O., Lisowski, W., Lisovytskiy, D., Kamińska, A., & Łomot, D. (2017). Dual functionality of TiO2/Biochar hybrid materials: Photocatalytic phenol degradation in the liquid phase and selective oxidation of methanol in the gas phase. *ACS Sustainable Chemistry & Engineering*, 5(7), 6274–6287. doi: 10.1021/acssuschemeng.7b01251

M. Sirija, G. B. Dharma Rao, Anjaneyulu Bendi*, Chandra Mohan*, and Prem Lata Meena

Chapter 6
Plastic waste management: a sustainable practice for green future

Abstract: Plastic is the common word for an extensive range of natural or synthetic solid materials. Plastics are defined as polymers of elevated molecular weight. The chemical and physical properties of plastic waste, combined with its non-degradable nature, contribute to a significant toxic waste problem that affects the atmosphere. As a result, plastics contaminating food supplies and pose serious health risks to both water animals and humans. On the other hand, improper waste management leads to plastic emissions into the environment, which adversely affects ecosystems. Burning of plastic waste generates too much greenhouse gas productions. At the same time, plastic as a substance is totally consumed and further cannot be reused. In Sweden, an effective and sustainable approach to managing plastic waste emphasizes recycling. The European Union (EU) is also taking steps to improve plastic recycling by setting higher recycling targets in its updated waste regulations. It's important to consider various factors that need focus while exploring innovative methods for managing plastic waste. Moreover, it is significant to extend a sympathetic of aspects that require to be emphasized even as allowing for unconventional advances and prospects to cash plastic waste. The book chapter offers readers an inclusive summary of the methodologies and advances to defeat the worldwide problem of plastic toxic waste and the viewpoint on the mistreatment of this squander as a resource.

Keywords: Plastic, sustain ability, waste management, polymer, nondegradable

6.1 Introduction

Plastics are the type of polymers that are used in our daily life for various things. After use, we must process, manage, and recycle in a better way. Otherwise, there will

*Corresponding author: **Anjaneyulu Bendi**, Department of Chemistry, Presidency University, Bengaluru, Karnataka, India, e-mail: anjaneyulu.bendi@gmail.com
*Corresponding author: **Chandra Mohan**, School of Basic and Applied Sciences, K. R. Mangalam University, Gurugram, India
M. Sirija, Department of Chemistry, Vardhaman College of Engineering, Hyderabad, India
G. B. Dharma Rao, Department of Chemistry, Kommuri Pratap Reddy Institute of Technology, Hyderabad, India
Prem Lata Meena, Department of Chemistry, Shyam Lal College, University of Delhi, Delhi, India

https://doi.org/10.1515/9783111618906-006

be more harm than good to the environment. There are thermoplastics and thermosetting plastics, which may include PVC, PS, PP, HDPE, and PETE. The methods we can use to recycle these plastics are Recycling, Incineration, Pyrolysis, Bioremediation, and Landfills. Plastic waste, medical waste, and E-waste should be recycled properly, as they are dangerous to humans, wildlife, and the environment. So, green technology and circular economy are the management strategies to follow to eradicate the buildup of waste [50]. During and after the Covid-19 pandemic, there is a rise in need to manage and minimize the use of plastic waste, including greenhouse gas (GHG) and footprint. Introduction of plastic waste footprint became a necessary need with the 6R initiatives (Figure 6.1) [32].

Figure 6.1: Different methods to minimize waste.

During the plastic production process, there is a need to add certain chemicals like additives that are toxic to the environment and they create harmful health problems. Microplastics are plastic fragments, which are less than 5 mm in length.

6.1.1 Harmful effects of plastics

- There are many more microplastics present in the agricultural land than in marine water. This is a big threat to food, human safety, and environment (*Assessment of Agricultural Plastics and Their Sustainability: A Call for Action*, 2021).

- Chemicals are released while storing food or water in plastic containers and plastic packaging, as also during plastic toys usage by children [10].
- The chemicals in plastics disrupt the endocrine system, lead to cancers, birth defects, and are even harmful to the reproductive system.
- Chemicals, for instance, phthalates, bisphenol-A, and brominated flame retardants are disease-causing agents and lead to hormonal imbalance in human beings. Microplastics that are present in cattle food can't be digested by them, leading to damage to their digestive system [15].
- Industrial waste contains microplastics that contaminate the water. Degradation of plastic waste releases different gases such as CO_2, which decreases the air quality, leading to climatic changes, global warming, acid rain, etc. Even if air quality is good enough, airborne diseases such as COVID-19 and pneumonia will easily spread.

6.1.2 Management strategies

Plastic consists of many numbers of monomer units undergoing polymerization at high temperature and pressure to form long-chain polymer units. We human beings use plastics because of their flexibility, durability, and resistance to moisture. For example, shopping bags, which are made up of polyethylene, consist of 20,000 monomer units of ethylene. Due to the presence of long-chain polymer units, they are hard to destroy. Some other examples are PE, PVC, PP, PS, PA, PET, PTFE, and so on. They take more than 100 years to decompose in the wet soil [63]. Due to the rapid industrial growth, population growth, and man-made activities, both biodegradable and nonbiodegradable wastes are being generated. So, there is a need to develop some management strategies such as recycling, incineration, landfills, pyrolysis, and bioremediation (Figure 6.2) [19, 77].

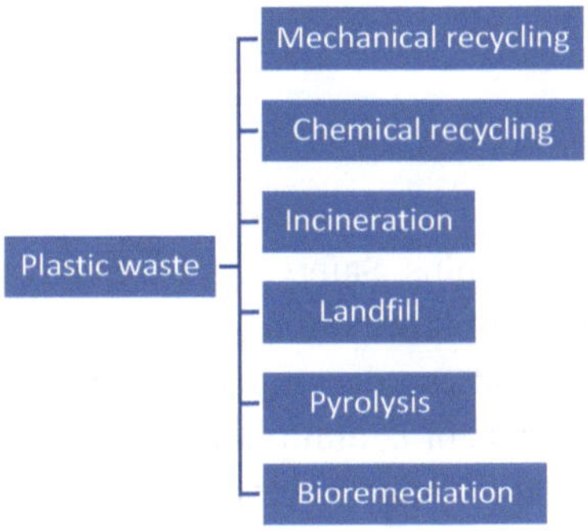

Figure 6.2: Different methods of management strategies of plastics.

6.1.2.1 Recycling

Renewing/recycling means converting the old plastic waste into raw materials, which can be used to make other materials. Like carbon-based products and other polymers, plastics are not degradable naturally. Bottles made of plastic and other materials, when melted, can be used to create other objects, such as chairs and tables. There are six processes involved in this procedure: aggregation of squander plastics, categorizing them, maintenance to acquire them free of pollutants, resizing and slicing, classifying and separating plastics, and molding. It promotes a sustainable lifestyle and conserves the national economy.

6.1.2.2 Incineration

The chemical process of burning waste in oxygen to release CO_2 and water molecules into the atmosphere is called Incineration. Fuel, usually referred to as energy, may be formed by the burning of organic compounds (Figure 6.3). The petroleum flues that are used in automobile and aircraft contamination may be found in an assortment of physical states, such as gas, solid, or liquid. To a certain extent, the energy formed by this ignition benefits the society in a number of ways. Furthermore, it makes a noteworthy contribution to squander reduction and waste-to-electricity formation, both of which are essential in today's contemporary world [15].

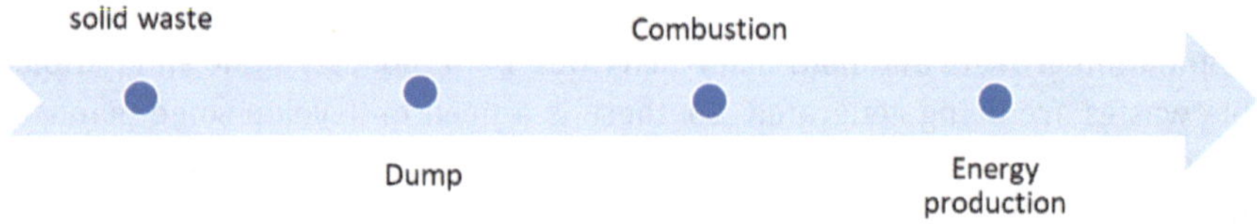

Figure 6.3: Procedure of conversion of solid waste to energy.

6.1.2.3 Landfills

After utilization, plastics can be disposed in different dustbins. Safety measures have to be taken to avoid soil degradation and ground water contamination [83]. So, a safe place is needed such as digging to a deeper depth, one that can decompose the plastic waste, which can take a long time, so as to avoid all kinds of contamination [38]. Because they create CO_2 and CH_4 gas throughout the biodegradation procedure, landfills are a great source of energy. It maintains sanitation, makes cities dirt-free, and separates the hazardous trash from other waste categories.

6.1.2.4 Pyrolysis

The procedure of converting gases and fatty oils to petroleum derivatives to produce renewable source of energy from plastic wastes is called pyrolysis [3] The three primary categories of the pyrolysis procedures are experienced by the heat energy essential to break the plastic linkages. There are media based on temperature: high, medium, and low temperatures. The range of temperatures necessary to destroy the plastic structure decides what constitutes medium and high temperatures. In the squander management region, these chemical treatment guidelines are established to make sure there is appropriate handling and gathering of plastic wastes. This scientific advance, although rather expensive up front, is an effective way to control waste.

6.1.2.5 Bioremediation

Bioremediation is the area of biotechnology that focuses on detoxification and refinement by the use of microorganisms to biodegrade naturally occurring substances that can be experienced by means of biodegradation of bacteria, fungus, algae, and plants (Figure 6.4) [5]. In certain circumstances, nutrients, culture medium, and specific enzymes may have to be added to maintain the microbial growth [6].

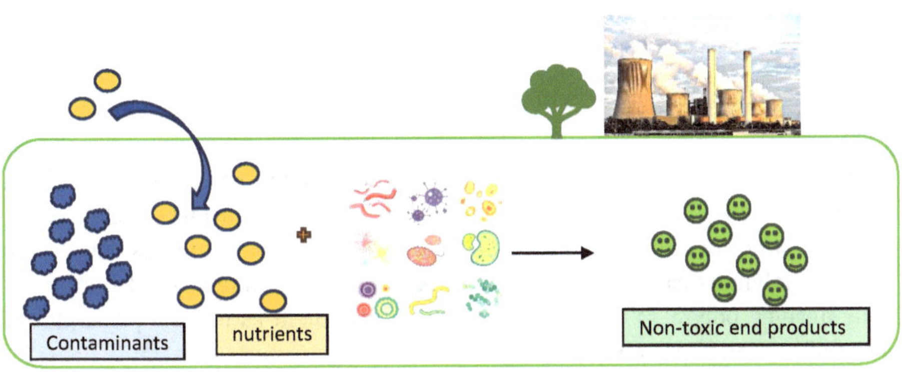

Figure 6.4: Process of conversion of toxic waste to nontoxic products.

6.2 Great alternative to nonbiodegradable plastic

Biodegradable bio-based polymers or green polymers are great alternatives for the green economy. Biomass, microorganisms, and synthetic polymers, consist of animal/plant proteins, cellulose, carbohydrates, lactic acid, and biogenic byproducts [50, 65]. Some examples of green polymers are PLA, TPS, PHAs, and cellulose, which are obtained from natural sources. PLA and TPS are obtained from corn starch, sugarcane, or sweet potato starch; PHAs are from microbial fermentation; and cellulose is from cotton, wood, and so on (Figure 6.5). These can be used for packaging, dustbin bags, and disposable food materials.

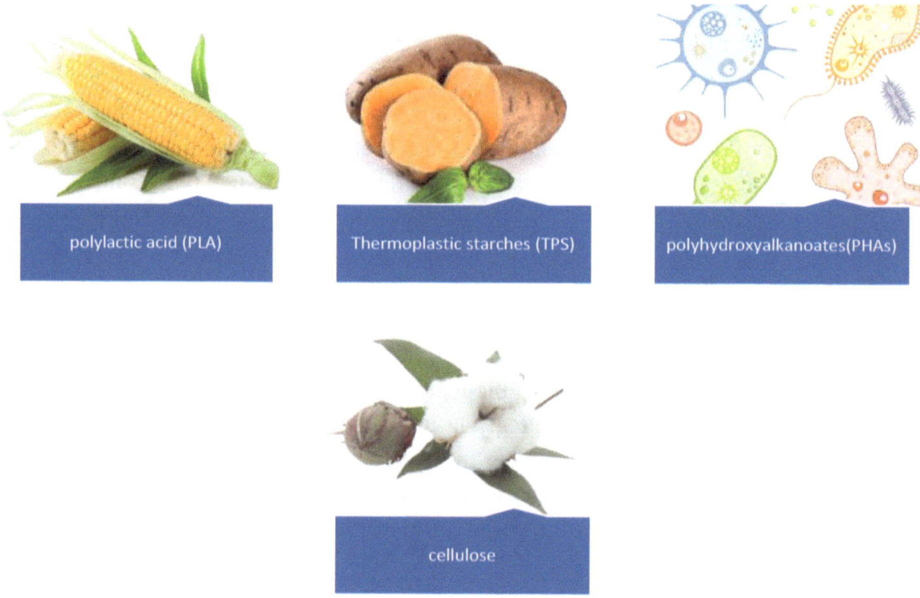

Figure 6.5: Different types of nonbiodegradable plastics.

Natural polymers can also be used instead of synthetic polymers such as plastics for a better environment. These natural polymers such as polysaccharides, nucleic acids, and proteins can be obtained from plant sources as well as animal sources (Figure 6.6) [26]. Natural polymers can be used in paper, textile, and food industries. Examples of naturally derived polymers that help in green environment as per the principles of green chemistry can be found in [23].

Cellulose is abundantly available in plant-based material, and cellulose derivatives like cellulose nanoforms (nanofibrils and nanorods) are used widely to make different materials because of its mechanical and good thermal stability properties. Methods such as solvent casting [21, 78], vacuum filtration [59], and layer-by-layer technology [20] extract cellulose from plants and algae.

Besides cellulose and its derivatives, there are many other natural polymers such as chitosan, gelatin, and many others. Applications of natural polymers are: (i) active food packaging; (ii) ion-exchange resigns for fuel cells; (iii) drug delivery and wound healing carried out by patches; (iv) microneedles for drug delivery and fluid uptake; and (v) sensors, smart coating, and many more [58].

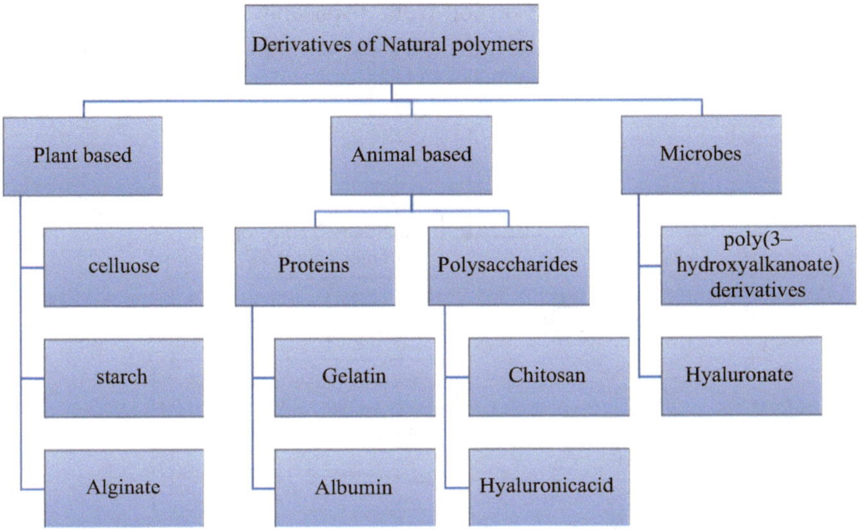

Figure 6.6: Classification of natural polymers.

As plastic trash is high in carbon content, it may be used as a starting material to create products with additional value that are based on carbon. These derived carbon products from plastic squander are called PWCMs (plastic waste-derived carbon materials). An extensive variety of carbon products, including graphene, carbon dots, CNTs (carbon nanotubes), porous carbon and carbon nanosheets, can currently be made from different kinds of plastic squander, as well as PVC, PS, PET, PE, PFR, LDPE, HDPE, PAN, and PP. Carbonization via anoxic pyrolysis, catalysis, hydrothermal carbonization, microwave heating, and flash Joule heating may all be used to do this [12, 66].

Advantages of PWCMS include elevated degrees of porosity, huge specific surface area, high conductivity and good stability [67], and use in sustainable environmental and green energy applications like water-splitting systems, supercapacitors and batteries, solar evaporation, and CO_2 capture [17]. By managing plastic waste, there will be economic and environmental benefits.

6.2.1 Role of PWCMs in batteries

PWCMs can be used as anode, cathode, and separators in the battery systems. Carbon materials without graphite can be used as anode in alkali-ion batteries [68]. Porous -based materials like activated carbon, soot carbon, turbostatic, disordered, and hard carbon can be effectively work as high-energy storage capacities for Na^+ and Li^+ [28].

[11] have prepared non-graphitic carbon from disposable PP masks treated with sulfuric acid, followed by pyrolysis to form hard carbon anode, which is shown as high reversible capacity for Na^+ (Figure 6.7).

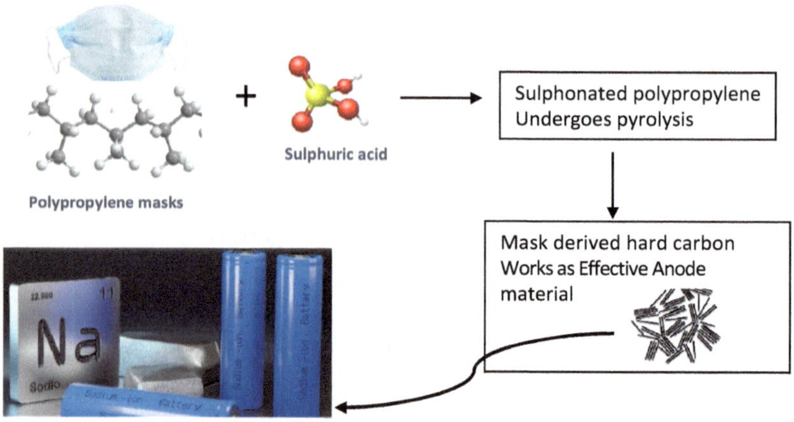

Figure 6.7: Process of converting disposable masks to hard carbon anode.

Kim et al. [30] has prepared Si/C nanofibers/C nanocomposites from HDPE waste, which is effectively, improving the anode performance in Li-ion batteries. CNTs with Fe/N-codoped materials are prepared by using squander plastics by pyrolysis maintained at 850°C. Hossain and Sahajwalla performed more efficiently than the Fe-doped CNTs and a commercial Pt/C catalyst in ORR performance [27]. The negatively charged (cathode) porous sulfonated carbon derived from LDPE inhibited the diffusion of S_n^{-2} (polysulfide) as a result of repulsive interactions throughout the discharge procedure in lithium–sulfur batteries [73].

6.2.2 Role of PWCMs in supercapacitors

Doping by means of heteroatoms, which include N, O, and S can improve the conductivity, wettability, and pseudo capacitance of PWCMs, which in turn improves the electrochemical experience of super capacitors [24].

One-dimensional CNTs can be prepared from plastic waste through catalytic carbonization. Karakoti et al. [29] proposed that CNTs prepared from PS, PE, and PP waste outperformed conventional CNTs used in supercapacitors due to the presence of oxygen-bearing functioning groups and large SSA in synthesized CNTs from PWCMs. Multiwalled CNTs, prepared from PP waste, showed elevated electrochemical ability, indicated by the specific capacitance of 59 F/g in 1 N in KOH [74].

Liu et al. [40] prepared the nanosheets of graphene from PS waste. Sahoo et al. [55] processed the graphene nanosheets to catalytic carbonization and KOH activation treatment. Due to the large SSA and pore volume, these porous carbon nanosheets have shown applicability in supercapacitive energy storing devices.

Min et al. [47] proposed that the sp^2/sp^3-hybridized carbon with co-etching throughout the CO_2 activation procedure promoted HPC (hierarchical porous carbon) production from PET waste. Composites with carbon base produced from plastic squander can also be used as electrodes in supercapacitors [37].

6.2.3 Role of PWCMs in water electrolysis

Green hydrogen production is achieved through water electrolysis. OER (oxygen evolution reaction) and HER (hydrogen evolution reaction) take place by the usage of effective catalysts. PWCMs and their derivatives have huge SSAs and elevated conductivities; appropriate porosities are used as OER and HER catalysts. Mir and Pandey [48] prepared carbon–metal carbide hybrid materials, obtained from plastic squander. They enhance the HER performance. The carbon matrix that contains nitrogen doping showed HER activity in Molybdenum-carbide/carbon composites [13]. MoS_2/CWM composite showed elevated carbon content, high nitrogen content, and high stability (stronger interaction between MoS_2 and CWM) during electrocatalysis [64].

Wei and Kamali [72] discussed an environmentally benign route for the synthesis of an MnO_x/S-doped composite of graphitized carbon. This composite has a mesoporous nature, where MnO_x, evenly circulated on the carbon sheets, showed good OER activity. [35] premeditated N-doped mesoporous carbon, functionalized with ZnO from PET squander, which acted as bifunctional electrocatalyst.

6.2.4 Role of PWCMs in pollutant adsorption

PWCMs such as porous carbon [45], CNTs [18], carbon nanosheets [34, 36], and graphene [35] have properties like electrostatic attraction, pore filling, π–π interactions, and hydrogen bonding [22], which can effectively work as adsorbents. Rai and Singh [52] synthesized porous carbon from waste PP and cyanobacteria, which effectively removed the MB dye. KOH can be used as an activator to produce a more porous nature of the activated carbon. Shen et al. [57] and Xu et al. [76] prepared the Fe-O groups in Fe_3O_4 for the absorption of cephalexin by magnetic PET-based activated carbon so that these magnetic PWCMs can be easily separated and reused. Wu et al. [75] prepared hydrochar from PVC waste, which consists of SO_3H groups that enhance the adsorption capability for Cu^{2+} and Cr^{4+} ions. It was discovered that ion exchange and electrostatic attraction mechanisms could effectively remove NO_3^- ions from solution using a polypyrrole-modified PET-derived composite of carbon [53]. Oxygen- or nitrogen-containing functional groups are observed to have high absorption capacity.

6.2.5 Role of PWCMs in pollutant degradation

PWCMs-based catalysts can be used to make active persulfate [$S_2O_8^{2-}$] and hydrogen peroxide [H_2O_2] to generate active radicals, for instance, OH· and SO_4· radicals [33]. These oxidizing groups have been experienced in the treatment of contaminated water. Using co-pyrolysis, [70] produced an iron/carbon composite. This substance showed the effective breakdown of amaranth dye by interactions with SO_4^{2-} and ˙OH radicals produced by $S_2O_8^{2-}$ activation. By the adsorption–degradation process, Cu/C hybrid prepared from PCBs can also degrade different dyes [51]. Carbon particles adsorb the orange 7 and can be further enhanced by UV irradiation; this shows high degradation efficiency.

The composite of PWCMs-metal oxide material can effectively work as photocatalysts for dye dilapidation in wastewater treatment [34, 36]. By passing the UV-IR light irradiation, electrons and holes are generated in the photocatalysts [39]. The produced free radicals are used to degrade the pollutants into nontoxic compounds [25]. Under visible light, NiO–MWCNT composite prepared from HDPE was found to degrade MB, which has high degradation efficiency.

6.2.6 Role of PWCMs in solar evaporators

Solar energy is used to convert polluted water to fresh water. Zhang et al. [80] developed HPC from PET waste for solar-driven water evaporation using carbonization technique. Song et al. [61] prepared Ni/C nanocomposite from PP and NiO, which enhanced the photothermal conversion process.

6.2.7 Role of PWCMs for CO_2 capture

CO_2 capture by PWCMs is environmentally sustainable and provides carbon neutrality. Algozeeb et al. [2] and Wang et al. [71] utilized potassium acetate as the initiator to fabricate a carbon product with porous character, which showed high CO_2-capture efficiency. Doping by means of heteroatoms, which includes N, S, and O, can improve the chemical absorption of CO_2. Yuan et al. [79] described a N-doped microporous carbon-based materials using squander PET bottles by the one-pot synthesis method. These materials showed better CO_2 absorption than the two-pot synthesis. An AI framework can be designed to segregate the waste automatically into biodegradable and nondegradable plastics by using an intelligent waste bin. The segregated nonbiodegradable plastic can be degraded using microbes. The biodegradable bins can be labeled clearly and transferred to the municipal landfill. This can decrease manpower, save time, and make the environment safe. Rajendran et al. described squander isolation using AI techniques like ML (machine learning) algorithms, arranging the squander products into non-squander and squander items with 80% accuracy

[41]. The recognition and degradation of squander plastic using UAV-supported multi-spectral data and deep learning along with degradation using microbial materials [1]. A drone-supported smart solid squander exposure based on deep learning and image processing tools during characteristic withdrawal that assists in knowing squander-spoiled places and dispatches the locality coordinates as a notification [60].

Reduction of packing materials reduces the production of harmful chemicals from industries and reducing the products that last longer are one of the primary ways to prevent pollution and waste. Buying reusable products, repairing broken things, and reducing new products are things every individual should be aware of. Every country has to contribute recycling funds for the proper recycling of plastic waste. Biomedical plastic waste also has to be monitored with proper controls. Phthalates and their metabolites, metals and their active xenobiotics leach liquid from the solid waste and need constant monitoring. Plastic contamination has become a major concern, especially in the marine ecosystem. Microplastics, small pieces of plastic materials, are increasing day by day in water. Aquatic organisms are consuming the polluted waste, and in turn these aquatic organisms are consumed by human beings, ingesting the microplastics. Therefore, proper disposal practices and guidelines have to be maintained [56].

Plastic waste can be used for construction purposes, which will improve environmental sustainability and be a useful resource as a component in cement composites. By this, we can solve the solid waste management problem as well as utilize as raw material for construction purpose. The employment of PW in construction provides a path for these wastes to be used in a continuing application than short-term ones, resembling as recycling into new-fangled products that become waste in a little period of time. PW's capability to be used as a binder, collective, and fiber enables them to be a feasible alternative for each and every cement composites, with very minor negative impacts on the composite's performance. The reuse of PW for different building applications will provide numerous income streams [4].

Translation of squander plastics into hydrocarbon fuel appears an outstanding alternative for partial substitution of petroleum components. Nevertheless, the economic value is to be measured, and the procedure has not been validated at this moment. The observations of the above illustrated various procedures specify that mechanical recycling is a capable procedure for squander plastic management in the current circumstance of Nepal [16].

Bhutan has taken significant measures to reduce plastic squander, for instance, banning plastic baggage and promoting waste categorization and recycling. Sustainable materials, storm water management, landscaping, energy effectiveness, and energetic mobility are all aspects of "green" path construction. It promotes traffic protection, protects animal surroundings, as well as reduces carbon productions. Illustrative instances of green road program implementation can be found in the Netherlands, Singapore, and India. To accomplish green path improvements in Bhutan, it is required to experience technical, fiscal, and policy concerns though incorporating stakeholders and the common public. Prospects for increasing the sustainability of green path systems require integrating renewable energy, developing self-healing materials, and using smart technology.

Green road development is a long-term approach for minimizing plastic waste and boosting eco-friendly road infrastructure. Bhutan may benefit from using green road development to alleviate road deterioration concerns, while minimizing environmental effect. Bhutan has engaged in drastic initiatives to control plastic squander, including prohibiting plastic baggage and encouraging garbage segregation and recycling. Green road development focuses on sustainable materials, storm water management, vegetation integration, energy effectiveness, and energetic mobility [9].

Most Croatian municipalities face significant challenges in improving current waste management owing to legal responsibilities imposed by various European Union (EU) waste management directives, for instance, reducing squander production and landfilling or increasing individually collected squander and recycling rates. Annual separate rubbish collection generates to some extent less than 5 kg per capita of a variety of squander components, which is much lower than the average number for the EU's [28] capital cities of 108 kg per capita. This can only be achieved by improved and more sustainable development of potential activities and infrastructure that takes into consideration the economic, environmental, and social features of squander management. It suggests that, in order to achieve the goals, the city of Zagreb will need to provide novel infrastructure while also raising public awareness about waste separation at the household point. The solution to the novel squander management problem described in this article is a means of applying the circular economy concept to existing squander management methods in Zagreb [14].

Dentistry is dedicated to preserving oral health and increasing function. However, it has a considerable environmental impact since it produces a large amount of rubbish, especially metallic waste, and consumes an excessive amount of water and electricity. Green dentistry conserves water and energy, generates less waste, and uses cutting-edge technology to decrease pollution. Oral health practitioners should recognize the value of contributing to long-term sustainability. Dental practitioners are becoming more concerned about their practice's environmental impact. Dental practitioners have a moral and ethical obligation to maintain an ecologically sustainable practice for the benefit of patients, dental staff, and the planet. This involves correct handling of solid materials, liquid waste, poisonous vapors, recycling, and lowering energy use. Thus, by focusing on prevention, caution, sustainability, and increasing awareness, a dental practitioner may make a significant contribution to improving the health of patients, the community, and the environment. Suggestions include adding required environmental sustainability education to undergraduate programs. Dental schools should make environmental awareness a priority in their curricula, give knowledge on efficient waste disposal, and provide an online tool for dentists to monitor carbon emissions for sustainability [54].

Sustainable squander management has arisen as a significant problem for the global higher education business. Over the past two decades, the UK higher education business has experienced increasing amounts of squander as an outcome of rapid extension, resulting in soaring costs that are worsened through the yearly raise in Landfill Tax. The industry has failed to address these and other atmospheric concerns for a number of causes, as well as a lack of extensive university collaboration, institutional bureaucracy, and implementation. The care observation described in this article gives a thorough evaluation of a few of the main obstacles, concerns, and achievements experienced throughout the development of the UoS four-phase squander management approach during the past 15 years. There is no question that the University of Sydney has extensively increased and enhanced its waste management operational methods and services, resulting in a further sustainable, cost-effective, and user-friendly system. The entire plan used, which recognizes the PESTLE components and the need of a coordinated ISB approach, serves as a realistic, effective, and useful instance for other institutions looking to control their squander efficiently and sustainably [49].

Plastic use increases in line with worldwide population expansion, resulting in increased plastic garbage output. Plastic waste, which is not biodegradable, causes numerous environmental and human health troubles. Several observations have recently been studied to identify secure and ecologically advantageous procedures of plastic squander disposal. This communication uses a computational approach to undertake a bibliographic analysis of the existing literature on plastic squander management. The bibliographic analysis looked at the most referenced papers and authors, as well as vigorously contributing countries and publishing resources. In addition, numerous plastic squander management techniques and their environmental benefits were discussed. It has been discovered that the two most successful techniques to plastic waste management are road building and tar and concrete manufacture. This is owing to the noteworthy features such as easier localization, fewer greenhouse gas discharges, and increased lifetime and sustainability of man-made items, structures, and roads. The use of landfills, on the other hand, is the least desirable technique, owing to environmental and human being health issues. Recycling has equivalent benefits and disadvantages. Pyrolysis and liquefaction, on the other hand, are favorable since they create char and fuel, but their benefits are restricted due to high energy requirements. Therefore, the employment of plastic squander for construction purposes is encouraged [81].

Plastic production on a global scale has consistently increased during the last five decades. This phenomenon might be attributed to the growing utilization of plastics, which is driven by their increasing range of uses and advantageous characteristics, including lightweight, robustness, longevity, cost-effectiveness, resistance to corrosion, and low production expenses. The manufacture of plastic materials contributes about 3–5% of global greenhouse gas emissions. The substitution of conventional plastics with biodegradable polymers has been proposed and implemented, resulting in diverse environmental consequences, such as the instability of the biodegradable plastics industry, along with recommendations for eco-friendly plastic alternatives that do not contribute to environmental issues. China is the primary recipient of imported waste plastic. Investments in the fabrication of bio-based or biodegradable polymers have led to significant growth in the market for these materials. The biodegradability of these materials is mostly governed by their chemical structure and the specific chemicals used in their production, rather than the carbon content of the polymers. Furthermore, it has been shown that the decomposition of bio-based end-products in various conditions is not consistently foreseeable. "Beyond Biodegradable" aims to develop innovative polymer bio-based materials and additives that may be effectively used in industrial applications via technological developments. Industrial companies will face the task of selecting bio-based plastic materials in a suitable manner. From a waste management perspective, it is desirable for biodegradable plastic items to decompose organically with little release of substances into the environment [8, 31].

The generation of solid waste has been significantly accelerated by the worldwide population growth, flourishing economy, rapid urbanization, and enhanced community living standards. Another recommended solution is the concept of zero squander. The conversation was on examining the challenges and potential of solid waste management, zero-squander practices, and zero-squander strategy, with the aim of transitioning from conventional squander management to a zero-squander vision. Implementing zero-squander techniques is a very effective strategy for mitigating the rising volume of solid squander. The concept of "zero waste" is an exceptionally effective method for tackling difficulties associated to solid waste. The notion of zero waste seeks to facilitate the restructuring of resource life cycles to attain full product reuse. In order to attain a state of zero waste, it is necessary to use a variety of crucial methods that span the whole lifespan of a product, beginning from the extraction of raw materials and culminating in its final disposal. These solutions include eco-labeling, clean manufacturing, closed-loop supply chain management, and eco-design, among other approaches. Attaining a state of zero waste is undoubtedly a praiseworthy goal [46].

A number of developing Asian countries are seeing difficulties in effectively handling solid waste as a result of problems such as inefficient garbage collection, inadequate waste disposal facilities, insufficient financial resources, and inadequate governmental activities. Unlike affluent nations, emerging Asian countries often transport their accumulated solid waste to the nearest appropriate site for final disposal. The sustainability of landfills in Asia is being hindered by several problems, including limited space, the need for modern gas capture and leachate treatment equipment, the tropical environment, and the characteristics of the trash. The proliferation of plastic and paper waste is escalating due to rapid economic expansion. Recognizing rubbish as a valuable resource is the first step toward attaining sustainable waste management and resource preservation. Composting and anaerobic digestion have shown to be effective strategies for treating biodegradable waste, resulting in positive results. The management of nonbiodegradable waste fractions necessitates the use of the 3Rs: reduce, reuse, and recycle. Hence, it is essential to rectify the deficiencies in this design to efficiently tackle waste management concerns and make progress toward the tenets of reduce, reuse, and recycle (3Rs) [62].

Currently, almost everyone recognizes the feasibility of embracing ecologically sustainable activities. Considering the well-known occurrence of global warming, it is crucial to swiftly find and execute remedies that will assist the Earth and expedite the shift toward a sustainable future. Global breakthroughs in plastics methodology, bioplastic manufacture, the recycling industry, and squander management are the focus of several international events. The objective is to convert plastic trash into a profitable trade opening for industrialists and manufacturers. The recycling process includes the examination of the reusing of thermoplastic, biopolymers, thermoset polymers, and their intricate composite analogues, such as nanocomposites and fiber-reinforced polymers. Furthermore, this research will highlight the need for enhancing squander management techniques to maximize the recycling capabilities of plastics and bioplastics. This may be accomplished by increasing the specificity of materials, maintaining cleanliness, and reducing the level of contamination in the gathered materials. To create a viable and easily accessible business opportunity in recycling, it is crucial to build a unified and complete market for secondary raw materials.

The prospects of improving polymer products, oligomers, and/or monomers, as well as micro-/nanoparticles, as also of the reusing process of (bio)plastics, which includes both thermosets, thermoplastics and (bio)plastic composites, need to be studied. For recycling thermoplastic materials, all recycling procedures, ranging from primary to quaternary recycling, may be used. Developing second-life goods must consider the degradation of material properties that happens with each reprocessing step. It is essential to examine the properties of the materials being collected, such as their homogeneity or heterogeneity, while evaluating mechanical recycling. This has a substantial influence on the categorization, cutting/grinding/shredding, and reprocessing procedures. For that reason, the environment of the collected materials is essential for chemical recycling as well [69].

6.2 Conclusion

In conclusion, the book chapter has systematically designed and acknowledged why sustainable plastic waste management has turned into a most important problem worldwide. The observations throughout the chapter are that insufficient communications and financial support are a few of the important difficulties to be overcome for waste plastic management exercises. In spite of the reality that plastic waste can be recycled to construct new stuff, these plastic wastes are presently littering every accessible open breathing space. Biodegradable plastic waste could be commonly reused as cutlery, food serving containers, and as disposable items. This chapter can demonstrate and tremendously be useful for research scholars or chemists in the understanding of waste plastic management for a sustainable world.

References

[1] Abu-Qdais, H., Shatnawi, N., & AL-Alamie, E. (2024). Intelligent system for solid waste classification using combination of image processing and machine learning models. *Journal of Experimental & Theoretical Artificial Intelligence*, 1(12). https://doi.org/10.1080/0952813x.2024.2323043

[2] Algozeeb, W. A., Savas, P. E., Yuan, Z., Wang, Z., Kittrell, C., Hall, J. N., Chen, W., Bollini, P., & Tour, J. M. (2022). Plastic waste product captures carbon dioxide in nanometer pores. *ACS Nano*, 16(5), 7284–7290. https://doi.org/10.1021/acsnano.2c00955

[3] Anuar Sharuddin, S. D., Abnisa, F., Wan Daud, W. M. A., & Aroua, M. K. (2016). A review on pyrolysis of plastic wastes. *Energy Conversion and Management*, 115, 308–326. https://doi.org/10.1016/j.enconman.2016.02.037

[4] Artificial Intelligence in Finance and Accounting (2019) *International Journal of Innovative Technology and Exploring Engineering*, 8(12S), 903–905. https://doi.org/10.35940/ijitee.l1203.10812s19

[5] Asgher, M., Rani, A., Khalid, N., Qamar, S. A., & Bilal, M. (2021). Bioconversion of sugarcane molasses waste to high-value exopolysaccharides by engineered *Bacillus licheniformis*. *Case Studies in Chemical and Environmental Engineering*, 3, 100084. https://doi.org/10.1016/j.cscee.2021.100084

[6] Asgher, M., Urooj, Y., Qamar, S. A., & Khalid, N. (2020). Improved exopolysaccharide production from *Bacillus licheniformis* MS3: Optimization and structural/functional characterization. *International Journal of Biological Macromolecules*, 151, 984–992. https://doi.org/10.1016/j.ijbiomac.2019.11.094

[7] Assessment of agricultural plastics and their sustainability: A call for action (2021). https://doi.org/10.4060/cb7856en

[8] Babita Singh, V., & Mohan, C. (2023). Plant-derived compounds and their green synthesis in pharmaceuticals and nutraceuticals. In *Book: Green Chemistry Approaches to Environmental Sustainability*, Elsevier, Amsterdam, pp. 149–163. Access date: 11/6/2025 https://doi.org/10.1016/B978-0-443-18959-3.00001-X

[9] Bhandari, N. L., Bhattarai, S., Bhandari, G., Subedi, S., & Dhakal, K. N. (2021). A review on current practices of plastics waste management and future prospects. *Journal of Institute of Science and Technology*, 26(1), 107–118. https://doi.org/10.3126/jist.v26i1.37837

[10] Bilal, M., Mehmood, S., & Iqbal, H. M. N. (2020). The beast of beauty: Environmental and health concerns of toxic components in cosmetics. *Cosmetics*, 7(1), 13. https://doi.org/10.3390/cosmetics7010013

[11] Cai, N., Xia, S., Zhang, X., Meng, Z., Bartocci, P., Fantozzi, F., Chen, Y., Chen, H., Williams, P. T., & Yang, H. (2020). Preparation of iron- and nitrogen-codoped carbon nanotubes from waste plastics pyrolysis for the oxygen reduction reaction. *ChemSusChem*, 13(5), 938–944. https://doi.org/10.1002/cssc.201903293

[12] Chen, J., Wu, J., Sherrell, P. C., Chen, J., Wang, H., Zhang, W., & Yang, J. (2022). How to build a microplastics-free environment: Strategies for microplastics degradation and plastics recycling. *Advanced Science*, 9(6). https://doi.org/10.1002/advs.202103764

[13] Chen, Z., Ibrahim, I., Hao, D., Liu, X., Wu, L., Wei, W., Su, D., & Ni, B. J. (2020). Controllable design of nanoworm-like nickel sulfides for efficient electrochemical water splitting in alkaline media. *Materials Today Energy*, 18, 100573. https://doi.org/10.1016/j.mtener.2020.100573

[14] Chojey, N. (2023). A review on green road development as an effective strategy for plastic waste reduction in Bhutan. *International Journal of Scientific Research in Engineering and Management*, 07 07. https://doi.org/10.55041/ijsrem24211

[15] De Weerdt, L., Sasao, T., Compernolle, T., Van Passel, S., & De Jaeger, S. (2020). The effect of waste incineration taxation on industrial plastic waste generation: A panel analysis. *Resources, Conservation and Recycling*, 157, 104717. https://doi.org/10.1016/j.resconrec.2020.104717

[16] Debnath, B., Bari, A. B. M. M., Ali, S. M., Ahmed, T., Ali, I., & Kabir, G. (2023). Modelling the barriers to sustainable waste management in the plastic-manufacturing industry: An emerging economy perspective. *Sustainability Analytics and Modeling*, 3, 100017. https://doi.org/10.1016/j.samod.2023.100017

[17] Deka, N., Barman, J., Kasthuri, S., Nutalapati, V., & Dutta, G. K. (2020). Transforming waste polystyrene foam into N-doped porous carbon for capacitive energy storage and deionization applications. *Applied Surface Science*, 511, 145576. https://doi.org/10.1016/j.apsusc.2020.145576

[18] El Essawy, N. A., Ali, S. M., Farag, H. A., Konsowa, A. H., Elnouby, M., & Hamad, H. A. (2017). Green synthesis of graphene from recycled PET bottle wastes for use in the adsorption of dyes in aqueous solution. *Ecotoxicology and Environmental Safety*, 145, 57–68. https://doi.org/10.1016/j.ecoenv.2017.07.014

[19] Evode, N., Qamar, S. A., Bilal, M., Barceló, D., & Iqbal, H. M. (2021). Plastic waste and its management strategies for environmental sustainability. *Case Studies in Chemical and Environmental Engineering*, 4, 100142. https://doi.org/10.1016/j.cscee.2021.100142

[20] Fonseca, D. F. S., Carvalho, J. P. F., Bastos, V., Oliveira, H., Moreirinha, C., Almeida, A., Silvestre, A. J. D., Vilela, C., & Freire, C. S. R. (2020). Antibacterial multi-layered nanocellulose-based patches loaded with dexpanthenol for wound healing applications. *Nanomaterials*, 10(12), 2469. https://doi.org/10.3390/nano10122469

[21] Gautam, R., Jindal, R., Mohan, C., & Mishra, A. K. (2024). Sustainable innovations as a solution to energy crisis and industrial pollution. *Book: Promoting Multi-Sector Sustainability with Policy and Innovation, IGI Global USA*, https://doi.org/10.4018/979-8-3693-2113-3.ch007

[22] Gómez-Serrano, V., Adame-Pereira, M., Alexandre-Franco, M., & Fernández-González, C. (2020). Adsorption of bisphenol A by activated carbon developed from PET waste by KOH activation. *Environmental Science and Pollution Research*, 28(19), 24342–24354. https://doi.org/10.1007/s11356-020-08428-6

[23] Anastas, P. T., & Warner, J. C. (2000). *Green Chemistry*. Oxford University Press, Oxford. 2000, paperback, 135 pp. £14.99. ISBN 0–19-850698-9., 2000. https://doi.org/10.1021/op000054t

[24] Ha, K. H., & Kim, M. S. (2012). Application to refrigerator plastics by mechanical recycling from polypropylene in waste-appliances. *Materials & Design*, 34, 252–257. https://doi.org/10.1016/j.matdes.2011.08.014

[25] Hao, D., Liu, C., Xu, X., Kianinia, M., Aharonovich, I., Bai, X., Liu, X., Chen, Z., Wei, W., Jia, G., & Ni, B. J. (2020). Surface defect-abundant one-dimensional graphitic carbon nitride nanorods boost photocatalytic nitrogen fixation. *New Journal of Chemistry*, 44(47), 20651–20658. https://doi.org/10.1039/d0nj04068a

[26] Herniou–Julien, C., Mendieta, J. R., & Gutiérrez, T. J. (2019). Characterization of biodegradable/non-compostable films made from cellulose acetate/corn starch blends processed under reactive extrusion conditions. *Food Hydrocolloids*, 89, 67–79. https://doi.org/10.1016/j.foodhyd.2018.10.024

[27] Hossain, R., & Sahajwalla, V. (2021). Microrecycling of waste flexible printed circuit boards for in-situ generation of O- and N-doped activated carbon with outstanding supercapacitance performance. *Resources, Conservation and Recycling*, 167, 105221. https://doi.org/10.1016/j.resconrec.2020.105221

[28] Kali, R., Padya, B., Rao, T., & Jain, P. (2019). Solid waste-derived carbon as anode for high performance lithium-ion batteries. *Diamond and Related Materials*, 98, 107517. https://doi.org/10.1016/j.diamond.2019.107517

[29] Karakoti, M., Pandey, S., Tatrari, G., Dhapola, P. S., Jangra, R., Dhali, S., Pathak, M., Mahendia, S., & Sahoo, N. G. (2022). A waste to energy approach for the effective conversion of solid waste plastics into graphene nanosheets using different catalysts for high performance supercapacitors: A comparative study. *Materials Advances*, 3(4), 2146–2157. https://doi.org/10.1039/d1ma01136g

[30] Kim, P. J., Fontecha, H. D., Kim, K., & Pol, V. G. (2018). Toward high-performance lithium–sulfur batteries: Upcycling of LDPE plastic into sulfonated carbon scaffold via microwave-promoted sulfonation. *ACS Applied Materials & Interfaces*, 10(17), 14827–14834. https://doi.org/10.1021/acsami.8b03959

[31] Kinuthia, J., & Nidzam, R. (2011). Towards zero industrial waste: Utilisation of brick dust waste in sustainable construction. *Waste Management*, 31(8), 1867–1878. https://doi.org/10.1016/j.wasman.2011.03.020

[32] Klemeš, J. J., Fan, Y. V., Tan, R. R., & Jiang, P. (2020). Minimising the present and future plastic waste, energy and environmental footprints related to COVID-19. *Renewable and Sustainable Energy Reviews*, 127, 109883. https://doi.org/10.1016/j.rser.2020.109883

[33] Kwon, G., Cho, D. W., Yoon, K., & Song, H. (2021). Valorization of plastics and goethite into iron-carbon composite as persulfate activator for amaranth oxidation. *Chemical Engineering Journal*, 407, 127188. https://doi.org/10.1016/j.cej.2020.127188

[34] Li, J., Pan, W., Liu, Q., Chen, Z., Chen, Z., Feng, X., & Chen, H. (2021). Interfacial engineering of $bi_{19}br_3s_{27}$ nanowires promotes metallic photocatalytic CO_2 reduction activity under near-infrared light irradiation. *Journal of the American Chemical Society*, 143(17), 6551–6559. https://doi.org/10.1021/jacs.1c01109

[35] Li, L., Lv, Y., Wang, J., Jia, C., Zhan, Z., Dong, Z., Liu, L., & Zhu, X. (2022). Enhance pore structure of cyanobacteria-based porous carbon by polypropylene to improve adsorption capacity of methylene blue. *Bioresource Technology*, 343, 126101. https://doi.org/10.1016/j.biortech.2021.126101

[36] Li, Z., Chen, K., Chen, Z., Li, W., Biney, B. W., Guo, A., & Liu, D. (2021). Removal of malachite green dye from aqueous solution by adsorbents derived from polyurethane plastic waste. *Journal of Environmental Chemical Engineering*, 9(1), 104704. https://doi.org/10.1016/j.jece.2020.104704

[37] Lian, Y. M., Utetiwabo, W., Zhou, Y., Huang, Z. H., Zhou, L., Faheem, M., Chen, R. J., & Yang, W. (2019). From upcycled waste polyethylene plastic to graphene/mesoporous carbon for high-voltage supercapacitors. *Journal of Colloid and Interface Science*, 557, 55–64. https://doi.org/10.1016/j.jcis.2019.09.003

[38] Liang, Y., Tan, Q., Song, Q., & Li, J. (2021). An analysis of the plastic waste trade and management in Asia. *Waste Management*, 119, 242–253. https://doi.org/10.1016/j.wasman.2020.09.049

[39] Liu, X., Chen, Z., Tian, K., Zhu, F., Hao, D., Cheng, D., Wei, W., Zhang, L., & Ni, B. J. (2021). Fe^{3+} promoted the photocatalytic defluorination of perfluorooctanoic acid (PFOA) over In_2O_3. *ACS ES&T Water*, 1(11), 2431–2439. https://doi.org/10.1021/acsestwater.1c00275

[40] Liu, X., Wen, Y., Chen, X., Tang, T., & Mijowska, E. (2020). Co-etching effect to convert waste polyethylene terephthalate into hierarchical porous carbon toward excellent capacitive energy storage. *Science of the Total Environment*, 723, 138055. https://doi.org/10.1016/j.scitotenv.2020.138055

[41] Maharjan, N., Miyazaki, H., Pati, B. M., Dailey, M. N., Shrestha, S., & Nakamura, T. (2022). Detection of river plastic using UAV sensor data and deep learning. *Remote Sensing*, 14(13), 3049. https://doi.org/10.3390/rs14133049

[42] Milios, L., Esmailzadeh Davani, A., & Yu, Y. (2018). Sustainability impact assessment of increased plastic recycling and future pathways of plastic waste management in Sweden. *Recycling*, 3(3), 33. https://doi.org/10.3390/recycling3030033

[43] Min, J., Zhang, S., Li, J., Klingeler, R., Wen, X., Chen, X., Zhao, X., Tang, T., & Mijowska, E. (2019). From polystyrene waste to porous carbon flake and potential application in supercapacitor. *Waste Management*, 85, 333–340. https://doi.org/10.1016/j.wasman.2019.01.002

[44] Kumar, M., Mohan, C., Kumar, S., Epifantsev, K., Singh, V. et al. (2022). Coordination behavior of Schiff base copper complexes and structural characterization. *MRS Advances*, 7, 939–943. doi: 10.1557/s43580-022-00348-6

[45] Mendoza-Carrasco, R., Cuerda-Correa, E. M., Alexandre-Franco, M. F., Fernández-González, C., & Gómez-Serrano, V. (2016). Preparation of high-quality activated carbon from polyethylene terephthalate (PET) bottle waste. Its use in the removal of pollutants in aqueous solution. *Journal of Environmental Management*, 181, 522–535. https://doi.org/10.1016/j.jenvman.2016.06.070

[46] Milios, L., Esmailzadeh Davani, A., & Yu, Y. (2018). Sustainability impact assessment of increased plastic recycling and future pathways of plastic waste management in Sweden. *Recycling*, 3(3), 33. https://doi.org/10.3390/recycling3030033

[47] Min, J., Zhang, S., Li, J., Klingeler, R., Wen, X., Chen, X., Zhao, X., Tang, T., & Mijowska, E. (2019). From polystyrene waste to porous carbon flake and potential application in supercapacitor. *Waste Management*, 85, 333–340. https://doi.org/10.1016/j.wasman.2019.01.002

[48] Mir, R. A., & Pandey, O. P. (2020). An ecofriendly route to synthesize C-Mo_2C and C/N-Mo_2C utilizing waste polyethene for efficient hydrogen evolution reaction (HER) activity and high-performance capacitors. *Sustainable Energy & Fuels*, 4(2), 655–669. https://doi.org/10.1039/c9se00516a

[49] Mittal, R., Maheshwari, R., Tripathi, S., & Pandey, S. (2020). Eco-friendly dentistry: Preventing pollution to promoting sustainability. *Indian Journal of Dental Sciences*, 12(4), 251. https://doi.org/10.4103/ijds.ijds_12_20

[50] Nandy, S., Fortunato, E., & Martins, R. (2022). Green economy and waste management: An inevitable plan for materials science. *Progress in Natural Science: Materials International*, 32(1), 1–9. https://doi.org/10.1016/j.pnsc.2022.01.001

[51] Nath, A., Shah, A., Singh, L. R., & Mahato, M. (2021). Waste plastic-derived NiO-MWCNT composite as visible light photocatalyst for degradation of methylene blue dye. *Nanotechnology for Environmental Engineering*, 6(3). https://doi.org/10.1007/s41204-021-00163-8

[52] Rai, P., & Singh, K. P. (2018). Valorization of Poly (ethylene) terephthalate (PET) wastes into magnetic carbon for adsorption of antibiotic from water: Characterization and application. *Journal of Environmental Management*, 207, 249–261. https://doi.org/10.1016/j.jenvman.2017.11.047

[53] Ribeiro, R. S., Vieira, O., Fernandes, R., Roman, F. F., Diaz de Tuesta, J. L., Silva, A. M., & Gomes, H. T. (2022). Synthesis of low-density polyethylene derived carbon nanotubes for activation of persulfate and degradation of water organic micropollutants in continuous mode. *Journal of Environmental Management*, 308, 114622. https://doi.org/10.1016/j.jenvman.2022.114622

[54] Ribić, B., Voća, N., & Ilakovac, B. (2016). Concept of sustainable waste management in the city of Zagreb: Towards the implementation of circular economy approach. *Journal of the Air & Waste Management Association*, 67(2), 241–259. https://doi.org/10.1080/10962247.2016.1229700

[55] Sahoo, N., Tatrari, G., Tewari, C., Karakoti, M., Bohra, B. S., & Danadapat, A. (2022). Vanadium pentaoxide-doped waste plastic-derived graphene nanocomposite for supercapacitors: A comparative electrochemical study of low and high metal oxide doping. *RSC Advances*, 12(9), 5118–5134. https://doi.org/10.1039/d1ra07458j

[56] Seema, H., Kemp, K. C., Le, N. H., Park, S. W., Chandra, V., Lee, J. W., & Kim, K. S. (2014). Highly selective CO_2 capture by S-doped microporous carbon materials. *Carbon*, 66, 320–326. https://doi.org/10.1016/j.carbon.2013.09.006

[57] Shen, Y., Chen, N., Feng, Z., Feng, C., & Deng, Y. (2022). Treatment of nitrate containing wastewater by adsorption process using polypyrrole-modified plastic-carbon: Characteristic and mechanism. *Chemosphere*, 297, 134107. https://doi.org/10.1016/j.chemosphere.2022.134107

[58] Silva, A. C. Q., Silvestre, A. J. D., Vilela, C., & Freire, C. S. R. (2021). Natural polymers-based materials: A contribution to a greener future. *Molecules*, 27(1), 94. https://doi.org/10.3390/molecules27010094

[59] Silva, N. H., Vilela, C., Almeida, A., Marrucho, I. M., & Freire, C. S. (2018). Pullulan-based nanocomposite films for functional food packaging: Exploiting lysozyme nanofibers as antibacterial

and antioxidant reinforcing additives. *Food Hydrocolloids*, 77, 921–930. https://doi.org/10.1016/j.foodhyd.2017.11.039

[60] Singh, P., & Sharma, V. (2016). Integrated plastic waste management: Environmental and improved health approaches. *Procedia Environmental Sciences*, 35, 692–700. https://doi.org/10.1016/j.proenv.2016.07.068

[61] Song, C., Hao, L., Zhang, B., Dong, Z., Tang, Q., Min, J., Zhao, Q., Niu, R., Gong, J., & Tang, T. (2020). High-performance solar vapor generation of Ni/carbon nanomaterials by controlled carbonization of waste polypropylene. *Science China Materials*, 63(5), 779–793. https://doi.org/10.1007/s40843-019-1243-8

[62] Song, Q., Li, J., & Zeng, X. (2015). Minimizing the increasing solid waste through zero waste strategy. *Journal of Cleaner Production*, 104, 199–210. https://doi.org/10.1016/j.jclepro.2014.08.027

[63] Szostak, E., Duda, P., Duda, A., Górska, N., Fenicki, A., & Molski, P. (2020). Characteristics of plastic waste processing in the modern recycling plant operating in Poland. *Energies*, 14(1), 35. https://doi.org/10.3390/en14010035

[64] Ubaidullah, M., Al-Enizi, A. M., Shaikh, S., Ghanem, M. A., & Mane, R. S. (2020). Waste PET plastic derived ZnO@NMC nanocomposite via MOF-5 construction for hydrogen and oxygen evolution reactions. *Journal of King Saud University – Science*, 32(4), 2397–2405. https://doi.org/10.1016/j.jksus.2020.03.025

[65] Vafaeva, K. M., Chhetri, A., Sudan, P., Mishra, M., Pakkiraiah, B., & Mohan, C., 2024, 'Polymer matrix nanocomposites for sustainable packaging: A green approach', *E3S Web of Conferences*, vol. 511, p. 01008. https://doi.org/10.1051/e3sconf/202451101008

[66] Vatin, N. I., Negi, G. S., Yellanki, S. V., Mohan, C., & Singla, N., 2024, 'Sustainability Measures: An Experimental Analysis of AI and Big Data Insights in Industry 5.0', *BIO Web of Conferences*, vol. 86, p. 01072. https://doi.org/10.1051/bioconf/20248601072

[67] Vieira, O., Ribeiro, R. S., Diaz de Tuesta, J. L., Gomes, H. T., & Silva, A. M. (2022). A systematic literature review on the conversion of plastic wastes into valuable 2D graphene-based materials. *Chemical Engineering Journal*, 428, 131399. https://doi.org/10.1016/j.cej.2021.131399

[68] Villagómez-Salas, S., Manikandan, P., Acuña Guzmán, S. F., & Pol, V. G. (2018). Amorphous carbon chips Li-ion battery anodes produced through polyethylene waste upcycling. *ACS Omega*, 3(12), 17520–17527. https://doi.org/10.1021/acsomega.8b02290

[69] Visvanathan, C., Adhikari, R., & Ananth, A. P. (2007). 3R practices for municipal solid waste management in Asia. *Linnaeus Eco-Tech*, 11, 11–22. https://doi.org/10.15626/eco-tech.2007.002

[70] Wang, C., Jiang, X., Huang, R., Cao, Y., Xu, J., & Han, Y. (2019). Copper/carbon composites from waste printed circuit boards as catalysts for Fenton-like degradation of Acid Orange 7 enhanced by ultrasound. *AIChE Journal*, 65(4), 1234–1244. https://doi.org/10.1002/aic.16519

[71] Wang, J., Yuan, X., Deng, S., Zeng, X., Yu, Z., Li, S., & Li, K. (2020). Waste polyethylene terephthalate (PET) plastics-derived activated carbon for CO_2 capture: A route to a closed carbon loop. *Green Chemistry*, 22(20), 6836–6845. https://doi.org/10.1039/d0gc01613f

[72] Wei, S., & Kamali, A. R. (2020). Dual-step air-thermal treatment for facile conversion of PET into porous carbon particles with enhanced dye adsorption performance. *Diamond and Related Materials*, 107, 107914. https://doi.org/10.1016/j.diamond.2020.107914

[73] Wen, X., Chen, X., Tian, N., Gong, J., Liu, J., Rümmeli, M. H., Chu, P. K., Mijiwska, E., & Tang, T. (2014). Nanosized carbon black combined with Ni2O3 as "universal" catalysts for synergistically catalyzing carbonization of polyolefin wastes to synthesize carbon nanotubes and application for supercapacitors. *Environmental Science & Technology*, 48(7), 4048–4055. https://doi.org/10.1021/es404646e

[74] Wen, Y., Kierzek, K., Min, J., Chen, X., Gong, J., Niu, R., Wen, X., Azadmanjiri, J., Mijowska, E., & Tang, T. (2019). Porous carbon nanosheet with high surface area derived from waste poly(ethylene terephthalate) for supercapacitor applications. *Journal of Applied Polymer Science*, 137(5), 48338. https://doi.org/10.1002/app.48338

[75] Wu, W., Zhu, S., Huang, X., Wei, W., & Ni, B. J. (2021). Mechanisms of persulfate activation on biochar derived from two different sludges: Dominance of their intrinsic compositions. *Journal of Hazardous Materials*, 408, 124454. https://doi.org/10.1016/j.jhazmat.2020.124454

[76] Xu, X., Zhu, D., Wang, X., Deng, L., Fan, X., Ding, Z., Zhang, A., Xue, G., Liu, Y., Xuan, W., Li, X., & Makinia, J. (2022). Transformation of polyvinyl chloride (PVC) into a versatile and efficient adsorbent of Cu(II) cations and Cr(VI) anions through hydrothermal treatment and sulfonation. *Journal of Hazardous Materials*, 423, 126973. https://doi.org/10.1016/j.jhazmat.2021.126973

[77] Yadav, S., Yadav, A., Mohan, C., Garg, V. K., & Kumari, N. (2023). Introduction to environmental and green chemistry. In: V. K. Garg (ed.), *Book: Green Chemistry Approaches to Environmental Sustainability*. Elsevier, Amsterdam 1–22. Access date: 11/6/2025 https://doi.org/10.1016/B978-0-443-18959-3.00005-7

[78] Yu, L. (Ed.) 2008. Biodegradable polymer blends and composites from renewable resources. https://doi.org/10.1002/9780470391501

[79] Yuan, X., Wang, J., Deng, S., Suvarna, M., Wang, X., Zhang, W., Hamilton, S. T., Alahmed, A., Jamal, A., Park, A. H. A., Bi, X., & Ok, Y. S. (2022). Recent advancements in sustainable upcycling of solid waste into porous carbons for carbon dioxide capture. *Renewable and Sustainable Energy Reviews*, 162, 112413. https://doi.org/10.1016/j.rser.2022.112413

[80] Zhang, B., Song, C., Liu, C., Min, J., Azadmanjiri, J., Ni, Y., Niu, R., Gong, J., Zhao, Q., & Tang, T. (2019). Molten salts promoting the "controlled carbonization" of waste polyesters into hierarchically porous carbon for high-performance solar steam evaporation. *Journal of Materials Chemistry A*, 7(40), 22912–22923. https://doi.org/10.1039/c9ta07663h

[81] Zhang, N., Williams, I., Kemp, S., & Smith, N. (2011). Greening academia: Developing sustainable waste management at higher education institutions. *Waste Management*, 31(7), 1606–1616. https://doi.org/10.1016/j.wasman.2011.03.006

[82] Zheng, Y., Yanful, E. K., & Bassi, A. S. (2005). A review of plastic waste biodegradation. *Critical Reviews in Biotechnology*, 25(4), 243–250. https://doi.org/10.1080/07388550500346359

Jyotsna Sharma*, Ansh Kumar, Chandra Mohan, and Lata Vodwal

Chapter 7
Data-driven approaches for aligning nanopackaging innovations with Sustainable Development Goals (SDGs)

Abstract: Nanotechnology is rapidly revolutionizing the packaging industry by introducing various advanced materials and smart functionalities that help in enhancing product safety, longevity, and sustainability. This chapter explores the intersection of nanotech-enabled packaging and the United Nations Sustainable Development Goals, highlighting the urgent need to align technological innovations with global sustainability objectives. While nanomaterials promise breakthroughs in biodegradability, antimicrobial activity, waste reduction, and environmental remediation, their deployment must be guided by ethical frameworks and ecological considerations. We will delve into how data-driven technologies – including artificial intelligence, Internet of Things, and machine learning – serve as powerful tools in this alignment process. These tools support life cycle assessments (LCAs), optimize material selection, improve efficiency, predict environmental impacts, and foster traceability across the supply chain – making nanotech-enabled packaging both scalable and sustainable. This convergence of nanotechnology and intelligent systems not only accelerates innovation but also reinforces circular economy principles and transparency.

The chapter also presents a critical analysis of the opportunities and risks associated with the widespread adoption of nanotech in packaging. Issues such as data privacy, nanotoxicity, and regulatory ambiguity are examined alongside the benefits of efficiency, safety, and SDGs compliance. Ultimately, the goal is to present a balanced roadmap that ensures innovation progresses responsibly and sustainably in the packaging sector.

Keywords: Sustainability, environmental remediation, machine learning, nanotoxicity

***Corresponding author: Jyotsna Sharma,** Department of Physics, Faculty of Physical Sciences, South Asian University, New Delhi, India, e-mail: plasmajyoti@gmail.com
Ansh Kumar, Department of Computer Science, South Asian University, New Delhi, India,
Chandra Mohan, School of Basic and Applied Sciences, K. R. Mangalam University, Gurugram, India
Lata Vodwal, Department of Chemistry, Maitreyi College, University of Delhi, Delhi, India

https://doi.org/10.1515/9783111618906-007

7.1 Introduction

Packaging has become both a necessity and a nemesis in the modern world. It is integral to the global economy – preserving products, streamlining logistics, and shaping consumer experiences. However, the rapid growth of single-use packaging, particularly plastics, has led to an alarming environmental crisis [1]. According to the United Nations Environment Program, the world produces approximately 460 million metric tons of plastic waste annually, with a large chunk of it coming from packaging alone. These plastics often end up in landfills or oceans, causing long-term ecological damage, microplastic contamination, increased greenhouse gas emissions, and contributing to climate change [2, 3]. Continued demand growth, along with the surge in e-commerce and globalization, has further intensified packaging demand, accelerating consumption and straining global waste management systems [4, 5]. In response, governments, organizations, and consumers are increasingly vouching for sustainable packaging solutions that balance functionality with environmental responsibility [6]. The current scenario calls for a fundamental shift – from traditional linear models of "take-make-dispose" to innovative, circular, and eco-conscious approaches. Amid this escalating packaging crisis, the spotlight now turns to sustainable packaging – not merely as a solution, but as a necessary evolution in how we design, use, and dispose of materials [7–9]. This global challenge sets the stage for exploring how nanotechnology and data-driven tools can redefine packaging for a more sustainable future.

Conventional packaging – especially single-use plastic – generates massive pollution and resource waste [10, 11]. For example, 9.2 billion tons of plastic have been produced to date, yet only about 9% has ever been recycled. Since plastic never truly biodegrades, every piece ever made still contributes to pollution (often as microplastics). Microplastics from discarded packaging harm organisms across the food web, destabilizing entire ecosystems [12]. Such packaging accumulates in landfills and waterways, threatening biodiversity and contaminating food chains. Sustainable packaging aims to mitigate these impacts by eliminating waste and pollution at source.

Advances in materials science, coupled with data-driven innovation, are enabling smarter packaging systems. For instance, using biodegradable or compostable materials (such as PLA or algae-based polymers) prevents persistent plastic debris. A notable example is *seaweed-based film packaging*: a swimwear company developed a 100% recycled cardboard box with a compostable seaweed film window. Unlike conventional plastics, seaweed film fully biodegrades into soil, eliminating the risk of microplastic pollution. Moreover, cultivating seaweed for packaging can improve ocean health (it adds habitat, cycles nutrients, and even sequesters carbon). Sustainable packaging is not only eco-friendly – it is also cost-effective. Reducing waste and optimizing materials often cuts costs across the supply chain. By eliminating pollution at the source, sustainable packaging breaks the traditional link between consumption and waste [13–15]. Environmentally, it prevents pollution and cuts emissions. Economically, it trims costs and meets market and regulatory demands. Technologically, it leverages data, connectivity, and materials science to innovate. Governments worldwide are also introducing stricter packaging regulations, making sustainable alternatives not just ideal, but inevitable. These efforts directly support UN SDGs: reducing plastic pollution (SDG 14), lowering carbon footprints (SDG 13), and promoting responsible production (SDG 12). More benefits of sustainable packaging will be highlighted as we dive further into the chapter. As sustainable packaging gains momentum, the integration of nanotechnology [16–18] marks a promising next step, offering powerful enhancements in safety, functionality, and environmental performance through innovations at the molecular scale.

Nanotechnology-enabled packaging refers to the application and integration of nanoscale materials and techniques – typically ranging from 1 to 100 nm – into packaging systems to enhance their strength, intelligence, safety, and sustainability [19–24]. At the nanoscale, materials exhibit unique physical, chemical, and biological properties that differ significantly from their bulk counterparts. In the context of packaging, nanotechnology is used to create advanced materials with superior barrier properties, antimicrobial functions, and mechanical strength, while also enabling smart packaging features. Some common nanomaterials include nanosilver [25, 26], nanoclays [27], nanocellulose [28, 29], and carbon nanotubes (CNTs) [30, 31], which are integrated into films, coatings, or structural layers in packaging materials. They extend shelf life, minimize waste, and enhance safety, addressing both consumer demands and sustainability goals. Moreover, innovations such as nanosensors and active packaging systems – like freshness-indicating films for perishable foods – allow for real-time monitoring of temperature, contamination, and spoilage, transforming conventional packaging into intelligent, responsive solutions. We will explore these nanomaterials in detail later. Yet beyond material enhancements, nanotechnology's full potential emerges when paired with intelligent digital systems.

Data-driven technologies [32] play a transformative role in advancing sustainable packaging — not just in how materials are engineered, but in how they're designed, optimized, and monitored at scale. At the material level, breakthroughs in nanotechnology have led to the development of smart, high-performance packaging films — materials that are lighter, stronger, biodegradable, and even antimicrobial. However, to unlock their full potential, these innovations rely on data-driven design, simulation, and computational modeling.

From artificial intelligence (AI) and machine learning (ML) [33, 34] optimizing nanomaterial formulations to Internet of Things (IoT) [35, 36] and embedded sensors enabling smart packaging [37, 38] that tracks freshness, humidity, and temperature, tech is revolutionizing both functionality and sustainability. Additionally, blockchain [39, 40] ensures transparent sourcing and traceability of advanced sustainable materials, helping verify eco-claims and align with regulatory frameworks. Digital twins and virtual simulations [41, 42] allow researchers to test nanomaterial performance in silico, minimizing trial and error in physical labs.

Nanotechnology provides the sustainable "what," and data-driven technologies provide the intelligent "how." Their synergy enables a new generation of intelligent, sustainable, and circular packaging systems, directly supporting SDGs through innovation that scales. In the chapters ahead, we will explore how this integration is reshaping packaging science from the molecular level to the global supply chain. Building on this synergy between nanotechnology and computational tools, data-driven decision-making [43, 44] emerges as the operational backbone, turning intelligent packaging concepts into actionable, real-time systems that enhance sustainability at every stage of the value chain.

Data-driven decisions in packaging refer to using real-time or historical data collected from packaging materials, supply chains, and consumer interactions to optimize packaging design, performance, logistics, and sustainability. Instead of relying on fixed schedules, estimates, or traditional best practices, businesses use empirical data – data based on real-life observations and experiments – often powered by sensors, IoT, and AI to guide every packaging-related decision.

Data-driven decision-making is revolutionizing packaging by transforming it from a passive protective layer into an intelligent, responsive asset. By integrating smart packaging systems, such as nanosensors [45], AI algorithms, and IoT connectivity, into packaging materials, companies can now capture and analyze real-time data on freshness, temperature, humidity, and contamination directly from the product environment. For instance, companies like IBM and Tetra Pak are leveraging AI and sensor data in milk cartons to monitor spoilage in real-time, enhancing shelf-life predictions and reducing waste. IoT connectivity enables this data to be transmitted to cloud platforms, where AI and ML models analyze it for spoilage prediction, shelf-life estimation, and quality assurance. Additionally, digital twins simulate various packaging systems and supply chain scenarios using real-time data, guiding continuous improvement in material design and logistics. These insights help refine nanomaterial selection, streamline cold chain logistics, and tailor marketing strategies based on packaging performance. This closed-loop system supports sustainability by reducing waste, increasing product traceability, and enabling smarter, faster decision-making across the value chain. Ultimately, the convergence of nanotechnology, AI/ML, and IoT is laying the groundwork for a circular, low-emission packaging economy.

The integration of data-driven technologies and nanotechnology into packaging opens up a world of possibilities. For businesses, these innovations provide opportunities to reduce waste, enhance supply chain efficiency, and improve consumer engagement. Real-time data collection allows for predictive analytics, enhancing shelf-life estimations and minimizing spoilage, while nanotech solutions offer the potential for more sustainable, high-performance packaging materials. Companies can harness these technologies to align with global sustainability goals and appeal to an increasingly eco-conscious consumer base, thus fostering a more circular and resource-efficient packaging ecosystem. Moreover, this data-driven transformation paves the way for more personalized packaging experiences, further boosting both user satisfaction and brand value.

However, these advancements are not without challenges. Integrating cutting-edge technologies such as AI, IoT, and nanomaterials can be resource-intensive and complex, especially for small and medium-sized enterprises. Data privacy and security are major concerns, given that the vast volumes of consumer and product data being collected could become vulnerable to breaches and misuse. Additionally, the long-term environmental impact of widespread nanotech use remains uncertain, and questions around disposal and recyclability require careful consideration. As the industry progresses, finding a balance between innovation and risk mitigation will be crucial for ensuring that the benefits of these technologies outweigh the potential drawbacks and for steering the packaging industry toward a truly regenerative future.

This chapter explores how nanotechnology and intelligent systems such as AI, IoT, and ML are revolutionizing sustainable packaging. It includes real-world case studies, highlights cutting-edge innovations, and maps their alignment with key United Nations Sustainable Development Goals (SDGs). Together, these insights aim to chart a path toward a smarter, cleaner, and more circular packaging economy.

7.2 Understanding nanotechnology in packaging

7.2.1 The science behind nanotechnology and nanomaterials

Nanotechnology involves manipulating matter at the nanoscale (1–100 nm), where materials exhibit properties radically different from their bulk counterparts. At this scale, an increased surface-area-to-volume ratio enhances strength, reactivity, and catalytic behavior. Quantum effects also emerge – such as size-dependent fluorescence in quantum dots. Nanomaterials may become more chemically active, offer superior electrical and thermal conductivity, and display unique optical properties – for example, gold nanoparticles appear red and silver yellow due to surface plasmon resonance. These nanoscale phenomena enable the development of smarter, lighter, and more functional materials, particularly in advanced packaging applications.

Nanomaterials are defined as materials possessing internal or external structures within the nanoscale range (~1–100 nm). They can be **naturally occurring**, such as:
- **Nanocellulose**, derived from plant cell walls, known for its strength and biodegradability,
- **Chitosan nanoparticles**, extracted from crustacean (crab and shrimp) shells, offer antimicrobial and film-forming properties.

Or **engineered/synthetic**, such as:

- **Silver and gold nanoparticles** for antimicrobial and sensor functionalities.
- **CNTs** for electrical conductivity and mechanical strength.
- **Nanoclays** (e.g., montmorillonite) – while naturally occurring and derived from layered silicate minerals found in nature, are often **engineered or surface-modified** (e.g., ion-exchange, polymer grafting) to enhance properties such as gas/moisture barrier effectiveness and thermal stability.

These nanomaterials – whether sourced from nature or designed in labs – form the backbone of nanotechnology-enabled packaging, offering transformative capabilities that traditional materials cannot match. In the next section, we delve into the various types of nanomaterials commonly used in packaging and their specific roles in creating safer, more durable, and sustainable solutions.

7.2.2 Types of nanomaterials used in packaging

The following nanomaterials are widely utilized in the development of nanotechnology-enabled packaging due to their unique functional properties.

Table 7.1: Common nanomaterials used in packaging and their functional roles.

Nanomaterial	Key functions in packaging	Use case
Nanoclay (e.g., MMT, i.e., montmorillonite and halloysite)	Enhances mechanical strength; **gas/ barrier protection** (O_2, moisture); can impart antimicrobial effect	Used in PE/PP films to stiffen material and block O_2/H_2O. Improves shelf life by slowing gas diffusion.
Silver (AgNP)	Broad-spectrum **antimicrobial** (bacteria, fungi, and viruses); odor control	Incorporated in plastic containers or coatings to kill microbes, e.g., in meat or dairy packaging. Helps prevent microbial spoilage during storage, thereby enhancing product safety.
Titanium dioxide (TiO_2)	Blocks UV light; **photocatalytic antimicrobial**; optical whitening	Embedded or coated in films to protect contents from UV light; also can degrade surface microbes under UV. For example, TiO_2 in PVA films has demonstrated significant improvement in UV shielding performance.
Zinc oxide (ZnO)	**Antimicrobial** (both bactericidal and fungicidal); UV-absorbing; mechanical reinforcement	Integrated into biodegradable packaging films (e.g., starch blends) to combat spoilage microbes like *S. aureus*, keeping food fresher for longer.

Table 7.1 (continued)

Nanomaterial	Key functions in packaging	Use case
Carbon nanotubes (CNTs)	Exceptional **strength** and **stiffness** (stronger than diamond); electrical/ thermal conductivity; enhance barrier properties when uniformly dispersed and well-aligned	Blended into polymers (e.g., polyolefins) to form high-performance nanocomposites with added strength and durability; researched for ultra-durable packaging, including aerospace applications.
Nanocellulose	Outstanding **oxygen/water vapor barrier** (when dry); high tensile strength; **renewable and biodegradable**	Used as coatings or fillers in biopolymer films to dramatically reduce gas permeability; ideal for sustainable packaging solutions.

Each of these nanomaterials is utilized in packaging in application-specific ways. For instance, nanoclay and CNTs are commonly mixed into polymer melts or coatings to enhance barrier properties and mechanical strength. In contrast, metal oxides and silver nanoparticles (AgNPs) are typically coated or blended into films to provide active functionalities such as antimicrobial or UV protection. Nanocellulose, owing to its biodegradable nature, is often processed into paper-like films or used as fillers in bio-based composites. These advanced functions stem from their nanoscale dimensions and high surface-area-to-volume ratios, which dramatically amplify their performance at minimal material usage.

7.2.3 Applications of nanotechnology in packaging enhancement

(a) **Barrier properties:**
Nanotechnology significantly enhances the ability of packaging materials to block the transmission of gases, moisture, and other environmental factors – essential for preserving product integrity.

- **Layered nanoclays** (e.g., montmorillonite) embedded in polymer matrices form *tortuous diffusion paths*, making it difficult for gas or vapor molecules to pass through. For instance, adding just ~3% MMT to polystyrene reduces oxygen permeability by ~30% and water vapor transmission by ~50%.
- **Metal oxide coatings**, such as thin layers of silica or alumina, are applied to polymer films to block gas ingress. Coca-Cola, for example, uses a silica-based barrier in PET (polyethylene terephthalate) bottles to double shelf life by minimizing CO_2 loss and O_2 ingress.
- **Nanocellulose films**, derived from plant matter, offer exceptional oxygen barrier performance. Their dense, hydrogen-bonded network creates a highly imperme-

able structure – rivaling or even surpassing conventional petroleum-based high-barrier films.

These nanoscale enhancements not only extend shelf life but also reduce the need for multilayer or composite packaging, supporting sustainability goals by simplifying material recovery and recycling.

(b) **Mechanical strength:**

- Nanofillers act as reinforcement agents within polymer matrices, significantly enhancing mechanical properties — allowing packaging to be lighter, thinner, and more resource-efficient without compromising integrity.
- For instance, incorporating just **1% montmorillonite** into a biodegradable PHBH (poly(3-hydroxybutyrate-co-3-hydroxyhexanoate)) matrix led to a **111% increase in tensile strength**, showcasing the potential of nanoclay–polymer synergy.
- Similarly, embedding **2% layered double hydroxide (LDH)** into a polymer boosted tensile strength by approximately **83%**, while simultaneously reducing **oxygen permeability by ~54%** – achieving both durability and barrier performance.

Moreover, nanocomposites often exhibit higher elastic **modulus, toughness, and impact resistance**, making them suitable for demanding packaging conditions such as stacking, shipping, and temperature variations. This enhanced performance can enable material downgauging, reducing raw material usage and supporting sustainability goals.

(c) **Antimicrobial activity:**

- **Silver (AgNPs), ZnO, TiO_2,** and some nanoclays exhibit strong **antibacterial and antifungal effects**.
- Films embedded with **AgNPs** can **inactivate microbes**, extending **shelf life** and preserving freshness.
- **ZnO and TiO_2** show broad-spectrum action — effective against pathogens like ***E. coli*** and ***S. aureus***.
- **Hybrid nanocomposites** (e.g., **Ag/TiO_2 blends in LDPE**) significantly reduce microbes like ***E. coli*** and ***Candida albicans***.
- These active materials also help **reduce chemical preservatives**, promoting cleaner label packaging.

(d) **UV and thermal resistance:**

- **TiO_2 and ZnO nanoparticles** absorb UV light effectively, shielding **light-sensitive foods** from degradation.
- A film with **5% TiO_2 + 5% cellulose nanocrystals** increased UV-blocking by **~70,000%**, making it virtually **UV-opaque**.
- **CNTs** enhance **thermal conductivity**, **flame retardancy**, and **overall heat resistance** in nanocomposites.
- These enhancements improve **product shelf life**, **material safety**, and **transport resilience**.

(e) **Smart/sensing capabilities:**
- **Nanosensors** can detect **spoilage gases (e.g., ammonia and H_2S)** or **changes in pH**, signaling food freshness in real-time.
- **Quantum dots, CNTs,** and **metal oxide nanoparticles** are used in **color-changing films** or **electronic tags** for smart packaging.
- **Printed nanosensors** on labels can alert consumers or supply chains to **temperature abuse, contamination,** or **leaks.**
- This enables **active monitoring**, better **inventory control**, and **less food waste** — aligning tech with sustainability.

(f) **Environmental sustainability:**
- **Nanocellulose** is renewable, biodegradable, and can rival petroleum-based plastics in strength and barrier performance.
- **Nanocomposites** enable **thinner, lighter packaging**, reducing raw material consumption.
- **Functional nanoadditives** (e.g., antimicrobials) extend shelf life, helping curb **food waste**.
- However, caution must be exercised regarding the environmental and health implications of nanomaterials, particularly with respect to their safe disposal and effective recycling in packaging applications.

The above functional benefits have already been realized across various commercial packaging solutions [46]. The next section explores key case studies and industry applications that illustrate how nanotechnology is reshaping sustainable packaging in practice.

7.2.4 Case studies and industry applications of nanotechnology-enabled packaging

Table 7.2 highlights key industry sectors leveraging nanotechnology in packaging, outlining their functional benefits and notable real-world applications.

These case studies exemplify the diverse potential of nanotechnology in enhancing packaging performance, while aligning with industry-specific sustainability and efficiency objectives.

Increasingly, nanotechnology-enabled packaging systems are being integrated with digital technologies to create smart, data-driven solutions. In practice, nanosensors and printed electronics are embedded into packaging and connected via **IoT networks**, while **AI/ML algorithms** analyze the resulting data to monitor freshness, authenticate products, and optimize logistics. Additionally, **blockchain technology** is being explored for secure, tamper-proof data sharing across the supply chain — enhancing traceability, enabling product authentication, and fostering consumer trust. These emerging integrations will be explored in greater detail in Section 7.4.

Table 7.2: Industry-wise applications of nanotechnology-enabled packaging and their functional benefits.

Industry	Key nanotechnology features and benefits	Industry examples/key players
Food and beverages	Gas/vapor-barrier nanocomposites (clay and nanocellulose); antimicrobial nanocoatings (Ag and ZnO); freshness sensors (e.g., CO_2 and volatile organic compounds); biodegradable nanopolymers for sustainability. Improved shelf life, reduced spoilage, and waste.	Nfinite Nanotech barrier coatings (Amcor, PepsiCo pilots); Innoscentia/BioTip colorimetric labels for spoilage; Apeel edible coatings; Sonoco "GreenCan" high-barrier paper packaging.
Pharmaceuticals	Extreme barrier (O_2/H_2O/UV) from nano-silica, clays; active scavengers (nanosilica and nanocellulose); tamper-evident nanostructured seals; RFID/NFC traceability tags. Enhanced drug stability and compliance, anti-counterfeit.	PackagingGateway IoT initiatives with RFID; Active Pouch tamper-proof nanolaminate (industry report); DuPont nanofiber blisters; specialized temperature-sensitive nanopackaging solutions for vaccine stability.
Cosmetics and personal care	UV-block and oxygen-barrier nanocoatings; nano-reinforced plastics for strength; nanoporous biodegradable cartons; hydrophobic nanolayers. Protects sensitive formulations while enabling lightweight and eco-friendly options.	L'Oréal, Shiseido R&D in nanocoatings; BASF biodegradable nanoplastics; cosmetic tubes with clay-nano reinforcement; "smart perfume" boxes with nano-LED freshness indicators (prototype).
Electronics	Anti-static nanofillers (CNTs, graphene) in films; moisture barrier nanocoatings; flame-retardant nanoadditives; thermally conductive nanocomposites. Protects circuits from ESD, humidity; facilitates thermal management by dissipating heat from electronic components.	ESD-shielding bags with CNTs (manufacturers of ESD packaging); graphene-infused power electronics enclosures; thin nanoceramic insulators for battery packs.
Logistics and cold chain	Nanofoam insulations (aerogels); phase-change microcapsules; antimicrobial/antifungal nanocoatings; integrated nanosensors. Longer freshness in transit, reduced losses; real-time shipment monitoring. Integrated nanosensors enable real-time shipment monitoring and traceability.	Sensorized cold chain shippers (e.g., Laerdal Nitroshippers); PCM nanocomposite insulated boxes; RFID + IoT platforms by DHL/Intel using nano-enabled labels; silver nanoparticle liners in produce crates.

7.3 Sustainable Development Goals (SGDs) and packaging

The UN SDGs serve as a comprehensive framework for promoting global equity, sustainability, and environmental responsibility by 2030. While often overlooked, packaging significantly influences these goals through its effects on resource efficiency, waste generation, pollution, and consumer health. As the demand for sustainable development accelerates, innovation in packaging materials and design has become essential—particularly in sectors such as food, healthcare, and logistics. Nanotechnology introduces advanced functionalities and environmentally friendly alternatives that can make packaging smarter, safer, and more sustainable, thereby contributing meaningfully to multiple SDGs.

7.3.1 SDGs addressed through nanotechnology-enabled packaging

a) **SDG 3: good health and well-being:**
Safe, hygienic packaging directly supports public health by preventing contamination and foodborne illness. **Antimicrobial surfaces** and **active barriers**, such as **nanoscale coatings**, can inhibit pathogens and block oxygen and moisture, keeping food fresh longer. **AgNPs** embedded in packaging kill bacteria, reducing spoilage and preventing contamination – an important contribution to food safety. Beyond passive protection, nanotechnology also enables **intelligent sensor tags** using **graphene or CNTs** that detect spoilage gases or toxins in real time, alerting consumers to quality issues. By extending shelf life, reducing waste, and enhancing safety, nano-enabled packaging plays a vital role in achieving SDG 3 (good health and well-being).

b) **SDG 6: clean water and sanitation:**
Nanotechnology-enabled packaging can contribute to clean water access through both direct purification tools and safer material design. For example, researchers have developed flexible sachets filled with **silver-based nanomaterials** that, when immersed in contaminated water, release disinfectant ions that kill microbes, enabling portable, low-cost purification without bulky infrastructure. Beyond direct purification, safe-by-design packaging materials – free from harmful leachables such as **PFAS** (persistent "forever chemicals" linked to serious health and environmental risks) – help prevent water contamination from packaging waste. Together, these nano-enabled solutions support SDG 6 by advancing point-of-use water treatment and reducing toxic runoff into water systems.

c) **SDG 9: industry, innovation, and infrastructure:**
Nanotechnology and data-driven tools are driving innovation in sustainable packaging by enabling the development of materials that are lighter, stronger, and more eco-

friendly. For instance, the International Iberian Nanotechnology Laboratory (INL) uses ultrasonic spray coating to apply nanolayered, bio-based polymers onto paper substrates, creating hydrophobic packaging that repels water and fats while remaining fully recyclable or compostable. These innovations offer high-performance alternatives to conventional plastics like polyethylene and polypropylene.

At the system level, intelligent systems and AI tools such as real-time sensors, predictive analytics, and automated quality control enhance manufacturing efficiency and foster smarter infrastructure. EU programs like FlexFunction2Sustain and Convert2Green demonstrate open-innovation ecosystems for circular, nano-enabled packaging. Together, these technological advancements support SDG 9 by strengthening industrial innovation and building resilient, sustainable infrastructure.

d) **SDG 12: responsible consumption and production:**
SDG 12 emphasizes reducing waste and resource consumption – areas where packaging design plays a critical role. Nano-enhanced, data-driven sustainable packaging supports a **circular economy** by promoting reusability, recyclability, and reduced material use. **Nanocomposite materials** (like nanoclays or nanocellulose in polymer films) improve barrier properties, so less packaging material is needed, reducing material consumption. Research projects are explicitly replacing fossil-fuel plastics with renewable nanomaterials.

e) **SDG 13: climate action:**
Packaging production and disposal contribute significantly to greenhouse gas emissions. Smart packaging solutions that combine nanotechnology and AI can reduce carbon footprints by optimizing material selection, supply chain logistics, and life cycle energy use. For example, lighter-weight or bio-based nanopackaging consumes less energy in production and transport. Additionally, by extending shelf life and minimizing food spoilage – a major source of methane emissions from landfills – nanopackaging helps lower overall greenhouse gas emissions. These innovations directly support SDG 13 by mitigating climate change through smarter, more sustainable packaging practices.

f) **SDGs 14 and 15: life below water and life on land:**
Packaging waste threatens terrestrial and aquatic ecosystems through litter, microplastics, and toxic leachates. SDGs 14 and 15 emphasize protecting biodiversity on land and below water, which is endangered by poorly managed packaging waste. Biodegradable nanopackaging and safe-by-design materials are vital solutions. For example, **chitosan and cellulose nanocomposites** (from renewable sources) are being developed as plastic alternatives that will fully degrade without harming soil or wildlife. In agriculture, **nanoparticle films** can even aid soil health by degrading agrochemicals and releasing benign residues. Together, nano-enabled packaging technologies and responsible end-of-life management help minimize ecosystem pollution and support biodiversity, fulfilling the goals of SDG 14 (life below water) and SDG 15 (life on land).

7.3.2 Circular economy

The **circular economy** is a systems-level economic model that aims to **eliminate waste** and **keep resources in use for as long as possible**:

Traditional linear model: Take → Make → Dispose

Circular model: Make → Use → Reuse → Recycle → Regenerate

In packaging, circular design principles promote resource efficiency and a low carbon footprint. Companies increasingly reduce material usage, adopt mono-materials or easily separable multilayer structures, and shift to renewable feedstocks. For instance, EU regulations now mandate design-for-recycling standards and enforce minimum recycled content targets –rising to 65% for plastic bottles by 2040.

Data-driven technologies like AI and nanotechnology are playing a transformative role in enabling circular economy principles within packaging systems. **AI-powered waste sorting systems** (e.g., by Tomra or AMP Robotics) use **computer vision** to accurately separate packaging waste by material, improving recycling rates and material recovery. Predictive analytics and ML optimize packaging design by simulating barrier performance or mechanical strength, allowing manufacturers to reduce material use without compromising quality. **Blockchain and digital product passports (DPPs)** embedded in smart packaging enable traceability, informing users about recyclability and composition, thus encouraging responsible reuse and recovery. On the nanotechnology front, **biodegradable nanocomposites such as nanocellulose and chitosan** enhance compostable packaging with better strength and barrier properties, making them viable alternatives to fossil-based plastics. **Nanotech-enabled sensors** integrated into packaging also reduce food waste by detecting spoilage or temperature anomalies. Together, these innovations help close material loops, reduce resource extraction, and advance the shift toward sustainable, circular packaging ecosystems.

Circular economy practices, especially when combined with nanotechnology and data-driven design, support sustainable packaging systems that minimize waste, extend product life cycles, and reduce environmental impact.

7.4 Data-driven approaches for advancing sustainable nanopackaging

Data-driven technologies are at the forefront of transforming nano-enabled packaging into smarter, safer, and more sustainable systems. As the demand grows for packaging that is both high-performing and eco-friendly, data-driven technologies like AI, ML, IoT, data analytics, digital twins, and blockchain are helping researchers and industries tackle challenges such as spoilage prediction, environmental impact, and supply chain transparency. By combining these digital tools with nanomaterials, we can better design, monitor, and optimize packaging systems to align with real-world sustainability goals. The following sections take a closer look at how each technology contributes to this transformation.

7.4.1 Artificial Intelligence (AI)

AI (especially ML and deep learning) can process large and complex datasets from packaging systems to improve design and operations. In nanopackaging, AI is widely applied for **predictive spoilage detection** and **smart material design**. For example, AI models can analyze real-time data from embedded sensors (e.g., VOC sensors, temperature/humidity logs, and imaging) to predict how long a food product will stay fresh and when it will spoil, enabling smarter inventory management and dynamic shelf-life labeling. Some startups have even developed AI-driven "smart films" that visually change color based on freshness indicators, cutting food waste by as much as 30% in pilot deployments.

Beyond freshness prediction, AI is also playing a role in **material and design optimization**. Using past performance data, AI models can suggest the best combination of nanomaterials and packaging structure – balancing strength, barrier function, cost, and sustainability. For instance, industry reports note that AI-driven data analysis can optimize box size and material thickness to reduce raw material use.

What is more, **generative AI** is emerging as a cutting-edge tool for designing novel materials. Models like **generative adversarial networks** and **variational autoencoders** can generate entirely new nanocomposite formulations with tailored properties, such as high oxygen resistance or biodegradability, before they are even tested in the lab.

- **Spoilage prediction:** Deep learning and predictive analytics use sensor data (gas emissions, pH changes, imaging) to forecast shelf life and spoilage time, potentially adapting labeling or pricing to minimize waste.
- **Design optimization:** AI optimizes package dimensions, layer structures, and nanofiller loadings to meet performance targets while minimizing material use. For example, packaging firms use AI tools to "right-size" packaging and simulate different designs for stress and durability.
- **Material discovery:** AI-driven material design (such as generative models) rapidly screens nanoreinforced polymers or biocomposites for desired functions (e.g., impermeability, strength, biodegradability).

Many of these applications are powered by ML – the predictive engine behind modern AI. Let us explore how ML contributes to performance modeling, safety assessment, and environmental forecasting in nanopackaging.

7.4.2 Machine learning (ML)

ML techniques (often overlapping with AI) enable performance modeling, safety assessment, and environmental prediction for nanomaterials. ML models can be trained on lab or simulation data to predict how a given nanocomposite will behave, without running exhaustive experiments. For example, ML regression or classification models have been developed to **predict nanomaterial toxicity**, a crucial concern for nanopackaging safety. A recent toxicology study demonstrated that ML classifiers (random forest, SVM, and neural nets) can predict cell death from nanoparticle exposure using inputs like particle size, composition, and dose. This "in silico" approach reduces both time and cost by virtually screening candidate nanoparticles, such as silver, TiO_2, and ZnO, for safety prior to conducting in vivo experiments.

ML is also used for **performance and barrier modeling** of packaging materials. For instance, regression models can forecast mechanical strength, oxygen/water permeability, or degradation rates from composition and process parameters. Similarly, ML can predict how packaging materials will release additives or degrade under environmental stress, aiding in life-cycle predictions.

- **Performance modeling:** ML algorithms (e.g., random forests and neural nets) model the relationship between nanoblend composition and material properties (strength, flexibility, and barrier) to accelerate formulation of sustainable nanocomposites.
- **Toxicity and safety:** Supervised ML predicts biological/toxicological outcomes of nanomaterials, enabling safer design. A random forest model, for example, achieved high accuracy in forecasting nanoparticle-induced cell death based on exposure data.
- **Environmental fate:** Emerging ML studies focus on predicting nanomaterial behavior in the environment (e.g., migration, degradation, and interaction with ecosystems), though literature is still growing.

While ML enables powerful predictions from data, the IoT serves as the bridge that collects that data in real time – from the packaging itself. Let us explore how IoT technologies are making nanopackaging smarter and more responsive.

7.4.3 Internet of Things (IoT)

IoT refers to networks of connected devices and sensors that monitor physical objects in real time. In sustainable nanopackaging, IoT enables smart tracking and condition monitoring throughout the supply chain. Tiny sensor labels or embedded RFID (radio frequency identification) tags can record a package's temperature, humidity, location, and shock history, all of which can be transmitted to the cloud. This real-time data helps maintain the cold chain and track product quality end to end.

For example, **Evigence** (a food-tech startup) embeds ultra-small time–temperature sticker sensors in produce packaging. These stickers measure cumulative temperature exposure and estimate current freshness; retailers and consumers can scan the sensor to see if a fruit or salad meets freshness thresholds. In trials, such IoT freshness monitors have been deployed in meal kit deliveries, and company data suggest ~20% longer shelf life and 33% less food waste from using real-time freshness data.

Similarly, IoT systems track packaging containers themselves (via GPS or RFID) to optimize logistics. Connected tags on cartons or pallets provide location and environment data, enabling dynamic routing or reordering when conditions change.

- **Supply chain tracking:** RFID/GPS and IoT platforms monitor package location, transit conditions, and handling, improving transparency and reducing loss.
- **Freshness sensors:** Embedded sensors (time–temperature indicators, humidity sensors, gas sensors, and pH indicators) send live data to AI models to detect spoilage, triggering alerts when quality falls. As one review observes, IoT in the food supply chain is nascent but can markedly extend shelf life by constant monitoring.
- **Cold chain management:** IoT-enabled cold boxes and refrigerated trucks keep logs of temperature history; AI analyzes these logs to signal risk (e.g., Time Temperature Indicator (TTI) sensors provide data that AI uses to optimize storage).

As IoT devices continuously generate real-time data from smart packaging systems, the next step is making sense of that data – this is where data analytics comes in, turning raw sensor inputs into actionable insights for sustainability and performance.

7.4.4 Data analytics

Data analytics is the hidden engine driving smarter, greener decisions in nano-enabled packaging – from design to disposal. Big data analytics and business intelligence underpin many sustainability efforts by **aggregating and interpreting large datasets** from packaging, manufacturing, and supply chains. In nanopackaging, analytic tools help quantify environmental impact (life cycle emissions, water use, and waste generation) and predict future trends. For example, LCA often relies on vast product and supply chain data; advanced analytics can streamline LCA by automating data collection and identifying hotspots of carbon or material use. Data platforms support **design for environment** by simulating material choices and estimating carbon footprint before full-scale implementation. An industry blog notes that analytics can refine LCA methodologies and make them more accessible to businesses.

Analytical platforms are also used to **monitor operational sustainability**. By collecting sensor and production data, companies can create dashboards tracking packaging waste rates, recycling rates, or energy use in real time. Predictive analytics can forecast a product's full carbon footprint or end-of-life impact based on material choices and volumes shipped. In practical terms, analytics can help a manufacturer *right-size* production and logistics: by analyzing historical sales and packaging data, analytics helps avoid overstocking or excessive transportation emissions:

- **Environmental monitoring:** Aggregate sensor/production data to quantify environmental KPIs (energy, water, and waste) and feed back into design decisions.
- **Life cycle prediction:** Use analytics platforms to model cradle-to-grave impacts and forecast carbon footprint or recyclability of packaging options.
- **Operational insights:** Analyze supply chain and sales data to optimize inventory and routing, preventing waste and minimizing carbon from transportation. For example, analytics can enable dynamic inventory management that aligns production with actual demand, cutting unnecessary packaging throughput.

These analytical insights form the foundation for advanced simulation technologies – especially digital twins, which bring real-time modeling and system-wide visibility to sustainable packaging.

7.4.5 Digital twins

A digital twin is a virtual replica of a physical object, system, or process. It simulates how a real-world system behaves, responds, and evolves –virtually, but in real time. Think of it like a simulation or a mirror model that reflects the state of the physical entity, but in a virtual space. These virtual models are powered by data from sensors, IoT devices, and other data-gathering technologies. In nanopackaging, digital twins can simulate packaging behavior throughout design, production, and use. For example, engineers create digital twins of a new packaging carton (including material properties and nanocomposite layers) to run simulated tests. These models can predict how the package withstands stacking, moisture, and dropping without building a physical prototype. This greatly accelerates design cycles and ensures durability with minimal resource waste [47, 48].

Digital twins also extend to the supply chain. By modeling logistics scenarios, companies can fine-tune packaging dimensions and stacking patterns to reduce transportation volume and lower emissions. Inside Packaging magazine describes how a digital twin of a production line or logistics network can analyze effects of different package sizes on truck loading and emissions. For sustainability, digital twins let designers virtually test **alternative materials** (e.g., biodegradable film vs plastic) before manufacturing, checking performance and environmental impact. Digital twins enable companies to evaluate recyclability, material efficiency, and carbon footprint of packaging options by simulating full life cycles:

- **Virtual prototyping:** Simulate packaging under stress (compression, vibration, temperature) to predict failure points and shelf life. For example, a digital twin can test carton rigidity under stacking pressure, reducing prototyping costs.

- **Supply chain modeling:** Model the impact of package design on transportation (truck space, fuel use) and on the circular economy (e.g., how many times a reusable container can be cycled). A digital twin can compute the ideal carton packing arrangement to maximize pallet efficiency and cut fuel use.
- **Life cycle simulation:** End to end, the twin can simulate material degradation, recyclability, or exposure to environmental conditions. As industry reports suggest, digital twins allow in silico testing of barrier properties, shelf life, and degradation, ensuring both performance and environmental compatibility.

While digital twins simulate and optimize packaging systems in real time, another emerging technology – blockchain – ensures the trust, traceability, and transparency of these systems across their life cycle.

7.4.6 Blockchain

Blockchain is a decentralized digital ledger that securely stores records across a network of computers in a way that is transparent, immutable, and resistant to tampering. It provides an unalterable record of transactions, which can ensure **transparency** in sourcing and life cycle of packaging materials. In sustainable nanopackaging, blockchain is primarily envisioned for enhancing **traceability and accountability** across the supply chain. For instance, IBM's blockchain technology has been utilized to trace food products from source to store, allowing consumers to scan QR codes on packaging to learn about the origins and journey of their food products. Blockchain along with AI are being piloted to verify authenticity: for example, **tamper-evident nanolaminate seals** – smart seals made using nanomaterials that change or break when someone tries to open the package – could log on-chain when a package is opened. In other words, blockchain combined with AI is being explored as a way to confirm if a product is genuine and has not been tampered with. Packaging industry analysts emphasize that blockchain's key benefit is transparency, which helps prevent **greenwashing** – a deceptive practice of conveying a false impression or providing misleading information about how a company's products are more environmentally sound than they truly are – and ensures compliance. For instance, a paper packaging manufacturer can use blockchain to prove its fibers come from certified forests and that nanocoatings are ethically sourced. In case of a product recall, blockchain enables **quick traceback** of affected lots. Importantly for sustainability, brands can append verifiable data on carbon footprint or recyclability to each product.

- **Supply chain transparency:** Blockchain logs each step from raw materials (e.g., nanoclay-mined) to manufacturing to end-of-life. Consumers or auditors can verify sustainability claims (certifications, carbon credits) by checking the chain.

- **Smart contracts:** Automated contracts on blockchain can enforce sustainable practices (e.g., only release payment if a material's certificate is validated) and streamline reporting.
- **Secure documentation:** Immutable records mean certifications (FSC for paper, eco-labels for bioplastics) cannot be counterfeited. This is crucial in a market wary of "greenwashing."
- **Integration with IoT:** Blockchain can ingest IoT sensor data (e.g., temperature logs) in tamper-proof form. For instance, a sensor reading from a smart package could be stored on-chain to prove cold chain compliance.

To better compare these technologies in the context of sustainable nanopackaging, the following table provides a side-by-side overview of their core roles, applications, benefits, and implementation challenges.

Table 7.3: Comparative analysis of digital technologies in sustainable nanopackaging.

Approach/ technology	Key roles in nanopackaging sustainability	Examples/ applications	Benefits	Challenges
AI	Spoilage prediction; packaging/material design optimization; adaptive logistics	Amazon's Package Decision Engine; FreshTrack spoilage-**AI film** (color-changing); AI-driven design tools in packaging firms	Automates complex analyses; reduces trial and error; can yield lighter, smarter packaging	Requires large datasets; "black-box" models; high compute cost
ML	Performance modeling of nanocomposites; toxicity and safety prediction; environmental fate modeling	Nanoparticle toxicity models (random forest on cell data); predictive models of mechanical/barrier properties	Accelerates R&D by predicting outcomes; handles nonlinear relationships; improves safety screening	Data quality and quantity; model interpretability; transferability
IoT	Smart tracking (RFID/ GPS) of packages; sensor-based freshness monitoring; cold chain control	Evigence freshness stickers (real-time produce monitoring); RFID/humidity sensors in smart cartons; GPS-tracked pallet monitoring	Provides real-time visibility; reduces spoilage and losses; enables dynamic supply chain optimization	Sensor cost and power; connectivity in transit; data privacy/security

Table 7.3 (continued)

Approach/ technology	Key roles in nanopackaging sustainability	Examples/ applications	Benefits	Challenges
Data analytics	Life cycle and carbon-footprint analysis; inventory and demand forecasting; material usage optimization	Analytics platforms for LCA; inventory systems predicting packaging demand; carbon-footprint tools integrated with supply data	Informs sustainable decisions; identifies waste/savings; quantifies impact metrics (CO_2, waste)	Garbage-in, garbage-out: needs quality data; complex modeling; integration of disparate datasets
Digital twins	Virtual prototyping of packaging; supply-chain and logistics simulation; lifecycle/ climate-impact simulation	Arafura (Aust. alumina co.) packaging digital twin; bi0SpaCE digital-twin product passports; Tetra Pak using twins for packaging lines	Tests new designs and materials without physical waste; optimizes transport and reusability; embeds sustainability early	Building accurate models requires detailed data; high setup complexity; continuous data sync needed
Blockchain	Provenance and traceability of raw materials; transparent sustainability claims; automated compliance	Packaging-traceability pilots (e.g., scanning blockchain QR codes); bioSpaCE digital passports on blockchain	Immutable audit trail; prevents greenwashing; empowers consumer trust; integrates with IoT data	Implementation cost and scalability; lack of standards; energy use of some chains; requires broad adoption

The integration of data-driven technologies into nano-enabled packaging is rapidly transforming how materials are designed, monitored, and optimized for sustainability. From AI-driven material modeling to blockchain-backed traceability, these technologies are helping the industry move toward smarter, more responsible packaging systems. Together, they enable data-informed decisions that reduce waste, improve performance, and ensure transparency across the product life cycle.

However, as we embrace these digital advancements, it is equally critical to evaluate their real-world environmental and health implications. The next section explores how nanopackaging materials interact with ecosystems and human health – key dimensions in assessing their overall sustainability.

7.5 Environmental and health impacts of nanomaterials in nanotech-enabled packaging

Nanotechnology has emerged as a powerful enabler of sustainable and smart packaging systems, offering enhanced material performance, extended product shelf life, and reduced environmental burden. However, alongside these advancements, the use of nanomaterials in packaging applications raises important concerns regarding their potential environmental and human health impacts. While certain nanomaterials contribute to reducing plastic waste and food spoilage – thus indirectly **mitigating climate change** – others may pose toxicity risks, present challenges in end-of-life disposal, and exert unforeseen effects on ecosystems. This section critically examines both the positive contributions and the emerging challenges associated with nanotech-enabled packaging, with particular focus on environmental sustainability, human health safety, and regulatory considerations.

7.5.1 Positive impacts of nanomaterials in packaging

a) **Mitigating climate change:** Climate change mitigation is increasingly being addressed through nanotech-enabled packaging, both directly and indirectly.

   i) **Reduction in carbon footprint**

   - Nanocomposites are stronger and lighter, allowing thinner packaging with equal or better performance, which reduces raw material use and transportation emissions.
   - Lighter packaging = reduced fuel consumption in logistics, thus reducing excessive emissions of GHGs (greenhouse gases) such as carbon dioxide, methane, nitrous oxide etc. due to packaging production and waste.

   ii) **Reducing plastic consumption and waste generation**

   - Nanomaterials such as nanoclay, nanocellulose, and graphene oxide can reinforce biodegradable films (e.g., PLA, starch, and chitosan), enabling thinner and lighter packaging that performs like thicker conventional plastic.
   - Such lightweight nanoscale packaging **requires fewer resources** and is **more easily recycled**, thereby cutting overall plastic consumption. By enabling thinner multilayer films and stronger plastics, nanotechnology can directly reduce packaging volume and waste.
   - Reduction in plastic production (fossil fuel-based), lowers landfill GHGs, and lessens ocean pollution that releases methane and ethylene.

iii) **Food waste reduction (methane prevention)**

- Nanotech packaging with **antimicrobial nanoparticles (Ag, ZnO)** or **smart spoilage sensors** reduces microbial spoilage, extending shelf life for perishable items (meat, dairy, fruits).
- Food waste generates **methane** in landfills – one of the most potent greenhouse gases. Preventing food loss **cuts GHGs significantly**.

iv) **Improved energy efficiency in storage and transport**

- Certain nanomaterials (e.g., **aerogels** and **nanoinsulators**) can be used in packaging for temperature-sensitive goods like vaccines, dairy, or seafood.
- This helps reduce the **energy needed for refrigeration** and **cold chains**, which are major GHG contributors.

v) **Smart recycling with nanosensors**

- Nanotech-enabled packaging can integrate **embedded RFID/NFC sensors or color-changing indicators** to sort waste more effectively, signal when packaging is safe to compost or recycle and identify contamination risks.
- This improves recycling rates, lowers incineration or landfill use, and supports circular economy – all of which help cut GHGs.

Integrating nanomaterials with intelligent digital technologies – such as AI-based life cycle analysis, IoT-enabled spoilage monitoring, and blockchain-powered recycling traceability – amplifies their climate benefits, reinforcing nanotech-enabled packaging as a vital tool in combating climate change.

b) **Improved food safety and shelf life:**

- Incorporating **antimicrobial and barrier nanomaterials** into packaging greatly extends food freshness. **Silver, zinc oxide, titanium dioxide** and other nanoparticles inhibit bacterial growth and oxidation.
- **AgNPs** embedded in cellulosic films significantly inhibit spoilage organisms (e.g., *E. coli* and molds) and extend produce life.
- In one study, a tomato stored in AgNP nanocomposite packaging remained visibly fresh and microbiologically stable for up to 30 days longer than the control sample, demonstrating the powerful antimicrobial and shelf-life-extending capabilities of nano-enabled active packaging systems.
- **Active films** can also block UV or scavenge ethylene: for instance, **TiO_2 nanoparticles** in packaging have ethylene-scavenging and UV-shielding properties that slow fruit ripening.
- Researchers at NIT, Andhra Pradesh created nanoparticle-reinforced biopackaging that boosts mechanical stability and imbues antimicrobial sensors to detect pathogens and spoilage.
- Intelligent "smart" packaging with embedded nanosensors can signal spoilage (e.g., color-changing freshness tags) or monitor gas levels, helping to prevent foodborne illness and reducing waste by giving producers and consumers real-time safety information.

By integrating these antimicrobial and smart nanomaterials with AI-driven spoilage prediction, IoT-based cold chain monitoring, and real-time data analytics, nano-enabled packaging systems can drastically reduce food waste, enhance consumer safety, and build a responsive, intelligent food supply chain.

c) **Enhanced recyclability and circularity:**
  - Nanotechnology is enabling smarter, more circular packaging. For example, bio-based nanocoatings are being developed to make recyclable materials viable for foods normally requiring plastic. INL researchers created a nanostructured coating that makes paper hydrophobic (water- and fat-repellent) so it can safely pack moist foods, yet the paper remains compostable/recyclable.
  - By replacing conventional plastics with bio-based polymers and nanoscale functional coatings, the resulting packaging is **recyclable** and **compostable**.
  - More broadly, nanotechnology is also used in sorting and tracking: embedded RFID tags and printed QR codes (enabled by conductive nano-inks) can label the material composition of packaging, aiding automated separation at recycling centers. Active and smart nano-enabled packaging can simplify recycling and cut material inputs.

  By integrating nanomaterials with data-driven technologies such as AI-powered material recognition, IoT-enabled sorting infrastructure, blockchain-based material passports, and data analytics for tracking recycling patterns, packaging systems can become more intelligent and efficient, enhancing recyclability, reducing contamination, and facilitating truly circular life cycle management.

d) **Environmental remediation:**
  Environmental remediation refers to the process of removing pollutants, contaminants, or hazardous substances from soil, water, air, or sediments — to restore the environment to a safe and usable condition.
  - Nanotechnology is increasingly being leveraged in packaging not just to prevent pollution, but also to actively remediate environmental contaminants. Certain nanomaterials – such as titanium dioxide (TiO_2), zinc oxide (ZnO), iron oxide (Fe_3O_4), and carbon-based nanostructures – exhibit photocatalytic, adsorptive, or redox properties that enable them to degrade pollutants, neutralize toxins, or capture heavy metals.
  - For instance, TiO_2 nanoparticles embedded in packaging films can degrade organic pollutants like pesticides or dyes upon exposure to UV light, contributing to cleaner disposal or even environmental cleanup when such materials enter waste streams.
  - Similarly, iron oxide nanoparticles can adsorb heavy metals like lead or cadmium from contaminated water sources, offering potential applications in agricultural or aquaculture packaging systems.
  - Some nanopackaging solutions are designed to release neutralizing agents in response to environmental triggers, enabling them to remediate harmful by-

products in industrial or marine environments. These functional materials act as passive or active agents of environmental detoxification – an emerging area at the intersection of packaging and green engineering.

When combined with AI and ML algorithms for real-time pollutant detection and IoT-based sensors for environmental monitoring, these systems become increasingly intelligent and responsive. Digital Twins can simulate contaminant dispersion, while Blockchain ensures secure and traceable documentation of remediation events. With LCAs powered by data analytics, nano-enabled packaging evolves into a dynamic platform for environmental surveillance and restoration – ultimately supporting ecological resilience and sustainable waste systems.

e) **Consumer safety:**
- **Microplastics** from degraded packaging are now found in human blood, placenta, and vital organs. These particles have been associated with hormonal disruption, immune system impairment, and increased cancer risks, raising serious public health concerns.
- Nanocomposites, such as nanoclay or nanocellulose-reinforced biopolymers, enhance the structural integrity of packaging films, making them more resistant to mechanical breakdown. This helps reduce the formation of micro- and nanoplastics during use or after disposal.
- Conventional plasticizers like phthalates, often linked to reproductive toxicity, can be replaced by nanofillers that preserve flexibility without harmful chemical leaching. This reduces health hazards associated with direct and indirect contact with packaging.
- Nanomaterials improve barrier performance, limiting oxygen, moisture, and chemical migration. This not only protects food better but also prevents the leaching of plastic fragments or additives into the environment or human body.

As we have seen above, nanotechnology-enabled packaging and nanomaterials offer significant positive impacts on both environmental sustainability and human health. However, these advancements also come with certain **negative implications** that must be carefully examined and addressed to ensure alignment with **enhanced and responsible SDGs.**

7.5.2 Challenges associated with using nanomaterials in packaging

a) **Toxicological risks to human and animal health:**
- Many nanoparticles raise health concerns. AgNPs, while excellent antimicrobials, readily dissolve in stomach acid and accumulate in organs. Animal studies show that ingested nanosilver is sequestered (30–99%) in the liver,

where it induces hepatocyte necrosis and apoptosis. Additionally, inhaled AgNPs linger in the lungs and cause oxidative injury.
- **CNTs** are extremely strong but fibrous, and inhalation can cause inflammation and fibrosis akin to asbestos. Aquatic studies find that CNT exposure in fish leads to oxidative stress and organ toxicity, which further impacts the food chain.
- **Nanoclays** (e.g., organo-modified montmorillonite) used for strength and barrier enhancement also pose risks: in vitro tests with lung cells showed that certain nanoclay particles significantly reduce cell viability and alter morphology.
- Moreover, even **TiO_2 nanoparticles**, long used as white pigments and UV blockers, have come under scrutiny. The EU's food safety agency (EFSA) recently concluded that TiO_2 (additive E171) **can no longer be considered safe** due to possible **genotoxic effects**. (The United States still permits TiO_2 as a food/color additive, but emerging data may prompt review.)

b) **Waste management and end-of-life concerns:**
- Packaging materials with nanofillers complicates end-of-life treatment. Many nano-enabled plastics cannot be safely composted or biodegraded; when incinerated, they may release nanoparticles into emissions or ash.
- Landfill leaching is also a concern: engineered nanoparticles can gradually migrate out of discarded plastics. For example, nanocoatings on cardboard might detach and release nanosilver or titania (i.e., titanium dioxide or TiO_2) into leachate.
- Recycling streams may be disrupted if nanomaterials interfere with reprocessing or contaminate batches.
- Overall, dedicated studies of nanopackaging waste are limited, but regulators warn that safe disposal guidelines (like specialized hazardous waste labeling for nanomaterials) are needed to avoid environmental release.

c) **Occupational exposure and worker safety:**
- Inhalation or skin exposure to nanoparticles (e.g., during packaging film production) can pose occupational hazards.
- Factories using dry nanopowders (like TiO_2 or CNTs) must use strict containment and personal protective equipment (PPE), which raises costs.

d) **Biodegradability limitations of some nanomaterials:**
- While the base packaging might be biodegradable, **embedded nanoparticles often are not**.
- For example, silver or CNTs do not degrade in compost – this undermines the "eco-friendly" claim of some packaging solutions.

This raises critical questions about the true environmental sustainability of nano-enabled packaging.

To address the challenges associated with nano-enabled packaging, researchers and industries are increasingly focusing on safer-by-design nanomaterials that balance functionality with reduced toxicity. Innovations in biodegradable nanocomposites, such as cellulose nanocrystals or bio-based nanoclays, offer promising alternatives that degrade more safely while maintaining barrier performance. Furthermore, standardized LCAs and risk evaluation frameworks are being developed to better understand environmental and human impacts at every stage – from production to disposal.

Additionally, data-driven technologies like ML and predictive modeling are playing a pivotal role in accelerating this shift. AI can simulate nanoparticle behavior in biological or environmental systems to forecast toxicity before real-world deployment. Blockchain and IoT-enabled systems offer transparency across the supply chain, helping to track material origin, handling conditions, and end-of-life pathways. By integrating nanotechnology with smart data systems, packaging solutions can be optimized for performance and for health, safety, and sustainability.

As nano-enabled packaging becomes more widespread, managing its end-of-life impacts is crucial. Beyond addressing toxicity and safety, attention is shifting toward waste handling, recycling, and environmental fate. Emerging digital technologies, such as AI, IoT, and smart sensors, now play a key role in improving traceability, optimizing recycling systems, and supporting sustainable material flows. The next section explores how such innovations intersect with nanotech-enabled packaging waste management.

7.6 Waste management and recycling technologies in nanotechnology-enabled packaging

As nanotechnology becomes increasingly integrated into packaging across sectors like food, pharmaceuticals, and electronics, effective waste management and recycling strategies are essential. Nano-enabled materials, while offering enhanced performance, pose new challenges to traditional recycling systems. Leveraging technologies such as AI, ML, IoT, blockchain, and digital twins can enable smarter identification, sorting, and circular use of nanopackaging waste. This section explores how these digital tools are reshaping waste management practices to support sustainable, data-driven packaging systems.

7.6.1 Challenges in managing waste from nanotechnology-enabled packaging

Nanotechnology-enabled packaging includes a broad spectrum of materials and functionalities – ranging from nanobiosensors for spoilage detection in food products to antimicrobial coatings using AgNPs, and nanoreinforced polymers with clay or CNTs for enhanced barrier strength. While these innovations improve packaging performance and shelf life, they introduce significant challenges in post-consumer waste management and recycling processes.

One of the foremost issues is the heterogeneity of nanomaterials used across different products. Each type – such as nanoclays, nanosilver, titanium dioxide (TiO_2), or quantum dots – requires distinct treatment and recycling protocols. For example, plastic films embedded with nanoclays do not behave similarly to conventional polymers during mechanical or thermal recycling, often leading to contamination or inefficiencies in material recovery.

Furthermore, identifying and quantifying nanomaterials in mixed waste streams remains technologically complex. Standard recycling labels and classification systems typically do not indicate the presence of nanomaterials, making segregation and safe handling nearly impossible at the consumer and municipal levels. The environmental persistence and potential toxicity of certain nanomaterials – particularly heavy metals or reactive oxides – are still not fully understood, raising concerns about long-term ecological and human health effects.

In the absence of well-defined regulatory frameworks and labeling norms, most nano-enabled packaging waste is funneled into general plastic or metal streams, undermining both recyclability and safety. Traceability throughout the packaging life cycle is critical for enabling circular use; however, there is a lack of standardized data protocols or digital infrastructure to support this. Without targeted strategies, the increasing use of nanomaterials in packaging may exacerbate environmental and health risks and poses a barrier to sustainable circular economy models.

7.6.2 Existing strategies for the disposal and recycling of nanopackaging waste

Current waste management of nanotechnology-enabled packaging still relies heavily on traditional disposal routes: landfilling, incineration, and mechanical recycling. However, these strategies face limitations when dealing with embedded nanomaterials.

a) **Landfilling** is the most common method worldwide. While simple, it raises concerns for nanowaste like coated plastics, CNT films, or AgNP-infused packaging. Engineered nanomaterials (ENMs) may leach into groundwater or persist indefinitely, especially under acidic landfill conditions. Regulatory frameworks rarely differentiate nanocontent, leaving risk assessment uncertain.

b) **Incineration** is used extensively in Europe, East Asia, and for hazardous waste. High temperatures can degrade organic matrices, but many nanomaterials survive or transform, ending up in fly ash or bottom ash. For example, CNTs and metal oxides like titanium dioxide (TiO_2) can persist post-incineration. Modern filters capture some nanoparticulates, but efficiency is variable. The U.S. EPA conservatively assumes **0% destruction efficiency** for nanomaterials.
c) **Mechanical recycling**, including re-pelletizing and remelting of plastics, is widely practiced but poorly adapted to nanocomposites. While nanomaterials can enhance recycled plastic strength or barrier properties, they also complicate sorting and processing. Recyclers often reject unknown composites. During reprocessing, nanoparticles may be released through abrasion or thermal degradation, raising occupational and environmental risks.
d) **Chemical recycling** (e.g., pyrolysis or solvolysis) is emerging as an alternative, breaking plastics into fuels or monomers. However, the fate of embedded nanomaterials during thermal breakdown remains poorly understood – some may persist in char or emit as aerosols. Other experimental options include immobilization (e.g., blending nanowaste into concrete) or vitrification, but these are still costly and unscalable.

Policy responses remain fragmented. The EU's upcoming packaging and packaging waste regulation (PPWR) encourages design-for-recycling and extended producer responsibility (EPR), indirectly pressuring manufacturers to avoid complex nanocomposites. However, specific disposal guidelines for nanopackaging are still lacking. The United States manages nanoscale materials under general chemical laws, with few waste-specific rules. In Asia, countries like Japan and South Korea enforce strong recycling mandates, while China's rapid waste generation outpaces current nano-specific policy. India, meanwhile, has introduced regulations targeting plastic waste (such as the Plastic Waste Management Rules, 2016, and EPR guidelines for plastic packaging), but these do not yet explicitly address nanomaterials.

Overall, while traditional disposal methods dominate, their compatibility with nano-enabled packaging remains limited. Research on nanowaste risk remains limited, and enforcement varies across countries and states, suggesting the need for more targeted frameworks, as nanotech use in packaging grows.

Building on the challenges and limitations of current disposal methods, emerging data-driven and smart technologies offer promising solutions to improve the management and tracking of nanotech packaging waste.

7.6.3 Data-driven and smart technologies for efficient nanowaste management

While traditional methods for managing packaging waste focus on physical and chemical recycling techniques, data-driven technologies open up new opportunities to make nanotechnology-enabled waste management more precise, predictive, and efficient. These approaches utilize the power of smart systems – including ML, IoT, data analytics, and blockchain – to enhance waste identification, sorting, prediction, and resource recovery. This section explores how data-driven tools can help address key challenges in nanotech packaging waste management and contribute to a more sustainable packaging ecosystem.

a) **Smart sorting and classification**
   One of the core challenges in nanopackaging waste management is accurately identifying and separating nanomaterial-based packaging from conventional waste. ML and computer vision models can be trained on large datasets of packaging types to classify materials based on physical features (e.g., transparency, color, and barcode) or spectral signatures from sensors. For example, hyperspectral imaging combined with AI models can distinguish between nanocomposite plastics and regular polymers in real time, improving the accuracy of recycling streams and reducing contamination rates.
b) **Predictive analytics for waste generation and flow**
   AI models can also be used to predict the volume and composition of nanopackaging waste generated across supply chains. By analyzing data from packaging production, sales, and disposal records, ML algorithms can forecast future waste patterns – helping municipalities and industries plan better for waste collection, treatment, and recycling capacity. This is particularly useful in identifying peak waste periods or potential bottlenecks in the handling of nanomaterials that require specialized processing.
c) **Digital twins for simulation and optimization**
   Digital twin technology – virtual replicas of physical packaging systems – can simulate the entire life cycle of nano-enabled packaging, from production to disposal. This enables researchers to test how different nanomaterials behave in the environment, how long they take to degrade, or how effectively they can be recovered using various recycling techniques. These simulations guide the design of more recyclable nanopackaging and optimize waste-handling systems to minimize resource loss.
d) **Blockchain for transparent tracking and accountability**
   Blockchain technology can play a significant role in traceability and accountability of nanotech packaging waste. It enables secure, tamper-proof tracking of nanomaterials throughout their life cycle – from manufacturing to post-consumer disposal. Such decentralized records increase transparency, support regulatory

compliance, and can incentivize companies to recover and recycle nanomaterials more responsibly.

e) **Real-time monitoring using IoT sensors**
IoT sensors embedded in waste bins, recycling facilities, or smart packaging can monitor real-time data such as fill levels, composition, temperature, or chemical signatures. This helps waste management systems detect hazardous nanowaste early, optimize collection schedules, and ensure safe handling of sensitive materials. Combined with edge computing, this data can be processed locally for faster response times and better decision-making.

Table 7.4: Data-driven approaches for efficient nanopackaging waste management.

Approach	Technologies involved	Application / function	Sustainability impact
Intelligent waste sorting	Machine learning, computer vision, and hyperspectral imaging	Automatically detects and classifies nanowaste materials in recycling facilities	Reduces contamination and increases recycling efficiency
Waste generation forecasting	Predictive analytics and big data	Anticipates volume and type of nanopackaging waste based on usage and trends	Enables proactive planning, reduces overflow and mismanagement
Life cycle simulation	Digital twins, cloud computing, and AI	Simulates product lifecycle to optimize design, reuse, and end-of-life management	Supports eco-design and minimizes environmental footprint
Traceability	Blockchain and smart contracts	Tracks nanomaterials from manufacturing to disposal with secure, tamper-proof records	Ensures responsible handling and regulatory compliance
Real-time waste monitoring	IoT sensors and edge computing	Continuously monitors waste bins for toxicity, fill levels, or leakage	Promotes timely intervention, improves worker safety and operational control

Despite growing awareness of nanopackaging waste, existing disposal and recycling strategies often fall short due to limited scalability, inefficiencies in detecting nano-contaminants, and a lack of real-time monitoring and traceability. These limitations underscore the urgent need to move beyond conventional waste management approaches. By incorporating advanced data-driven technologies – such as AI-powered waste sorting, IoT-enabled tracking systems, and predictive analytics as discussed above – we can develop smarter, more adaptive, and resource-efficient recycling ecosystems. These innovations not only enhance the feasibility and effectiveness of nanowaste management but also act as transformative tools in achieving circularity, reducing environmental harm, and aligning nano-enabled packaging with long-term sustainability goals.

7.7 Challenges, gaps, and future opportunities in nanotech-enabled packaging

Throughout this chapter, we have navigated the transformative potential of nanotechnology in redefining the packaging industry to align with the imperatives of sustainability, safety, and circularity. Beginning with an in-depth understanding of nanomaterials – such as AgNPs, CNTs, and nanoclays – we explored their structural and functional roles in enhancing packaging performance. From antimicrobial properties to barrier enhancement and smart sensing capabilities, these innovations clearly demonstrate how nanotech-enabled packaging is not just an improvement over conventional materials but a necessary evolution in light of mounting environmental and industrial challenges.

However, technological advancement alone is insufficient. We delved into how the SDGs – particularly SDG 9 (industry, innovation, and infrastructure), SDG 12 (responsible consumption and production), SDG 13 (climate action), and SDG 3 (good health and well-being) – serve as guiding frameworks. Achieving these goals requires not only adopting cutting-edge materials but also ensuring that their life cycles are ethically, environmentally, and economically sustainable.

A core contribution of this chapter is the integration of nanotechnology with data-driven technologies. Section 7.4 highlighted how AI, ML, IoT, blockchain, digital twins, and data analytics can significantly enhance the efficiency, traceability, predictive capabilities, and sustainability of nano-enabled packaging. These tools present novel solutions to longstanding problems such as contamination detection, real-time condition monitoring, and waste stream optimization, paving the way for intelligent, adaptive packaging systems.

Despite these promising directions, serious challenges and gaps remain. Section 7.5 underscored critical concerns regarding the environmental and health impacts of nanomaterials, many of which remain under-researched. The toxicity, bioaccumulation, and long-term fate of nanomaterials in ecosystems and human bodies are areas that demand urgent and rigorous study. Likewise, Section 7.6 revealed that current recycling and waste management frameworks are ill-equipped to handle the complexity and scale of nanowaste, lacking in both real-time precision and economic feasibility.

To bridge these gaps, systemic innovation is essential. This includes:

- Technological innovations that integrate nanomaterials with intelligent digital systems for full life cycle monitoring.
- Regulatory frameworks that are dynamic and adaptive to the evolving science of nanomaterials.
- Social and industrial reforms that embrace sustainability as a design principle, not just an afterthought.

What emerges is a clear call for interdisciplinary and multi-stakeholder collaboration. Material scientists must work alongside AI and data experts, environmental researchers, industry leaders, and policymakers. Only by bringing together these diverse perspectives can we hope to build resilient, scalable, and ethically responsible packaging systems.

Looking forward, nanotechnology is poised to become the backbone of next-generation packaging. Its potential to address pressing global issues – such as climate change, food security, supply chain transparency, and material efficiency – is unmatched. But realizing this potential requires bold action: more open data-sharing frameworks, investment in sustainable research, development of green nanomaterials, and public–private partnerships that scale pilot innovations into global standards.

In conclusion, nanotech-enabled packaging – when combined with intelligent computing technologies – holds the key to a future that is not only more sustainable but also more secure, efficient, and aligned with planetary boundaries. As we confront the dual crises of environmental degradation and unsustainable consumption, this convergence of disciplines offers a powerful roadmap for innovation and hope. The next generation of packaging must be not only smart and sustainable but also systemic – capable of reshaping industry standards and contributing meaningfully to a healthier, climate-resilient planet.

References

[1] Ibrahim, I. D., Hamam, Y., Sadiku, E. R., Ndambuki, J. M., Kupolati, W. K., Jamiru, T., Eze, A. A., & Snyman, J. (2022). Need for sustainable packaging: An overview. *Polymers*, 14, 4430. doi: 10.3390/polym14204430

[2] Conti, G. O., Rapisarda, P., & Ferrante, M. (2024). Relationship between climate change and environmental microplastics: a one health vision for the platysphere health. *One Health Advances*, 2, 17. doi: 10.1186/s44280-024-00049-9

[3] Zheng, L., & Suh, S. (2019). Strategies to reduce the global carbon footprint of plastics. *Nature Climate Change*, 9, 374–378. doi: 10.1038/s41558-019-0459-z

[4] Geyer, R., Jambeck, J. R., & Law, K. L. (2017). Production, use, and fate of all plastics ever made. *Science Advances*, 3, e1700782. doi: 10.1126/sciadv.1700782

[5] Jiao, H., Ali, S. S., Alsorbany, M. H. M., Ebnalwaly, T., Abdelkarim, E., Schagerl, M., Al-Tohamy, R., & Sun, J. (2024). A critical review on plastic waste life cycle assessment and management: Challenges, research gaps, and future perspectives. *Ecotoxicology and Environmental Safety*, 271, 115942. doi: 10.1016/j.ecoenv.2024.115942

[6] Gavrilescu, D., Seto, B. C., & Teodosiu, C. (2023). Sustainability analysis of packaging waste management systems: A case study in the Romanian context. *Journal of Cleaner Production*, 417, 138578. doi: 10.1016/j.jclepro.2023.138578

[7] Ciawi, Y., Tonyes, S. G., & Dwipayanti, N. M. U. (2025). Multilayer packaging recycling: challenges, current practices, and future prospects. *Academia Environmental Sciences and Sustainability*, 2, 1–12. doi: 10.20935/AcadEnvSci7521

[8] Purushothaman, R., Alamelu, R., Selvabaskar, S., & Sudha, M. (2025). Theories, techniques and strategies of sustainable circular economy: a systematic literature review. *Circular Economy and Sustainability*, 6, 1–26. doi: 10.1007/s43621-025-01161-5

[9] Zhu, Z., Liu, W., Ye, S., & Batista, L. (2022). Packaging design for the circular economy: A systematic review. *Sustainable Production and Consumption*, 30, 1–15. doi: 10.1016/j.spc.2022.06.005

[10] Chen, Y., Awasthi, A. K., Wei, F., Tan, Q., & Li, J. (2021). Single-use plastics: Production, usage, disposal, and adverse impacts. *Science of The Total Environment*, 752, 141772. doi: 10.1016/j.scitotenv.2020.141772

[11] Nayanathara, P. G. C. T. P., & Ratnayake, A. S. (2024). The world of plastic waste: A review. *Cleaner Materials*, 4, 100220. doi: 10.1016/j.clema.2024.100220

[12] Lehel, J., & Murphy, S. (2021). Microplastics in the Food Chain: Food Safety and Environmental Impact. In: *Environmental Contaminants and Human Health*. Springer, pp. 1–20. doi: 10.1007/398_2021_77

[13] Cozzolino, A., & De Giovanni, P. (2023). Portfolios of sustainable practices for packaging in the circular economy: An analysis of Italian firms. *The International Journal of Logistics Management*, 34, 24–49. doi: 10.1108/IJLM-03-2022-0132

[14] Liu, W., Zhu, Z., & Ye, S. (2023). A Framework Towards Design for Circular Packaging (DfCP): Design Knowledge, Challenges and Opportunities. *Circular Economy and Sustainability*, 3, 2109–2125. doi: 10.1007/s43615-023-00264-3

[15] Wang, M. (2023). The role of sustainable packaging in reducing environmental pollution. *GeoJournal*, 5, 1–10. doi: 10.22271/27067483.2023.v5.i1c.268

[16] Kagdada, H. L., Bhojani, A. K., & Singh, D. K. (2023). An Overview of Nanomaterials: History, Fundamentals, and Applications. In: *Nanomaterials: Fundamentals and Applications*. Springer, pp. 1–26. doi: 10.1007/978-981-19-7963-7_1

[17] Reshma, R., ., Shaik, H., Venkatesan, K., Venugopal, A., Kadeer, M., & Puppala, V. S. (2022). A review on nanomaterials: introduction, synthesis, characterization, and applications. *Chemical Papers*, 79, 1234–1256. doi: 10.1007/s11696-025-04107-3

[18] Singh, A. K., Sharma, R. K., & Kumar, M. (2021). Nanotechnology: Applications, challenges, and prospects. In: *Green Energy and Technology*. Springer, pp. 1–20. doi: 10.1007/978-3-030-95820-6_1

[19] Ahmad, K., Li, Y., Tu, C., Guo, Y., Yang, X., Xia, C., & Hou, H. (2024). Nanotechnology in food packaging with implications for sustainable outlook and safety concerns. *Food Bioscience*, 58, 103625. doi: 10.1016/j.fbio.2024.103625

[20] Bendas, M. (2022). Pharmaceutical nanotechnology: from the bench to the market. *Future Journal of Pharmaceutical Sciences*, 8, 12. doi: 10.1186/s43094-022-00400-0

[21] Ghosh, S., Mandal, R. K., Mukherjee, A., & Roy, S. (2025). Nanotechnology in the manufacturing of sustainable food packaging: A review. *Discover Nano*, 20(36), doi: 10.1186/s11671-025-04213-x

[22] Goosey, M. (2009). Nanopackaging: Nanotechnologies and Electronics Packaging. *Circuit World*, 35, 5–8. doi: 10.1108/cw.2009.21735cae.001

[23] Kausar, A. (2020). A review of high performance polymer nanocomposites for packaging applications in electronics and food industries. *Journal of Plastic Film & Sheeting*, 36, 94–112. doi: 10.1177/8756087919849459

[24] Usha, T., Middha, S. K., Hemavathi, K. N., Hemalatha, P., Harini, R., Afreen, Z., Dwivedi, N., Akshaya, K. A., Goyal, A. K., & Baliga, M. S. (2025). Innovations in nanotechnology for food production and packaging: A review. *Current Nutrition & Food Science*, 21, 559–572. doi: 10.2174/0115734013297685240820043042

[25] Khan, M. A. (2024). Silver Nanoparticles: Synthesis, Structure, Properties and Applications. *Nanomaterials*, 14, 1425. doi: 10.3390/nano14171425

[26] Rai, A., Yadav, A., & Gade, A. (2009). Silver nanoparticles as a new generation of antimicrobials. *Biotechnology Advances*, *27*, 76–83. doi: 10.1016/j.biotechadv.2008.09.002

[27] Gobena, S. T., & Woldeyonnes, A. D. (2021). An Investigative Study on the Progress of Nanoclay-Reinforced Polymer Nanocomposites: A Review. *Polymers*, 13, 4401. doi: 10.3390/polym13244401

[28] Dasgupta, A., Bajpai, P., & Mitra, J. (2024). An assessment of extracted nanocellulose from agricultural wastes and its effect on the properties of developed biodegradable packaging film: a meta-analysis. *Journal of Food Science and Technology*, 62, 1234–1245. doi: 10.1007/s13197-024-06079-w

[29] Thakur, V., Guleria, A., Kumar, S., Sharma, S., & Singh, K. (2021). Recent advances in nanocellulose processing, functionalization and applications: a review. *Materials Advances*, 2, 1872–1895. doi: 10.1039/D1MA00049G

[30] Eatemadi, A., Daraee, H., Karimkhanloo, H., Kouhi, M., Zarghami, N., Akbarzadeh, A., Abasi, M., Hanifehpour, Y., & Joo, S. W. (2014). Carbon nanotubes: properties, synthesis, purification, and medical applications. *Nanoscale Research Letters*, 9, 393. doi: 10.1186/1556-276X-9-393

[31] Ramachandrarao, M., Khan, S. H., & Abdullah, K. (2021). Carbon nanotubes and nanofibers – reinforcement to carbon fiber composites: synthesis, characterizations and applications: A review. *Composites Part C: Open Access*, 16, 100551. doi: 10.1016/j.jcomc.2024.100551

[32] Badshah, A., Daud, A., Alharbey, R., Banjar, A., Bukhari, A., & Alshemaimri, B. (2021). Big data applications: overview, challenges and future. *Artificial Intelligence Review*, 57, 290. doi: 10.1007/s10462-024-10938-5

[33] Adeyi, T. A., Jaiyeola, C. O., Oresegun, O. I., Olorunfemi, A. D., Abdul-Rahman, M. I., Babalola, K. O., & Babalola, O. T. (2025). AI and nanotechnology integration in the development of biodegradable food packaging materials. *International Journal of Science and Research Archive*, 14, 726–736. doi: 10.30574/ijsra.2025.14.1.0119

[34] Khan, M. A., & Khan, S. A. (2021). Application of AI/ML techniques in achieving SDGs: a systematic review. *Environment, Development and Sustainability*, *25*, 1–25. doi: 10.1007/s10668-023-03935-1

[35] Vinuesa, R., et al. (2020). The role of artificial intelligence in achieving the Sustainable Development Goals. *Nature Communications*, 11, 233. doi: 10.1038/s41467-019-14108-y.

[36] Werran, R. (2021). Pack (ag)ing an AI punch. *Food Science and Technology*, *39*, 48–51. doi: 10.1093/foodst/vwaf012

[37] Bungar, S. P., Sharma, N., Trif, M., & Adeel, M. (2023). Editorial: Emerging active, smart and intelligent packaging solutions in the fourth phase of the industrial revolution (Industry 4.0). *Frontiers in Nutrition*, 10, 1278143. doi: 10.3389/fnut.2023.1278143

[38] Nižetić, S., Šolić, P., López-de-ipiña gonzález-de-artaza, D., & Patrono, L. (2020). Internet of Things (IoT): Opportunities, issues and challenges towards a smart and sustainable future. *Journal of Cleaner Production*, 274, 122877. doi: 10.1016/j.jclepro.2020.122877

[39] Bibri, S. E. (2018). The IoT for smart sustainable cities of the future: An analytical framework for sensor-based big data applications for environmental sustainability. *Sustainable Cities and Society*, 38, 230–253. doi: 10.1016/j.scs.2017.12.034

[40] Kerry, J. P., O'Grady, M. N., & Hogan, S. A. (2021). Past, current and potential utilisation of active and intelligent packaging systems for meat and muscle-based products: A review. *Meat Science*, 74, 113–130. doi: 10.1016/j.meatsci.2006.04.024

[41] Christidis, K., & Devetsikiotis, M. (2016). Blockchains and smart contracts for the internet of things. *IEEE Access*, 4, 2292–2303. doi: 10.1109/ACCESS.2016.2566339

[42] Prabhu, B. P. A., Dani, R., Abdullah, K. H., Sharma, T., Bhatt, C., & Chauhan, R. (2021). A comprehensive review of sensor-based smart packaging technology. *Power Engineering and Intelligent Systems*, 1097, 39–52. doi: 10.1007/978-981-99-7216-6_4

[43] Francisco, S., & Swanson, S. (2021). The supply chain has no clothes: Technology adoption of blockchain for supply chain transparency. *Logistics*, 2, 2–13. doi: 10.3390/logistics2010002

[44] Yli-Huumo, J., Ko, D., Choi, S., Park, S., & Smolander, K. (2021). Where is current research on blockchain technology?—a systematic review. *PLOS ONE*, 11, e0163477. doi: 10.1371/journal.pone.0163477

[45] Saberi, S., Kouhizadeh, M., Sarkis, J., & Shen, L. (2019). Blockchain technology and its relationships to sustainable supply chain management. *International Journal of Production Research*, 57, 2117–2135. doi: 10.1080/00207543.2018.1533261

[46] Kalra, G., Kishore, K., Kumar, A., Rajoria, Y. K., Yadav, A., & Boadh, R. (2022). FIS-based prediction and estimation of health insurance for workers in the manufacturing sector. *Journal of Pharmaceutical Negative Results*, 13(8), 2636–2645.

[47] Kumar, A., Singh, A. P., Chauhan, A., Rajoria, Y. K., & Kaur, N. (2024). Investigation of a Fuzzy Production Inventory Model with Carbon Emission using Sign Distance Method. *E3S Web of Conferences*, 511, 01005.

[48] Singh, A. P., Chauhan, A., Aarya, D. D., & Rajoria, Y. K. (2025). Fuzzy Economic Order Quantity Models with Carbon Emission Considerations and Planned Discounts. In: *Recent Trends In Engineering and Science for Resource Optimization and Sustainable Development*. CRC Press.pp. 255–258.

Sarika, Karan Gehlot, Prawar, Chintan Singh*, Dayawati

Chapter 8
Value-added materials: Solar energy applications

Abstract: Green nanomaterials, which provide sustainable alternatives in line with environmentally friendly methods, mark a revolutionary breakthrough in material science. The synthesis, use, and recovery of green nanomaterials are covered in this chapter, with an emphasis on their applications in several industries, such as biopolymers, fuel cells, biofuels, and biorefineries. This chapter highlights the ecologically friendly synthesis techniques that maximize the efficiency of producing nanomaterials while minimizing harmful consequences, such as green chemistry and biosynthesis. These novel materials improve the efficiency of traditional systems while also promoting sustainability by utilizing renewable resources. Apart from their manufacturing, the chapter delves into the various uses of green nanomaterials in augmenting the performance of biopolymers, boosting fuel cell efficiency, and streamlining the procedures involved in producing biofuel. We go over how adding or acting as catalysts with nanomaterials can boost energy output and reaction rates while cutting down on waste. Case studies show how these materials can close the gap between innovative technology and environmental stewardship, promoting a circular economy in the production of sustainable materials. They also display successful implementations of these materials. The possibility of repurposing these materials in new uses is explored, as is the strategy for efficiently recovering them from industrial processes and end-of-life items. This chapter highlights the role of green nanoparticles in accomplishing sustainable development goals by highlighting the significance of life cycle management. This helps to pave the way for a future in which environmental factors are incorporated into material innovation.

Keywords: Green nanomaterials, sustainable synthesis, biopolymers, resource recovery, circular economy

***Corresponding author: Chintan Singh**, Amity Institute of Forensic Sciences, Amity University, Noida, Uttar Pradesh, India, e-mail: csingh5@amity.edu
Sarika, Amity Institute of Applied Sciences, Amity University, Noida, India
Karan Gehlot, School of Basic and Applied Sciences, K. R. Mangalam University, Gurugram, India
Prawar, Amity Institute of Forensic Sciences, Amity University, Noida, India,
Dayawati, Department of Chemistry, PT NRS Government College, Rohtak, India

https://doi.org/10.1515/9783111618906-008

8.1 Introduction

In the late nineteenth century, the discovery of a variety of fossil fuels such as coal, petroleum, and natural gas as an energy source changed the shape of global developments. The rapid growth of population and industrialization has led to the development of high demand of energy. The extensive use of fossil fuels is one of the major reasons of climate change and greenhouse emissions; moreover, the rapid depletion of fossil fuels is also concerning. The polluting and nonrenewable nature of these exhaustible resources has developed the need for renewable and clean energy resources [1]. The exploration for sustainable and clean energy solutions has led the way to significant reliance on solar energy. Today, the emission of greenhouse gases on a large scale as also the availability of limited fossil fuel makes the efficient utilization of energy a crucial matter. For overcoming this global concern, solar energy is a reliable source.

Solar energy is one of the most abundant and clean energy sources that is widely known. Solar energy is an environment-friendly technology and a crucial renewable energy source. With the rise in global energy consumption, demand for solar energy systems is growing, and so is the requirement of innovative materials that magnify their performance. The rapid advancement in technology has led the way to develop energy-efficient conservation and conversion systems/devices such as rechargeable batteries, solar cells, and supercapacitors, which can be used in electric vehicles, portable devices, aerospace, and in other fields. Solar energy has immense potential in addressing the global energy requirement, and for harnessing this potential, value-added products come in the frame. Value-added materials for solar energy applications are those materials that enhance the performance, longevity, and efficiency of the solar technologies. Along with all these highly sought qualities, they are cost-effective as well. The most evident purpose served by these materials is the conversion of sunlight into electricity, enhanced lifetime of the solar panels, and to make solar energy panels or systems that are easily accessible and affordable. As industries are evolving day by day and for adapting to the markets' changing dynamics, the importance of value-added materials will only grow. Their ability to significantly enhance product offerings along with promoting sustainability proves them as vital components in the future of manufacturing and economic development.

This chapter dives into a variety of value-added materials used in solar energy applications, their synthesis, applications, challenges, and their potential to drive technology and innovation in revolutionizing solar energy systems. It aims to give insight and provide a comprehensive view of the application of the green nanomaterials in solar cells as value-added materials, such as their source, properties, and the mechanism by which they enhance the performance of the solar cells. The incorporation of green nanomaterials into solar energy systems and technologies holds great promise for the enhancement of efficiency while promoting sustainability. By exploring more innovative and diverse applications and going through the current challenges, these materials can play a significant role in transforming the solar energy landscape and contribute to a sustainable future.

8.2 Understanding value-added materials

8.2.1 Definition and importance

Value-added materials are crucial substances or materials that are modified or engineered to serve enhanced functionalities, marketability, and performance, further than their basic properties. The importance of value-added materials is that they are more than just individual products; they play a special role in driving innovation and economic growth. These materials are essential for enhancing the performance metrics of solar technologies along with energy conversion efficiency such as increased lifespan and thermal storage capacity [4]. First and the foremost, the benefits of value-added materials are enhanced performance. For example, these advanced materials lead to improved energy conversion efficiencies and greater durability of the solar cells. Secondly, these materials contribute to sustainability. For instance, green nanomaterials synthesized from plant extracts such as nanocellulose can enhance solar cell efficiency while reducing the ecological carbon footprint associated with the traditional manufacturing processes. This approach toward sustainable materials aligns with the collective global efforts to tackle climate change and promote safer and greener technologies. The significance of value-added materials is highlighted by their potential to lower the costs while increasing the accessibility of solar energy. Last but not the least, investing in value-added materials can enhance customer loyalty and attraction, increasing the profit. By providing products that align with sustainability and exceptional performance, manufacturers can build a healthy relationship with consumers who prioritize environmental health. In summary, the importance of value-added materials lies in their effectiveness, sustainability, and customer loyalty, which makes them essential in the advancement of the technology and driving economic growth. In the manufacturing sector, value-added materials can show multiple functionalities, such as advanced composites, nanomaterials, and conductive inks. For instance, integration of nanomaterials into conventional products can boost their strength, durability, and thermal characteristics. This enhancement not only elevates the performance of the final product but also prolongs its life-span, delivering increased value to consumers along with lesser impact on the environment. The concept of value addition is just not limited to the physical material but also the process, service, and technological advancements, which contribute to the overall value of a product.

8.3 Types of value-added materials

8.3.1 Photovoltaic materials

Photovoltaic (PV) materials are crucial for the conversion of sunlight into electricity through PV effect. PV materials can be categorized into silicones, copper indium gallium selenide (CIGS), cadmium telluride (CdTe), and perovskite solar cells (PSCs). Silicon is the most widely used material for PV cells. Due to their high efficiency, monocrystalline and polycrystalline silicon are widely used. The efficiency of commercial crystalline silicone solar cells ranges from 12% to 19% and is expected to reach 20–23% by 2025 [5, 6]. Thin-film materials such as CdTe and CIGS are widely used for their flexibility and affordability. Organic PV materials (OPVs) are used for conversion of sunlight into electricity and for that purpose, organic compounds are used [7, 8]. PSCs are highly efficient, with efficiency exceedingly more than 25%. Another highly appreciative factor is affordability and high scalability. PSCs have comparably low cost with high energy conversion efficiency and can be used in wearable electronics. PSCs are fabricated using petroleum-based polymer substrates, which are also a major source of the pollution [9, 10]. A comparison of PV materials is illustrated in Table 8.1.

Table 8.1: Comparison of the types of photovoltaic materials.

Types of materials	Efficiency	Cost	Advantages	Disadvantages
Monocrystalline silicon	~25%	High	High efficiency and longevity	Higher manufacturing cost
Polycrystalline silicon	15–20%	Moderate	Cost-effective	Lower efficiency as compared to mono
CdTe	~22%	Low	Low production cost	Cadmium causes environmental concerns
CIGS	~22%	Moderate	Lightweight and flexible	Complex manufacturing
Perovskite	>25%	Low	High efficiency	Stability issues
Organic PV	10–15%	Low	Lightweight and flexible	Lower efficiency than silicon

8.3.2 Enhanced conductive materials

Enhanced conductivity materials improve the electrical conductivity of solar cells and significantly reduce energy loss. Transparent conductive oxides are widely used in solar cells as transparent electrodes. They allow light to pass through the electrodes while conducting electricity. Conductive polymers are used in organic PVs. These polymers improve conductivity, are flexible, cost-effective, and used in multiple applications.

One of the categories of the enhanced conductive materials is metal–organic frameworks (MOFs). MOFs are composed of metal nodes connected by organic linkers, which provide a higher surface area as they are porous coordinate polymers. Recent studies show that modification of the composition and structure of the MOFs increases their electrical conductance properties. Another such category includes is the carbon-based substances such as carbon nanotubes, biochar, and graphene. Due to its 2D structure, graphene shows high electron mobility, so, it can be easily incorporated to increase the power conversion efficiency. Enhanced conductive materials not only increase the efficiency and functionality of solar energy devices but also contribute to achieving the goal of sustainability and cost-effectiveness [11, 12]. The other applications of enhanced conductive materials also include energy storage device, where they facilitate faster charge or discharge cycles and improve the efficiencies of batteries and supercapacitors.

8.3.3 Thermal energy storage materials

Thermal energy storage (TES) materials are vital for enhancing the utility of solar thermal systems. TES can be made by heating, melting, cooling, solidification, and vaporization of materials, with the energy becoming available as heat when the process is reversed. Phase-change materials (PCMs) absorb and release thermal energy during phase transitions, which leads to latent heat storage. This property of PCMs allows for effective temperature regulation by storing more heat produced during sunshine or daytime for use during the hours of least sunlight or nighttime. PCMs can increase the efficiency of solar thermal systems. Based on phase transition, they can be classified by their mode of phase transition as gas–liquid, solid–liquid, solid–solid, and solid–gas systems. Organic PCMs derived from fatty acids or paraffin wax are highly utilized for their properties such as nontoxicity and biodegradability [13, 14]. High thermal conductivity materials such as graphite or metal foams also increase the efficiency of solar thermal systems [15]. One of the encouraging fields of research is the application of thermochemical storage materials, which capture thermal energy via reversible chemical reactions. These materials can attain greater energy densities than traditional PCMs, making them ideal for large-scale uses. For example, metal hydroxides or salts can go through dehydration reactions to accumulate heat and later release it during rehydration. This technique not only boosts storage capacity but also offers a more durable solution for seasonal TES. The incorporation of TES material into renewable resources increases their role as value-added material by increasing their sustainability and performance. TES reduces the reliance on fossil fuels and lowers the carbon footprint and greenhouse gas emission.

8.3.4 Light management materials

Value-added materials enhance the efficiency of solar cells by increasing light absorption and minimizing reflection. Light-trapping structures are those materials that are micro- or nano-structured, for example, pyramidal textures and nanowires. They trap light inside solar cell systems, resulting in increased absorption efficiency. Antireflective coatings help in absorbing more light on the surface of solar panels and in better conversion into electricity. As the demand of efficient solar energy is increasing day by day, the incorporation of light management materials has become the focus point in solar energy technology development. Materials like silicon nitride are used so that lesser amount of light is reflected off the surface of solar panels. Reflection loss is one of the drawbacks in PV technology, which can be minimized by antireflective coatings. Thin-film coatings diminish the reflection occurring at the different films through the destructive interference principle [6, 16]. Bifacial solar panels are another example of light management materials and advanced technology. This type of panels are designed to capture sunlight from both faces. It utilizes a reflective surface to enhance performance. By using this technology, solar installations can visibly surpass the traditional single-sided panels' efficiency. As research continues to advance in this field, innovative light management solutions will play a vital role in optimizing solar energy capturing and utilization.

8.3.5 Light-absorbing and reflective materials for concentrated solar power (CSP)

Concentrated solar power (CSP) systems concentrate sunlight onto a small area with the help of mirrors or lenses. Usually, they are used to drive steam turbines or heat engines. At the high temperature, specialized ceramics, silver, and aluminum are used to reflect and focus sunlight efficiently without degrading. These materials can be concluded as high-temperature reflectors. Silvered polymer reflectors are light in weight and show greater flexibility. But silver polymer reflectors have lower lifetime due to poor adhesion with silver when exposed to water. Aluminum is one of the most abundant metals. It is comparatively inexpensive, and an extensively used nonferrous metal. Solar reflectance of aluminum reflector is in the range of 85–91%. Usually, aluminum reflector materials are used for domestic and industrial lamps to focus light in a small area. Silvered glass reflectors are comparatively more sensitive than aluminum reflectors to acidic rain [17]. Apart from these, molten salts or PCMs are used to store heat generated by CSP systems. The most promising property of these materials is that they regularly store energy from the sun and utilize it during the nonavailability of sunlight. PCMs having a melting point greater than 300° Celsius are categorized as high-temperature PCMs that store energy to be used in the solar thermal systems along with waste heat recovery from industrial applications. Even in medical field, to maintain temperature of medicines and vaccines during the transportation, PCMs are used [1].

8.3.6 Encapsulation and protective materials

Encapsulated and protective materials increase the longevity of solar panels by safeguarding them from various environmental factors such as moisture, temperature, and mechanical stress. Polymer encapsulation shields the devices from oxygen and water vapor, while improving their resistance to humidity, temperature, and degradation by solar irradiation. EVA (ethylene vinyl acetate) is prominently used as a protective layer for solar cells. It helps in encapsulating cells and preventing moisture. EVA copolymers also behave as backsheet laminates for PV modules [18, 19]. Materials like PET and TPT are called backsheet materials. They provide resistance to environmental degradation along with mechanical strength. These materials are used as outermost protective layers of a solar panel. Backsheets protect the PV systems from weathering issues throughout their lifetime, so the long-term performance of the backsheet material is vital. Due to backsheet cracking or degradation, a PV system may lose its electrical insulation, and moisture may affect the device and can cause irreversible or permanent damage to the material and its functioning [20]. Anti-soiling coatings reduce the accumulation of dirt, dust, and debris on the solar panel's surface, resulting in increase in efficiency for a long time. Atmospheric dust can be caused by several factors such as volcanic eruption, wind, and environmental contamination. Till date, for dust cleaning or removal from PV systems, manual cleaning using water is practiced. But this kind of manual treatment needs an intensive water supply and leads to high operational and maintenance cost. Chemical agents such as polyvinyl chloride, acrylics, and polyethylene have been used to form films on glass substrates, but these coatings are found to be not so stable. Silica and metal oxide composition are part of the soiling chemistry, apart from salts, sulfuric acid molecules, and soot – where sulfuric acid molecules arise by transforming from gaseous to solid state phase. In addition to all these, biofilms can be grown on the solar glasses [21].

8.3.7 Advanced manufacturing materials

Advanced manufacturing materials play a significant role in enhancing scalability, reducing costs, and improving the manufacturing processes. 3D printing can also be used for the fabrication of solar panels by creating new and advanced designs. The primary advantage of 3D printing is that it can efficiently produce complex geometries that are impossible to achieve with conventional manufacturing methods. Along with design flexibility, 3D printing is more sustainable; it significantly reduces waste as it precisely uses material according to the requirements. Conductive inks and flexible polymers are the materials used in 3D printing for solar applications. Conductive inks are particularly useful in the production of flexible electronics, printed circuit (PCB), and sensors. Flexible materials and lightweight substrates enable solar cells to have a variety of applications such as wearables, portable electronics, and integration into building materials as well [22, 36]. The use of conductive polymers aligns with the goal of achieving sustainability by minimizing waste material and enabling the recycling of electronic components. Conductive inks offer sustainable pathways that are highly sought after by industries. 3D printing and conductive polymers are innovative approaches that are the need of the hour.

8.3.8 Nanomaterials

Nanomaterials are the most vital part of the nanotechnology and nanoscience. Nanotechnology has a vast range of applications, and it is considered for the manufacturing of exceedingly small devices and materials. It is widely used in the field of health, biomedicine, agriculture, environment, and energy as well. Key advantages of these nanomaterials are their high surface area-to-volume ratio, which directly enhances the interaction and reactivity with other substances. Nanotechnology promises improvements in solar cell efficiency and manufacturing techniques. Quantum dots are nanorange semiconductor particles that can absorb a broad spectrum of light and enhance the efficiency of solar cells by harvesting more sunlight [23, 24]. Nanorods or nanowires are also used to significantly improve light absorption and conversion efficiency in the solar cells, as they possess a larger surface area and trap comparably more sunlight [14].

Nanomaterials are essential for increasing the efficiency of solar cells and batteries in the context of energy applications. For example, adding carbon-based nanomaterials, such as graphene, to solar cells may boost light absorption and charge transport, increasing the efficiency of energy conversion. Lithium-ion batteries also use nanostructured materials to lessen the charging periods and boost charge capacity. Their usefulness in energy storage systems is further increased by the ability to modify these nanomaterials' characteristics by using controlled synthesis and manufacturing techniques. Nanomaterials are rapidly being used in biomedical applications, especially in imaging and medication delivery systems. By functionalizing biocompatible nanomaterials to target cells or tissues, regulated medication release and improved therapeutic efficacy are made possible. Nanomaterials serve as an essential value-added material that enhances product performance in multiple sectors, as shown in Figure 8.1. Their unique properties enable them to improve efficiency, sustainability, and use in applications ranging from solar energy systems to biomedical fields. As per research that is still going on, nanomaterials can drive technological progress and address global challenges as well [25, 26].

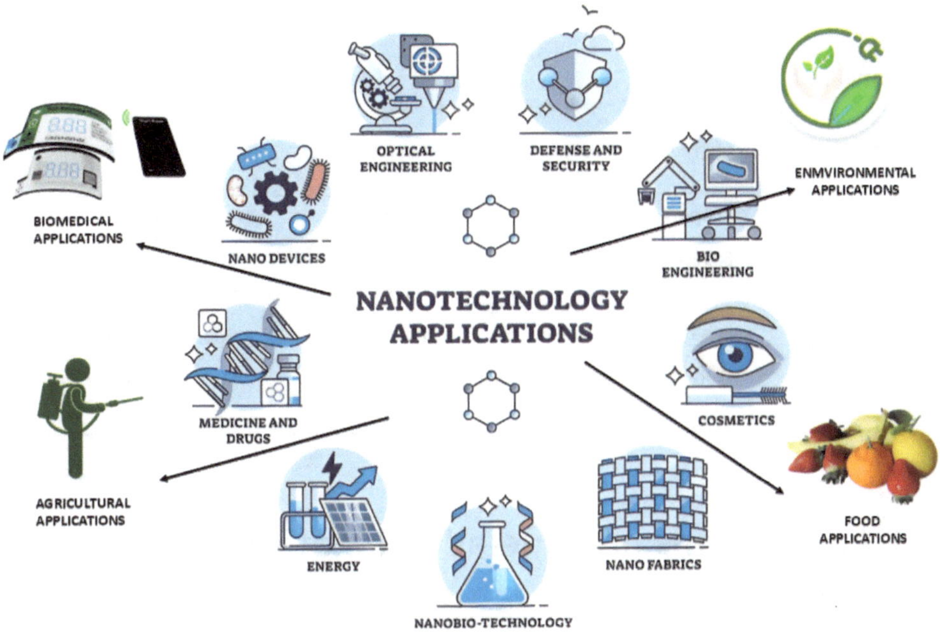

Figure 8.1: Application of nanotechnology in various fields.

8.3.9 Green nanomaterials

Green nanomaterials can be described as nanomaterials that are synthesized or manufactured in an environment-friendly process by limiting the toxic byproducts and utilizing renewable resources. Two main goals of green nanotechnology are to produce nanomaterials that are less harmful to the environment and still address the environmental challenges. Biodegradable products are composed of biomass. They include those derived from anaerobic digestate and are becoming more popular for their potential applications in the solar cell systems. Along with their contribution to waste reduction and environmental conservation, they align with the principle of sustainability [27]. Metal nanoparticles that are synthesized through green methods, such as copper oxide (CuO) and silver (Ag), show improvement in the efficiency of solar cells. In dye-sensitized solar cells (DSSCs), CuO nanoparticles enhance the charge transfer processes, resulting in increased overall efficiency [28]. Nanocellulose are derived from the cellulose-rich biomass, which have unique properties such as high surface area and high mechanical strength. Nanocellulose integration into solar cell designs enhances structural integrity along with providing functionalities such as improved light scattering [29]. In addition to enhancing performance metrics, green nanomaterials also show the way for sustainable coatings that protect solar cells from environmental factors while maintaining transparency. Some of the exemplary benefits of green chemistry are that it prevents pollution, reduces cost, uses less hazardous chemical synthesis, increases energy efficiency, and uses renewable feedstocks, as tabulated in Table 8.2.

Table 8.2: Comparison of green nanomaterials for solar cell systems.

Types	Properties	Sources	Applications
Biodegradable polymers	Eco-friendly and lightweight	Anaerobic digestate	Coatings for photovoltaic cells
Metal NPs	Enhanced conductivity	Plant extracts	Charge transport in DSSCs
Nanocellulose	More surface area and high strength	Cellulose-rich biomass	Structural reinforcement

8.4 Applications of value-added materials in solar energy

8.4.1 Enhances photovoltaic performance

Nanocellulose is observed to be a significant organic, sustainable, and environment-friendly nanomaterial. The sources to obtain nanocellulose are bacteria, trees, herbs, and even some animals. It can be used in multiple devices such as in solar cells, lithium-ion battery, and piezoelectric materials. Nanocellulose can be incorporated to improve light scattering within solar cells and as a result, enhancing overall efficiency and light absorption. Its high surface area permits for better interaction with incident light; this phenomenon is highly valuable in thin-film solar cells. Nanocellulose is used in energy conversion devices such as organic PV devices, PSCs, DSSCs, nanocomposite-based lithium-ion batteries, separators, electrodes, and even in piezoelectric devices. Even in energy conservation devices, nanocellulose-based supercapacitors, 2D supercapacitors, 3D supercapacitors, electrodes, and aerogels are used along with nanocellulose-based composites for CO_2 capture [30, 31].

8.4.2 Charge transfer enhancement

Charge transfer enhancement is a vital aspect of improving the efficiency of multiple electronic devices, most commonly in solar cells and OPVs. This phenomenon includes the motion of charge carriers (electrons and holes) across the materials, which is part of energy conversion. Metal nanoparticles such as CuO nanoparticles are synthesized by green method and show improved charge transfer processes in solar cells. For instance, the incorporation of CuO nanoparticles into nanoporous carbon counter electrodes results in a consequential enhancement in the efficiency of monolythic DSSCs.

Due to their high surface area and advantageous electrical characteristics, metal oxides like CuO, TiO_2, and ZnO are often used as photoanodes in DSSCs. The most researched metal oxide for DSSCs is TiO_2 because of its better light absorption abilities and capacity to promote effective charge separation. Increased dye loading is made possible by TiO_2's large surface area, which serves as essential for optimizing light harvesting and improving DSSC PV performance. Metal oxides serve as crucial value-added materials in DSSC, visibly enhancing their efficiency through the improved charge transfer mechanism [32, 33].

8.4.3 Sustainable coatings for solar cells

In solar cell technology, the development of sustainable coverings with biodegradable polymers from anaerobic digestate is an important milestone. By recycling waste materials, these coatings lessen their negative effects on the environment while simultaneously increasing the durability and efficiency of solar cells. Eco-friendly coatings that are derived from biomass waste display several advantages such as durability, cost-effectiveness, and environmental impact. These coatings significantly contribute toward the goal of sustainability by minimizing waste and reducing carbon footprint. Utilization and application of waste materials reduce production costs incorporated with traditional coatings. Along with that, all sustainable coatings also safeguard against environmental factors such as dust and dirt, while maintaining transparency.

Using biodegradable polymers made from renewable resources is one of the most farsighted strategies for sustainable coatings. The possibility of the polymers generated from anaerobic digestates as solar cell coatings has been studied recently. By recycling waste materials, these materials not only increase the longevity and efficiency of PV cells but also really help in environmental preservation. For example, research has demonstrated that applying biodegradable coatings can prolong the life of solar panels while drastically lowering the carbon footprint related to their manufacture and synthesis. By encouraging circular economy principles and reducing waste, the use of such polymers in the solar sector supports the aim of attaining the sustainability objectives.

8.5 Synthesis methods for green materials

8.5.1 Green synthesis techniques

Green synthesis includes synthesis of environment friendly methods by using plant extracts or biomass. Green synthesis is better than the traditional physical and chemical methods, as conventional/traditional method involve hazardous chemicals and toxic by-products. Traditional methods often require toxic reagents in harsh conditions. Green synthesis, on the other hand, can work with mild conditions. All these benefits make green synthesis environmentally safer. These techniques create nanomaterials without the use of hazardous chemicals or large energy inputs by using ecologically harmless procedures, which frequently involve biological agents, including plants, fungus, and microbes. In addition to lowering the impact on the environment, this environmentally friendly method improves the qualities of the materials produced. It makes them useful in a variety of applications, such as environmental remediation, energy production, and medicine. Plant-mediated synthesis is among the most popularly used research technique for green synthesis. This approach takes advantage of inherent reducing and stabilizing capabilities of phytochemicals that are present in plant extracts. For instance, various plant components, including leaves, roots, and seeds, have shown optimistic effect in converting metal ions into nanoparticles. Research has shown that nanoparticles produced using plant extracts possess distinct characteristics, such as increased antimicrobial activity and enhanced catalytic performance. Furthermore, plant-based synthesis methods are scalable and economical, making them appropriate for industrial uses [34]. There are two imminent methods for green synthesis; one of them is biological synthesis and other being physical/chemical synthesis. In biological synthesis, nanoparticles microorganisms are synthesized from plant extracts such as biomolecules, algae, fungus, and bacteria, as shown in Figure 8.2. Even for the chemical pathway of green synthesis, nontoxic solvents and reducing agents derived from natural sources are used for the synthesis of nanoparticles. To summarize, green synthesis techniques show significant advancement in the synthesis of value-added materials. By using biological resources for nanoparticle's synthesis, the methods or synthesis continues to unfold in this field. A comparison is depicted in Table 8.3 [35].

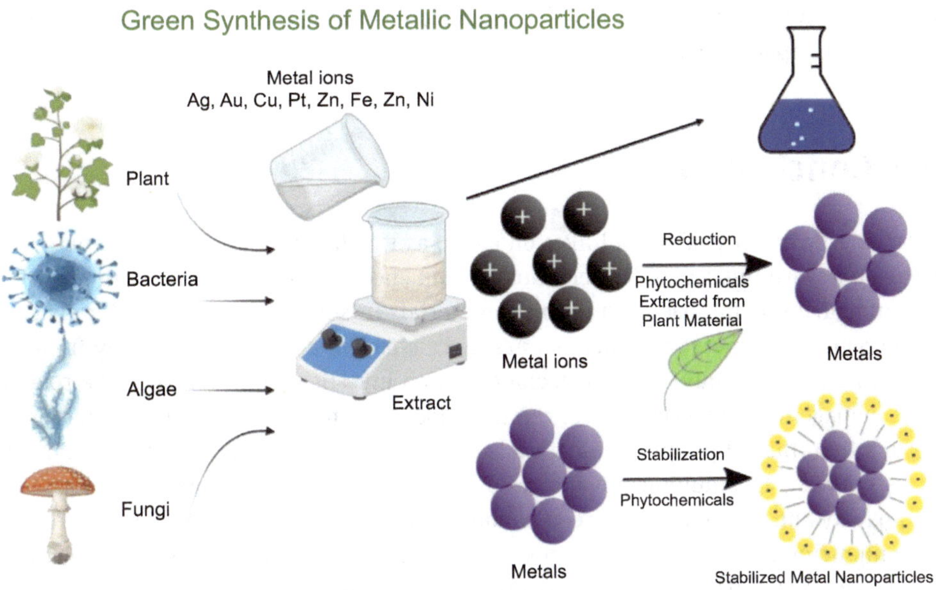

Figure 8.2: Sources and green synthesis of metallic nanoparticles.

Table 8.3: Comparison of the synthesis methods for green nanomaterials.

Synthesis	Advantages	Disadvantages
Biological synthesis	Eco- friendly	Slower reaction rates
Chemical synthesis	Faster synthesis	Toxicity concern

8.6 Challenges and future directions

The scalability of green nanomaterials for commercial manufacturing is a significant hurdle. Although results from laboratory-scale synthesis have been encouraging, consistency and cost-effectiveness concerns must be resolved prior moving on to large-scale manufacturing. It is difficult for green nanomaterials to transition from lab to industry [36]. More research is required to assess the long-term stability and endurance of green nanomaterials in solar cell applications. There is a high requirement for long-term performance assessments to ensure the durability of the material and its ability to withstand the operational stresses without unrepairable damages or degradation. The goal of research should be to make these materials more resilient against various environmental factors. The future scope of green nanomaterials in solar cell is optimistic. Further approaches in this field are needed to explore innovative synthesis methods that can enhance scalability while minimizing the environmental impact [37], [38].

8.7 Conclusion

Solar energy directly affects all lives on earth, as it is the cleanest source of energy. A key step toward improving PV performance and encouraging sustainability in material manufacturing is the use of green nanomaterials in solar cells. These materials can help create solar energy systems that are more effective and environmentally friendly by utilizing eco-friendly synthesis techniques and investigating novel uses. The role of light management materials, advanced manufacturing materials, and TES materials is no less than that of green nanomaterials in achieving sustainability. Green nanoparticles synthesis has multiple advantages over traditional chemical methods along with nontoxicity and sustainability. Green nanomaterials show a transformative opportunity for enhancing solar cells technologies along with sustainability. Integration of these green nanomaterials shows that environmentally friendly approaches can pave a way toward significant advancements in solar cell efficiency. For solving current scalability and stability issues and to light the path for widespread acceptance in the renewable energy industry, more collaborative research between academia, regulatory bodies, and industries is necessary. Researchers are already trying to not only improve energy conversion and conservation rates but also on minimizing the ecological footprint associated with the conventional/traditional synthesis and manufacturing. This alignment with the sustainability principles is crucial in addressing global energy demands. In conclusion, value-added materials represent a foundation of future advancement in solar technology. As all these technologies evolve, they without any doubt will shape the future generation of renewable energy solutions.

References

[1] Goel, V., Saxena, A., Kumar, M., Thakur, A., Sharma, A., & Bianco, V. (2023). Potential of phase change materials and their effective use in solar thermal applications: A critical review. *Applied Thermal Engineering*, 219. https://doi.org/10.1016/j.applthermaleng.2022.119417

[2] Seroka, N. S., Taziwa, R., & Khotseng, L. (2022). Solar energy materials-evolution and niche applications: A literature review. *Materials*, 15(15*)*. https://doi.org/10.3390/ma15155338

[3] Maka, A. O. M., & Alabid, J. M. (2022). Solar energy technology and its roles in sustainable development. *Clean Energy*, 6(3*), 476–483.* https://doi.org/10.1093/ce/zkac023

[4] Chowdhury, M. S., Rahman, K. S., Chowdhury, T., Nuthammachot, N., Techato, K., Akhtaruzzaman, M., Tiong, S. K., Sopian, K., & Amin, N. (2020). An overview of solar photovoltaic panels' end-of-life material recycling. *Energy Strategy Reviews*, 27, 100431. https://doi.org/10.1016/j.esr.2019.100431

[5] Seroka, N. S. (2022). Green synthesis of novel nanostructured silicon semiconductor thin film material from sugarcane bagasse ash for solar energy materials. Applied Sciences, 12(5), 2310 https://doi.org/10.3390/app12052310

[6] Shanmugam, N., Pugazhendhi, R., Elavarasan, R. M., Kasiviswanathan, P., & Das, N. (2020). Anti-reflective coating materials: A holistic review from PV perspective. *Energies*, 13(10*)*. https://doi.org/10.3390/en13102631

[7] Cui, Y., Hong, L., & Hou, J. (2020). Organic photovoltaic cells for indoor applications: Opportunities and challenges. *ACS Applied Materials & Interfaces*, 12(35*), 38815–38828.* https://doi.org/10.1021/acsami.0c10444

[8] Dada, M., & Popoola, P. (2023). Recent advances in solar photovoltaic materials and systems for energy storage applications: A review. *BMC Chemistry*, 17(1*)*. https://doi.org/10.1186/s43088-023-00405-5

[9] Bati, A. S. R., Zhong, Y. L., Burn, P. L., Nazeeruddin, M. K., Shaw, P. E., & Batmunkh, M. (2023). Next-generation applications for integrated perovskite solar cells. *Communications Materials*, 4(1*)*. https://doi.org/10.1038/s43246-022-00325-4

[10] Pitaro, M., Tekelenburg, E. K., Shao, S., & Loi, M. A. (2022). Tin halide perovskites: From fundamental properties to solar cells. *Advanced Materials*, 34(1*)*. https://doi.org/10.1002/adma.202105844

[11] Johnson, E. M., Ilic, S., & Morris, A. J. (2021). Design strategies for enhanced conductivity in metal-organic frameworks. *ACS Central Science*, 7(3), 445–453. https://doi.org/10.1021/acscentsci.1c00047

[12] Kong, T., & Zhang, W. (2023). Enhanced anaerobic digestion using conductive materials through mediation of direct microbial interspecies electron transfer: A review. *Fermentation*, 9(10*)*. https://doi.org/10.3390/fermentation9100884

[13] Pielichowska, K., & Pielichowski, K. (2014). Phase change materials for thermal energy storage. *Progress in Materials Science*, 65, 67–123. https://doi.org/10.1016/j.pmatsci.2014.03.005

[14] Wu, S., Yan, T., Kuai, Z., & Pan, W. (2020b). Thermal conductivity enhancement on phase change materials for thermal energy storage: A review. *Energy Storage Materials*, 24, 411–437. https://doi.org/10.1016/j.ensm.2019.10.010

[15] Hasan, M. M., Hossain, S., Mofijur, M., Kabir, Z., Badruddin, I. A., Yunus Khan, T. M., & Jassim, E. (2023). Harnessing solar power: A review of photovoltaic innovations, solar thermal systems, and the dawn of energy storage solutions. *Energies*, 16(18), 6456. https://doi.org/10.3390/en16186456

[16] Mavlonov, A., et al. (2020). *A Review of Sb2se3 Photovoltaic Absorber Materials and Thin-Film Solar Cells*. Elsevier Ltd. https://doi.org/10.1016/j.solener.2020.03.009

[17] Malwad, D., & Tungikar, V. (2021). Development and performance testing of reflector materials for concentrated solar power: A review. *Materials Today: Proceedings, Elsevier Ltd.*, 539–544. https://doi.org/10.1016/j.matpr.2020.10.744

[18] Ocaña, L., Montes, C., González-Díaz, B., González-Pérez, S., & Llarena, E. (2024). Evaluation of Ethylene-Vinyl Acetate, Methyl Methacrylate, and Polyvinylidene Fluoride as Encapsulating Materials for Perovskite-Based Solar Cells, Using the Low-Temperature Encapsulation Method in a Cleanroom Environment. *Energies (Basel)*, 17(1). https://doi.org/10.3390/en17010060

[19] Säckl, G., Wallner, G. M., Duchoslav, J., Tiefenthaler, M., & Stifter, D. (2024). Advanced analysis of ethylene vinyl acetate copolymer materials for photovoltaic modules. *Polymer Testing*, 132. https://doi.org/10.1016/j.polymertesting.2024.108381

[20] Abdallah, A. A., Abdelrahim, M., Elgaili, M., Pasha, M., Mroue, K., & Abutaha, A. (2024). Degradation of photovoltaic module backsheet materials in desert climate. *Solar Energy Materials and Solar Cells*, 277. https://doi.org/10.1016/j.solmat.2024.113118

[21] Hossain, M. I., Ali, A., Bermudez Benito, V., Figgis, B., & Aïssa, B. (2022). Anti-Soiling Coatings for Enhancement of PV Panel Performance in Desert Environment: A Critical Review and Market Overview. *MDPI.* https://doi.org/10.3390/ma15207139

[22] Fernandes, I. J., Aroche, A. F., Schuck, A., Lamberty, P., Peter, C. R., Hasenkamp, W., & Rocha, T. L. (2020). Silver nanoparticle conductive inks: Synthesis, characterization, and fabrication of inkjet-printed flexible electrodes. *Scientific Reports*, 10(1). https://doi.org/10.1038/s41598-020-65698-3

[23] Liu, L., Najar, A., Wang, K., Du, M., & Liu, S. (2022). Perovskite quantum dots in solar cells. *Advanced Science*, 9(7), 2104577. https://doi.org/10.1002/advs.202104577

[24] Rahman, M. M., Karim, M. R., Alharbi, H. F., Aldokhayel, B., Uzzaman, T., & Zahir, H. (2021). Cadmium Selenide Quantum Dots for Solar Cell Applications: A Review. *John Wiley and Sons Ltd*. https://doi.org/10.1002/asia.202001369

[25] Baig, N., Kammakakam, I., Falath, W., & Kammakakam, I. (2021). Nanomaterials: A review of synthesis methods, properties, recent progress, and challenges. *Royal Society of Chemistry*.

[26] Zhang, S., Jiang, S., Huang, B., Shen, X., Chen, W., Zhou, T., Cheng, H., Cheng, B., Wu, C., Li, W., Jiang, H., & Yu, H. (2020). Sustainable production of value-added carbon nanomaterials from biomass pyrolysis. *Nature Sustainability*, 3(9), 753–760. https://doi.org/10.1038/s41893-020-0538-1

[27] Li, W., Liu, Q., Zhang, Y., Li, C., He, Z., Choy, W. C., Low, P. J., Sonar, P., & Kyaw, A. K. (2020). Biodegradable materials and green processing for green electronics. *Advanced Materials*, 32(33). https://doi.org/10.1002/adma.202001591

[28] Abiodun, A. J., Alamu, G. A., Daramola, O. O., Adedokun, O., & Sanusi, Y. K. (2024). Green synthesis of copper oxide nanoparticles for improved performance in monolithic dye sensitized solar cells. *Fountain Journal of Natural and Applied Sciences*, 13(2). https://doi.org/10.53704/fujnas.v13i2.545

[29] Zinge, C., & Kandasubramanian, B. (2020). Nanocellulose-based biodegradable polymers. *Elsevier Ltd.* https://doi.org/10.1016/j.eurpolymj.2020.109758

[30] Lasrado, D., Ahankari, S., & Kar, K. (2020). *Nanocellulose-Based Polymer Composites For Energy Applications—A Review*. John Wiley and Sons Inc. https://doi.org/10.1002/app.48959

[31] Shoyiga, H. O., Akpasi, S. O., Akpan, J., Amune, U. O., & Kiambi, S. L. (2024). Novel Photoactive Material and Fabrication Techniques for Solar Cells Application: Nanocellulose-Based Graphene Oxide CdS Composite. Oxford University Press. https://doi.org/10.1093/ce/zkae010

[32] Bashir, M. B. A., Rajpar, A. H., Salih, E. Y., & Ahmed, E. M. (2023). Preparation and Photovoltaic Evaluation of CuO@Zn(Al)O-Mixed Metal Oxides for Dye Sensitized Solar Cell. *Nanomaterials*, 13(5). https://doi.org/10.3390/nano13050802

[33] Bykkam, S., Prasad, D. N., Maurya, M. R., Sadasivuni, K. K., & Cabibihan, J. J. (2021). Comparison study of metal oxides (CeO2, CuO, SnO2, CdO, ZnO and TiO2) decked few layered graphene nanocomposites for Dye-Sensitized solar cells. *Sustainability (Switzerland)*, 13(14). https://doi.org/10.3390/su13147685

[34] Abuzeid, H. M., Julien, C. M., Zhu, L., & Hashem, A. M. (2023). Green Synthesis of Nanoparticles and Their Energy Storage, Environmental, and Biomedical Applications. *Multidisciplinary Digital Publishing Institute (MDPI).* https://doi.org/10.3390/cryst13111576

[35] Osman, A. I., Zhang, Y., Farghali, M., Rashwan, A. K., Eltaweil, A. S., Abd El-Monaem, E. M., Mohamed, I. M., Badr, M. M., Ihara, I., Rooney, D. W., & Yap, P. (2024). Synthesis of green nanoparticles for energy, biomedical, environmental, agricultural, and food applications: A review. *Environmental Chemistry Letters*, 22(2), 841–887. https://doi.org/10.1007/s10311-023-01682-3

[36] De Oliveira Cândido, T. C., Pereira, A. C., & Da silva, D. N. (2023). Development and Characterization of Conductive Ink Composed of Graphite and Carbon Black for Application in Printed Electrodes. *Analytica*, 4(4), 513–526. https://doi.org/10.3390/analytica4040035

[37] Singh, C., Khajuria, H., & Nayak, B. P. (2023). A Study of Implementing a Blockchain- Based Forensic Model Integration (BBFMI) for IoT Devices in Digital Forensics. In International Conference on Computer Science, Engineering and Education Applications (pp. 318–327). Cham: Springer Nature Switzerland. https://doi.org/10.1007/978-3-031-36118-0_28

[38] Rani, M.S.S.; Singh, C.; Mishra, A (2021). Development of an Analytical Method for Detection of Imidacloprid Insecticide from Biological Matrix Using LC-MS/MS. Journal of Punjab Academy of Forensic Medicine & Toxicology, 21, 220–225. https://doi.org/10.5958/0974-083x.2021.00042.x.

Sayan Chatterjee*, Deepak Rathod, Shivangi Yadav,
Md. Muktasir Rohman, Annu Yadav, and Chandra Mohan

Chapter 9
Case studies demonstrating sustainable development for green chemistry approaches

Abstract: Sustainable development requires energy-efficient processes and technologies, particularly when nonrenewable energy sources such as coal and petroleum are limited and plays as a major currency in geopolitics. Energy is regarded as a criterion in assessing green chemistry, as defined by the 12 principles proposed by Paul Anastas. Inappropriate energy use has significant environmental and economic repercussions.

In recent decades, synthetic methodologies have transitioned towards ecologically benign and green reactions. Traditional methods emphasized on enhanced production and multiple output processes, neglecting safety, health, and environmental considerations. Solvents are essential in chemical processes, enabling solubilization, mass and heat transport, and influencing reaction kinetics. Hydrocarbons, being organic solvents, are favored in organic synthesis. Nonetheless, they also substantially contribute to waste production and elevate E-factor value. Catalysis is crucial for organic synthesis, with 90% of large-scale compounds interacting with a catalyst. It converts substrates tens of millions of times, promoting efficient and economical methods for various sectors. As an alternative, organic catalysis is a nontoxic, environmentally benign, and manageable catalyst that offers diverse applications and is ideal for pharmaceutical and academic sectors. Reactivity and selectivity are essential criteria in chemical synthesis.

Green chemistry encompasses procedures or products that substitute harmful compounds with environmentally benign substitutes. The Pollution Prevention Act of 1990 in the USA recognized green chemistry as a genuine scientific discipline, and it became the official emphasis of the EPA in 1991.Chemistry has substantially enhanced human life, increasing life expectancy via advanced healthcare facilities, rapid diagnostic methods, and effective medications. It has made water and food safe, enlarged and enhanced TVs, advanced efficient computers, swifter and less polluting vehicles, as well as alternate power sources such as batteries and inverters.

The case studies presented here show that green chemistry is a more practical need of the hour than just being a normal concept. It also highlights the role of collab-

---

***Corresponding author: Sayan Chatterjee**, University School of Biotechnology, Guru Gobind Singh Indraprastha University, New Delhi, India, e-mail: sayan@ipu.ac.in
Deepak Rathod, Shivangi Yadav, Md. Muktasir Rohman, Annu Yadav, University School of Biotechnology, Guru Gobind Singh Indraprastha University, New Delhi, India
Chandra Mohan, School of Basic and Applied Sciences, K. R. Mangalam University, Gurugram, India

https://doi.org/10.1515/9783111618906-009

oration between government, academia, and industry in promoting tenable practices. The chapter provides valuable insights for businesses, researchers, as well as policymakers in adopting greener methods.

Keywords: Green chemistry, sustainable development, waste minimization, atom economy, hazardous chemicals, nontoxic chemicals, renewable feedstocks, energy efficiency, catalysis, biodegradable products, pollution prevention, safe chemical practices

9.1 Introduction

Green chemistry is a solution to environmental issues like deforestation, pollution, and fossil fuel overexploitation by designing products and processes to minimize hazardous compounds and enhance resource efficiency. This approach reduces material use and utilizes renewable resources for environmental preservation. Advancements in technology and science have led to the production of chemicals and materials impacting our lives, but also resulting in the discharge of toxic substances and pollutants.

The direct discharge of industrial waste was stopped after Rachel Carson's 1962 study on synthetic pesticides' harmful effects on ecosystems, wildlife, and human health. The Environmental Protection Agency (EPA) and international environmental protection regulations aim to achieve sustainable development by limiting pollution emissions. The EPA's focus on pollution prevention has influenced the development of innovative chemical processes and product manufacturing.

Green chemistry benefits sectors like pharmaceuticals, agriculture, and medicine, encompassing atom economy, energy efficiency, renewable feedstocks, derivative minimization, catalysis, degradation design, pollution prevention, safer chemistry, and minimally hazardous chemical synthesis. This book chapter discusses the principles of green chemistry, emphasizing waste prevention as a cost-effective and environmentally friendly solution. It emphasizes the importance of optimizing materials, creating minimally hazardous chemical synthesis, using safer solvents, reducing energy demands, and using renewable feedstocks. The chapter also discusses the use of degradation products and real-time analysis for pollution prevention. It provides a comprehensive understanding of green chemistry, highlighting its benefits for humanity and the environment (Figure 9.1).

The atom economy chapter provides strategies for optimizing resource utilization and minimizing waste generation. It emphasizes waste minimization and green chemistry principles, such as pyroligneous acid, designing safer chemicals, principle 5, and the use of plants as solvents. The design for energy efficiency framework promotes renewable energy sources in the synthesis process. The seventh principal advocates for renewable feedstocks like soybean hull residues for valuable compounds. Reduced derivatives outline methods to minimize by-product production during synthesis. Catalysis studies chemicals that produce higher yields and exhibit more transparent reactions. Microbial biodegradation is emphasized in the design for degradation segment, addressing environmental pollution from pesticide use. Real-time analysis for pollution control emphasizes the importance of reducing waste and generating fewer hazardous materials. The chapter also highlights the use of polysaccharide precipitation for enzyme recovery, demonstrating its environmental sustainability and cost-effectiveness.

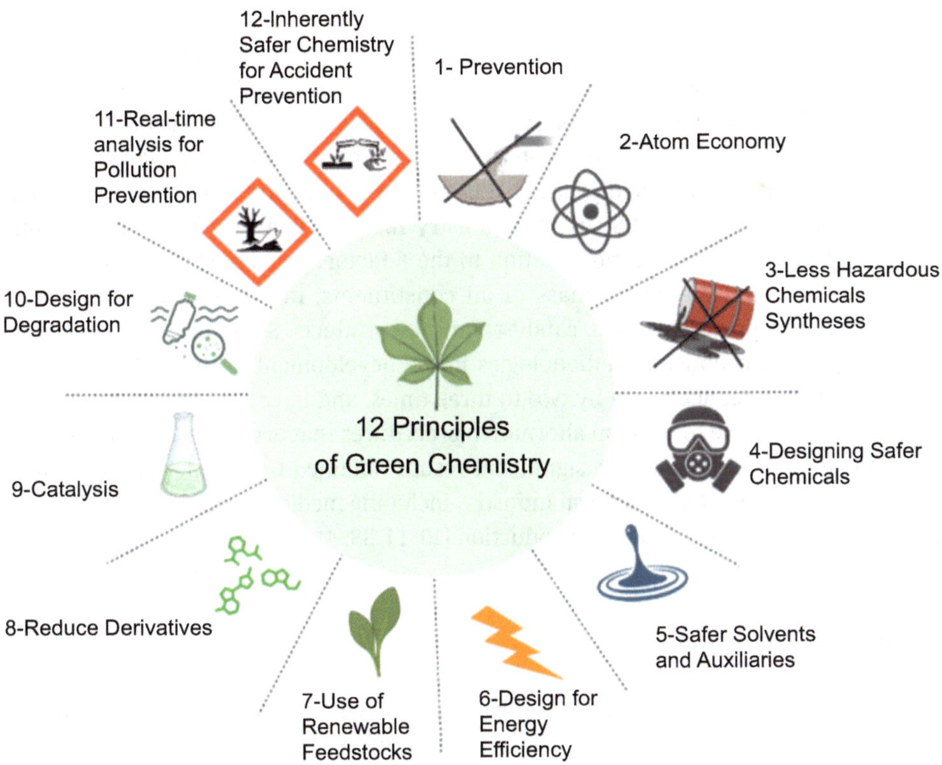

Figure 9.1: Twelve principles of green chemistry.

9.2 Prevention: waste minimization strategies

9.2.1 Introduction

9.2.1.1 Importance of waste prevention in green chemistry: source reduction or waste prevention – the base of the green chemistry pyramid

As industries expand worldwide, a growing global urgency to minimize waste, both to limit environmental damage and to define sustainable use of resources, has emerged.

Environmental diseases are caused by pollution from transportation, agriculture, industrial operations, home effluents, and waste management. Addressing these issues requires a multifaceted approach, often focusing on prevention.

9.2.2 Prevention of waste

Waste is any unvalued or created material, especially after extracting beneficial substances, with its environmental impact influenced by its chemical composition, toxicity, volume, and disposal method. Roger Sheldon developed the E-factor to oversee and regulate recycling. This scenario demonstrates a correlation between the weight of trash and the weight of the intended product. It calculates the amount of trash produced per kilogram of product. The ACS Green Chemistry Institute Pharmaceutical Roundtable (ACS GCIPR) has proposed a modification to the E-factor. This revised form quantifies the API concerning the entire mass of all constituents, including organic solvents, water, reagents, processing aids, catalysts, and by-products. Some firms are currently employing green chemistry methodologies in the development of their API processes. This strategy can reduce waste by two to three times, and in certain instances, even by 10 times. It is possible to develop alternative procedures that are more environmentally sustainable, as evidenced by the significant results obtained from using the preventive principle across the whole chemical industry, including medicines and other sectors utilizing synthetic chemistry in their production [10, 14, 38, 41, 45].

9.2.3 E-factor

The mass and volume of reactants and products (solvents and by-products) produced by chemical reactions in labs are measured in grams and milliliters, respectively. But these proportions are multiplied by a factor of a thousand at the industrial scale, with solvents and chemicals used in quantities ranging from kilograms to tons and liters. Industrial organic synthesis mostly involves halogenation, oxidation, alkylation, nitration, and sulfonation as its fundamental processes – also known as "dirty reaction" because of the undesirable by-products they produce and the hazardous and poisonous reagents and solvents that are necessary for them [11, 46].

9.2.4 The waste hierarchy

The inability to utilize raw materials completely renders efficient waste generation avoidance nearly impossible. Additionally, it is important to consider that post-usage, the product is classified as waste and requires disposal. Therefore, significant ecological repercussions may arise from managing this waste. The "waste hierarchy" was es-

E Factors in the Chemical Industry

Oil Refining
- **Tonnes per annum**: 10^6–10^8
- **E Factor**: < 0.1

Bulk Chemicals
- **Tonnes per annum**: 10^4–10^6
- **E Factor**: < 1–5

Fine Chemicals
- **Tonnes per annum**: 10^2–10^4
- **E Factor**: 5–50

Pharmaceuticals
- **Tonnes per annum**: 10–10^3
- **E Factor**: 25 – >100

Scheme 9.1: E-factors in chemical industry [46].

tablished to evaluate the environmental impact of different waste treatment methods. Although various waste management strategies may appear to function effectively in conjunction during the "transition" phase, they ought to be seen as alternatives in the long term, in accordance with the EU's waste hierarchy principle [24, 62].

Prevention: Biodegradable trash is one option here since it does not build up over time and can be used for prevention purposes. A prime example is the extensive study of biodegradable plastics.

Reduction of waste: The growing emphasis on design, package optimization, and material selection facilitates waste reduction.

Reuse: A highly effective method for managing on-site process scrap is re-extrusion, also referred to as primary reuse. This procedure entails utilizing various thermal cycles to manufacture distinct plastic items.

Recycling: A well-structured supply chain that includes recycling, sorting, and separate collecting stages repurposes materials that are considered waste commodities. This procedure is carried out by companies such as Pfizer and polymer makers, among others. One way that ETH Zurich helps is by reducing emissions and reusing solvents through redistillation.

Energy recovery: Biodigesters, waste-to-energy plants, and sophisticated pyrolytic methods can convert trash into usable energy rather than discarding it. Energy recovery from waste presents a more environmentally sustainable alternative to landfilling inside waste management systems. When evaluating the ecological footprint of a product, process, or activity, it is essential to monitor and quantify all resources (energy, materials, and waste) utilized in its production. Innovative assessment techniques for waste management have facilitated the comparison of material recovery, particularly recycling, with energy recovery.

Landfill disposal: It is the final alternative, poised for extinction in the future or, at the very least, to see a significant decline.

9.2.5 Case study of NatureWorks PLA production

The paper provides a detailed case study on the production of NatureWorks® polylactide (PLA), focusing on its life cycle assessment (LCA) and the integration of renewable energy sources. Here are the key points from the case study:

Background of NatureWorks: NatureWorks LLC, originally Cargill Dow LLC, has been a pioneer in PLA production since 2003. The company has developed technology for producing PLA from annually renewable resources, aiming to reduce environmental impacts compared to petroleum-based polymers.

LCA: The paper emphasizes the importance of LCA in evaluating the environmental performance of PLA. It provides cradle-to-factory-gate inventory data, which helps in benchmarking PLA against traditional fossil fuel-based polymers. The LCA data has been updated with actual production data from NatureWorks' facilities, enhancing the accuracy of the eco-profiles.

Renewable energy certificates (RECs): NatureWorks has committed to purchasing wind power-derived RECs, which allows the company to offset its electricity use with renewable energy. This program is part of their strategy to mitigate climate change by promoting the development of renewable energy sources.

Environmental impact: The study underscores the necessity of substantially decreasing greenhouse gas emissions and fossil energy consumption in PLA production. For instance, the 2006 PLA production system emitted only 0.27 kg CO_2 equivalent per kg of PLA, representing an 85% reduction compared to 2003 data.

Future developments: The use of wind energy in the production process is expected to further decrease reliance on nonrenewable fuels, with projections indicating a reduction of up to 67% in nonrenewable fuel use in future production systems.

Market position: PLA is positioned as a sustainable alternative to conventional plastics like polypropylene, polystyrene, and polyethylene terephthalate. Nonetheless, in comparison to these proven materials, it remains in the nascent phases of market development and optimization.

9.2.6 Conclusion

A reduction of 85% in emissions and a reduction of 50% in fossil energy usage were observed in 2006 PLA manufacturing, with 0.27 kg CO_2 equivalent per kg of PLA and 27.2 MJ of fossil energy consumed per kg of PLA, respectively, compared to the PLA eco-profile data from 2003. The main reason for the decrease in fossil fuel consumption and emissions of greenhouse gases is the use of wind power in PLA production. Using RECs generated by wind power, NatureWorks stated in 2004 their goal to lessen the environmental effects of producing PLA. Due to the substantial growth in PLA manufacture in January 2006, NatureWorks was forced to get certifications that corresponded to the electricity consumption in the PLA production system. According to the results, using wind energy significantly reduces the environmental impact of the PLA production system. Additional large reductions can be achieved by combining the use of renewable energy with innovative process technology [60].

9.3 Atom economy: maximizing material efficiency

9.3.1 Introduction

In 1991, Barry M. Trost proposed the idea of an atom economy. To accomplish the goal of organic synthesis, a number of reagents, catalysts, and solvents are required. After the reaction is completed, everything that is not the expected product or recyclable reagents (e.g., catalysts and solvents) will be thrown out as waste, mostly hazardous waste. Using the atom economy concept, we can find synthesis methods that use the least amount of reactant atoms feasible while yet producing a high-quality end product. Atom economy allows us to determine a reaction's efficiency by tallying the number of atoms moved from the reactants to the end product. New norms for pollution management can be developed with the help of atom economy, which is relevant to all synthetic processes [49, 56, 57].

One way to determine if a chemical reaction is environmentally friendly is to use the atom economy calculation. Most commonly employed in organic synthesis, the percentage yield does not take mass flow into consideration when assessing the efficiency of a reaction. As a percentage yield, a synthetic chemist records the yield of a certain reaction. A favorable yield is one that exceeds 90%. The calculations indicate that a one-to-one mole ratio of reactant to product results in a 100% yield. Nevertheless, the quantity of garbage generated is not included in the aforementioned estimate. A synthesis cannot be defined by its greenness, even if it achieves a 100% yield [13, 23]:

$$\%\,\text{Yield} = \frac{\text{Actual yield of the product}}{\text{Theoretical yeild of the product}} \times 100$$

The idea of an atom economy was put into numbers by Roger A. Sheldon, a professor at Delft University of Technology. One of the metrics he proposed for atom economy is the percentage atom use, which is the number of atoms from the reactants that end up in the product [29]:

$$\%\,\text{Atom utilization} = \frac{\text{Molecular weight of the desired product}}{\text{Molecular weight of the formed product} + \text{waste product}} \times 100$$

An example of a reaction where every atom is used is the synthesis of sodium thiosulfate from sodium sulfite and sulfur granules:

$$Na_2SO_3 + \text{Sulfur} \rightarrow Na_2S_2O_3$$

9.3.2 Economic aspects of atomic reactions

The atom consumption is significantly reduced because the final products of addition reactions usually consist of all reactant atoms. Substitution and elimination reactions are worse for the environment than addition processes, according to atom economy. Think about what happens when you add hydrogen bromide to methyl propene later on. All of the green atoms representing the reactants (9 and 11) make up the end product (7) in this example (Figure 9.2). The calculation of 100% atom economy and Table 9.1 further illustrate the reaction's superior atom economy.

Figure 9.2: Isobutyl bromide.

Table 9.1: Atom economy aspect for isobutyl bromide.

Reagents formula	Reagents FW	Utilized atoms	Weight of utilized atoms	Unutilized atoms	Weight of unutilized atoms
(9) C_4H_8	56	4C, 8H	56	-	0
(11) HBr	81	HBr	81	-	0
Total 4C, 9H, Br	137	4C, 9H, Br	137	-	0

The atom economy percentage is equal to (FW of atoms used/FW of all reactants) × 100, which is equal to (137/137) × 100 = 100%.

9.3.3 Economic aspects of atomic rearrangements

Elements in a molecule can rearrange themselves in rearrangement reactions. Since the molecule going through the rearrangement process does not lose, gain, or exchange any atoms, the atom economy of the reaction is 100%. So, from the standpoint of the atom economy, they are good for the environment. As an example, think about the conversion of 3,3-dimethyl-1-butene (12) to 2,3-dimethyl-2-butene (13), which is catalyzed by acid. Because they are an integral part of the final product, the atoms of reactant (12) are shown in green here (13) (Figure 9.3). Table 9.2 can be used to build an atom economy, and the proportion of atom economy can be determined in the same way as the other cases. Not included in the atom economy table or calculations is the acid (H^+), which is only utilized in catalytic quantities in this reaction and does not end up in the final product. The aforementioned rearrangements show that this reaction has a 100% atom economy.

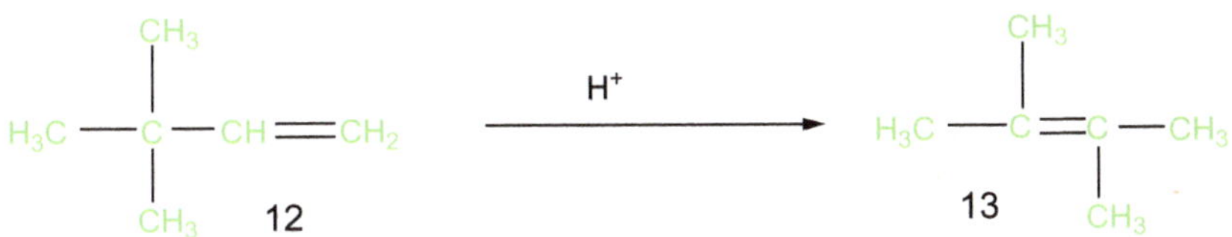

Figure 9.3: 2,3-Dimethyl-2-butene.

Table 9.2: Atom economy aspect for 2,3-dimethyl-2-butene.

Reagents' formula	Reagents' FW	Utilized atoms	Weight of utilized atoms	Unutilized atoms	Weight of unutilized atoms
(12) C_6H_{12}	84	6C, 12H	84	–	0
Total 6C, 12H	84	6C, 12H	84	–	0

Atom economy as a percentage = (FW of atoms used/FW of all reactants) ×100 = (84/84) × 100 = 100%.

Almost every chemical reaction can be categorized according to its nature: addition, rearrangement, substitution, or elimination. From an atom-economy perspective, addition and rearrangement, then substitution reactions, are the most eco-friendly processes. In terms of environmental impact, elimination responses rank last. Both the yield and the atom economy of a reaction must be considered when analyzing it in chemistry (atom economy).

9.3.4 Case study: atom economy of Ibuprofen process (BHC Company)

Ibuprofen became available without a prescription after the patent owned by the Boots Company ran out in the 1980s. Several businesses started creating new ways to synthesize ibuprofen and set up shop after realizing the lucrative potential of this medicine. The Hoechst Celanese Corporation (Somerville, NJ; now called the Celanese Corporation) created a new three-step process to synthesize ibuprofen. In an effort to manufacture and sell ibuprofen in conjunction with the Boots Company, the BHC Company was set up. The BHC synthesis is shown in Figure 9.4, where the green atoms represent the

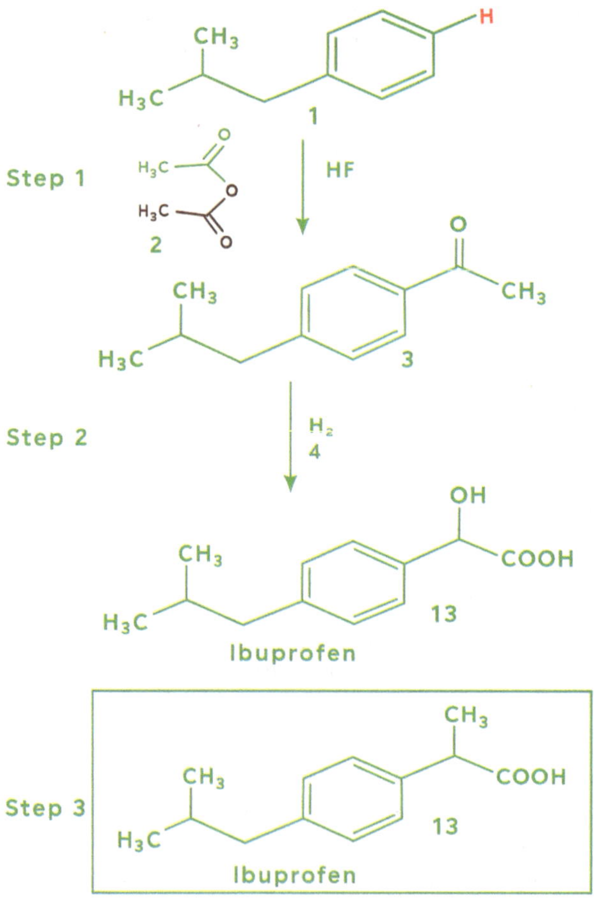

Figure 9.4: Ibuprofen.

utilized atoms and the brown atoms represent the unused atoms. Table 9.3 provides more evidence of the atom economy; here, the computed % atom economy reaches 77%, which is a substantial improvement over the 40% in Figure 9.4.

Table 9.3: Atom economy and the BHC's ibuprofen synthesis.

Reagents formula	Reagents FW	Utilized atoms	Weight of utilized atoms	Unutilized atoms	Weight of unutilized atoms
(1) $C_{10}H_{14}$	134	10C, 13H	133	H	1
(2) $C_4H_6O_3$	102	2C, 3H, O	43	2C, 3H, 2O	59
(4) H_2	2	2H	2	-	0
(6) CO	28	CO	28	-	0
Total 15C, 22H, 4O	266	Ibuprofen 13C, 18H, 2O	206	Waste products 2C, 3H, 2O	60

The formula for calculating % atom economy is (FW of atoms utilized/FW of all reactants) × 100, which is (206/266) × 100% or 77%.

9.3.5 Atom economy in catalysis

Catalysis shortens the time it takes for chemical reactions to begin by making them more efficient and decreasing the activation energy. Since catalysts do not play a role in the reaction, they can be omitted from the atom economy calculation. Atom economies are improved by reducing reaction steps, stoichiometric components, and energy utilization through the use of homogeneous, heterogeneous, or biocatalysts. The combined use of homogeneous and heterogeneous catalysts enhances reaction selectivity, efficiency, and sustainability by leveraging the advantages of both catalytic systems [4].

9.3.6 Conclusion

In addition to basic yield estimations, atom economy and experimental atom economy offer a novel approach to assess reaction efficiency by considering both utilized and discarded atoms in the reactants. Quantifying atom economy using the percentage experimental atom economy and the % atom economy facilitates quantitative comparisons among the atom economies of different reactions or syntheses. The optimal method for assessing reaction efficiency is to analyze both atom economy and yield by computing the product of percentage yield and experimental atom economy (PE × EAE).

9.4 Designing safer chemicals

9.4.1 Creating chemicals with minimal environmental impact

This is the most underdeveloped of the 12 principles of green chemistry, which have been established since the inception of the concept. The sluggish progress is mostly attributable to its involvement in various fields, including organic chemistry, toxicology, biology, and environmental science, among others. Over 20 years have elapsed since the principle was developed, certain guidelines or design requirements for the synthesis of safe chemicals have been devised to facilitate the speedy, efficient, and reliable reduction of compound toxicity; yet, these measures remain insufficient. Less hazardous chemical substances pertain not only to those that adversely affect human health but also to their environmental impacts. Before progressing with the creation of "safer chemicals," it is crucial to delineate the characteristics of a safe chemical and ascertain the attributes that need to be collected [15, 21].

9.4.2 A safer chemical must possess adequate potency

A chemical compound is considered more potent when its active ingredient is used in a lower proportion. The development of more potent pharmaceuticals offers environmental benefits by necessitating a diminished quantity of active ingredients, which subsequently reduces the need for raw materials, solvents, and other chemicals in the manufacturing process, hence decreasing waste generation.

9.4.3 High efficacy

It means that the treatment has the potential to provide the intended pharmacological effect with minimal side effects. Tolbutamide, which was not available until 1955, has a pharmacological effect that lasts for 6–12 h and requires three 500-mg doses per day. Glipizide, which has a pharmacological effect that lasts for up to 24 h and requires a single 5-mg dose daily, was introduced in 1984. Example: instances of therapeutic interventions exhibiting similar potency and efficacy that belong to the same structural class and pharmacological mechanism. Administering prednisone tablets the release in 1988 signified the commencement of the proton pump inhibitor era. Esomeprazole, the *S*-enantiomer of omeprazole, was initially introduced in 2001 (Figure 9.5). The fundamental structure of these persistent proton pump inhibitors consists of pyridinylmethylsulfinyl benzimidazoles. There are minor variations in the pharmacokinetics and pharmacodynamics regarding the rate and extent of stomach acid reduction. The drug can be generated effortlessly, efficiently, economically, and sustainably: Because of the high number of chemical components needed for medication formulation and the variety of reaction processes involved in chemical synthesis, this quality is readily apparent, driving up manufacturing costs due to the substantial by-products and waste produced. Minimizing solvent use, increasing yields, and shortening reaction times are all critical goals [27, 39, 40].

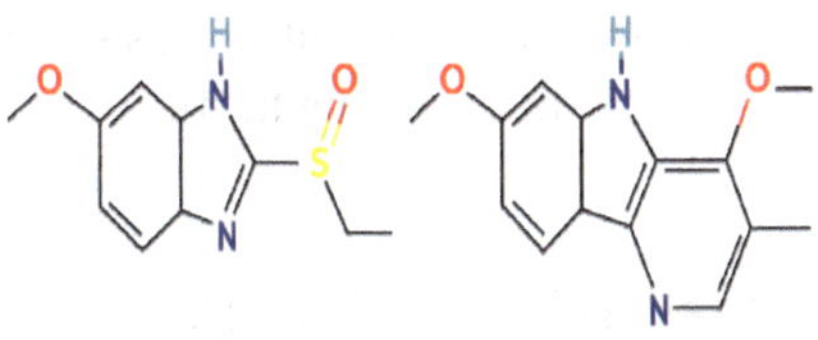

A) B)

Figure 9.5: (A) Omeprazole and (B) esomeprazole.

9.4.4 Minimal risks are associated with a harmless chemical

It is imperative that chemical compounds be developed with minimal toxicity in order to reduce potential harm to both humans and the environment. An example of this is the FDA's stringent monitoring of chemicals that may have an effect on human development during or after pregnancy. The 1957 over-the-counter "safe" sedative/ tranquilizer Thalidomide is a prime example of this. The drug's teratogenic effects were eventually disregarded when it was marketed to alleviate morning sickness in pregnant women. Pregnant women who were given this medication quickly discovered that their babies had serious congenital defects. A number of abnormalities, such as phocomelia (limbs that look like seals) and amelia (no limbs at all), were noted in children who had been exposed to thalidomide in 1960 [35].

9.4.5 Minimal worldwide and physical danger

In order to be considered safe, a chemical must not have any characteristics that could cause an explosion or fire, and it must also pose little threat to the environment on a global scale, meaning that it will not have any detrimental effects on the environment when exposed to it for long periods of time. Its chemical components should break down into harmless by-products in the environment when it biodegrades quickly. Toxic substances that are not harmful to humans should not build up in plants and lower-trophic animals like fish and algae before making their way up the food chain.

One example is radiographic procedures that make use of X-ray contrast agents based on iodine.

Since large doses are needed to achieve enough radiocontrast, the administration is done intravenously. As a potential X-ray contrast agent with minimal toxicity and high biodegradability, iodine compounds derived from sugars have been studied to lessen the likelihood of bioaccumulation. When compared to existing or potential alternatives, a safe chemical should have positive effects on human health, the environment, and the bottom line. It is expected that the process of synthesizing a novel chemical, as opposed to a commercialized chemical with the same function, will result in less waste and less or no solvent use, which can have positive effects on the environment and the economy. If the chemical molecule is less toxic, it will be easier to absorb, more biodegradable, and more effective than a commercial drug [2, 4].

9.4.6 Case study investigation into a biocatalytic method for cosmetic and personal care ingredients devoid of solvents

A major development in the manufacturing of esters, an essential component of personal care and cosmetic goods, is covered in the article.

9.4.6.1 Traditional manufacturing methods

- Chemical processes using strong acids and elevated temperatures are commonly used to create esters. This process not only consumes a lot of energy but also generates undesirable by-products that require energy-intensive purification in order to be removed.
- The conventional methods often utilize organic solvents, which pose risks to both workers and the environment due to their hazardous nature.

9.4.6.2 Eastman's innovative approach

- Eastman has developed a new method that employs immobilized enzymes for ester production.
- The enzyme-based method significantly reduces energy consumption and eliminates the need for strong acids and organic solvents, making it a more environmentally friendly option.

9.4.6.3 Market demand and trends

- Emollients, emulsifiers, and specialist performance ingredients, esters play an essential part in the enormous personal care and cosmetics industry. About 50,000 metric tons of esters were used in North America in 2006.
- The cosmetics business is seeing a shift toward more eco-friendly production methods and all-natural components.

9.4.6.4 Benefits of the new method

- By using immobilized enzymes, Eastman's method not only addresses the environmental concerns associated with traditional ester production but also enhances the quality and variety of esters available for cosmetic formulations [28].

9.5 Use of renewable feedstocks

9.5.1 Transitioning to sustainable raw materials

9.5.1.1 Use of renewable feedstocks: transitioning to sustainable raw materials

Feedstocks include a wide range of raw materials required for production and industrial processes. Renewable feedstocks, especially the ones which are used in flow processes are more natural replenishable materials. Renewable feedstock is part of the moiety upon which modern green chemistry stands. It is renewable fossil resources that constitute 86% of our energy and 96% of all organic compounds. By 2025, our energy appetite is likely to grow by 50%, which will consequently create imminent excess pressure on the nonrenewable sources. Thus, renewable feedstocks are expected to reduce the dependence on fossil fuels and to sustain future generations. It reduces effects on nature replacement by nonrenewable resources such as coal, oil, and gas, which can give items like biodegradable plastics (PHBV), medicines, solvents, enzymes, reagents, and catalysts. Large industrial machines can also potentially be alternative energy sources. Bio-based chemicals manufactured from renewable feedstocks show better carbon footprints than those produced from petrochemicals. We deplete resources at an unsustainable rhythm, with worldwide problems like climate change, environmental degradation, outdated and inadequate energy infrastructures limiting energy flows, pollution, deforestation and social inequality (Figure 9.6) [50].

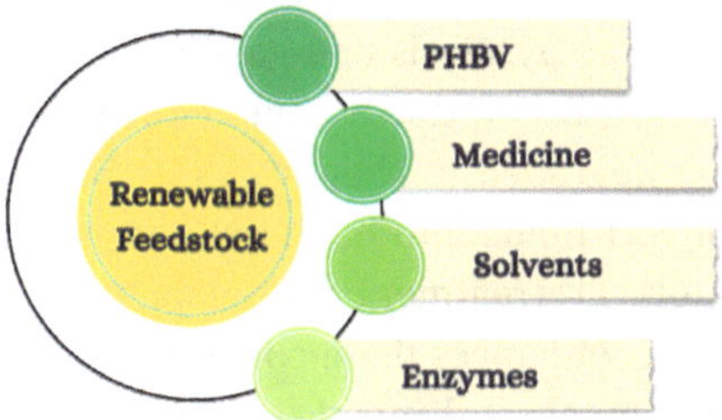

Figure 9.6: Renewable feedstock.

9.5.1.2 Economic and ecological significance

Renewable feedstocks comprise a vast category of material that, based on sustainability, provide alternative feedstocks to fossil feedstocks besides aiding in waste management and protection of the environment. Most of the biomass from the lingocellulosic sources comes from agricultural residues like crop by-products such as rice straw, wheat husks, stalks, and sugarcane bagasse. These residues are converted into biofuels, bioplastics, and chemicals such as lactic acid through fermentation and green chemistry processes, simultaneously reducing air pollution from biomass burning. Algae, particularly microalgae, are emerging as a highly efficient feedstock because they can grow on nonarable land (even in wasteland) and absorb CO_2. They are used for biodiesel production, bioethanol, and high-value compounds like pigments and omega-3 fatty acids. Forest waste, wood chips, sawdust, and bark from logging operations become a vital resource for bio-oil and adhesives by promoting circular economy practices in forestry. Organic municipal waste, comprising food scraps and garden waste, both solves the problems of urban waste and provides raw materials for biogas through anaerobic digestion and composting, improving soil health (left over slurry as fertilizer) and energy production. Industrial by-products such as dairy production brewery spent grains are converted into bioethanol, chemicals such as succinic acid, humic acid, and other valuable products, thus minimizing industrial waste streams. All these feedstocks put together provide a versatile and sustainable portfolio for the bio-economy, addressing the world's challenges of waste, pollution, and depletion of finite resources.

9.5.1.3 Procedure and processes

The conversion of renewable feedstocks into usable products occurs through several scientific methods involving biochemical, thermochemical, and catalytic processes. Biochemical techniques include fermentation, anaerobic digestion, which uses microbes or enzymes to break the biomass into simpler sugars, gases, or biofuels such as ethanol and biogas. Thermochemical processes such as pyrolysis and gasification involve applying heat to controlled environments that break down complex organic matter into bio-oil and syngas. These can then be upgraded to chemicals or energy. Catalytic processes such as transesterification convert lipids into biodiesel and other bio-based chemicals through reaction with alcohols. Pretreatment of feedstocks, for example steam-treating to break down lignocellulose, enhances these conversions. Innovations in enzyme engineering, catalyst development, and integrated LCAs have been major plays in enhancing yield and sustainability in converting renewable feedstock, hence promising the conversion path in reducing dependence on fossil fuel and propelling a circular economy. More specifically, renewable resource utilization helps in curbing greenhouse gas emissions where it stands so that fossil-derived materials will contribute to climate change. For example, biofuels from agricultural residues and algae have a much lower life-cycle carbon footprint. In addition, feedstocks such as algae and forest waste absorb CO_2 during their growth, thus supporting a carbon-neutral cycle that mitigates emissions. This transition aids in meeting global climate targets and reduces reliance on polluting energy sources. These feedstocks align with circular economy principles by promoting resource efficiency and waste reduction. Utilizing agricultural by-products, forest residues, and industrial waste transforms potential waste into valuable resources, contributing to a closed-loop system. This process minimizes reliance on finite resources and encourages regeneration. Adopting renewable feedstocks helps maintain economic growth while safeguarding the environment by creating sustainable production and consumption cycles that embody circular economy goals [52].

9.5.1.4 Future perspectives

Emerging technologies and innovations in renewable feedstocks are redrawing the landscape of sustainable production. Advanced genetic engineering and molecular biology are unlocking the potential of feedstocks like algae and agricultural waste for more efficient production of biofuels, bio-plastics (nylon 2–nylon 6), and special chemicals. For example, improvements in metabolic engineering enable optimization of microbial pathways to improve the yield and quality of bio-based products. Further, integrated bio-refineries that co-produce multiple products from the same biomass can maximize resource utilization and economic feasibility. New pretreatment technologies, such as steam explosion and enzymatic hydrolysis, are improving the extraction of fermentable sugars from lingo-cellulosic biomass, making it a viable feedstock for large-scale biofuel production.

9.5.1.5 Case study

A notable example of sustainable bio-based chemical production is the use of microalgae and agricultural waste as feedstocks. Microalgae, such as *Chlorella* and *Scenedesmus*, are very efficient in converting nutrients and CO_2 into biomass. This biomass can be further processed into chemicals, such as biofuels and biodegradable plastics like polyhydroxyalkanoates (PHAs), showing great promise for environmental and economic advantages. Moreover, agricultural residues, such as straw and bagasse, are reused to obtain fermentable sugars that can be fermented with microbes to yield valuable chemicals like lactic acid and ethanol. In this manner, waste can be both managed in terms of disposal and, at the same time, contribute to a circular economy by avoiding dependence on fossil resources. In 2023, the global bio-based chemicals market was valued at USD 73.16 billion and is anticipated to expand from USD

99.86 billion in 2024 to USD 207.95 billion by 2032, with a compound annual growth rate (CAGR) of 9.6% over the forecast period (2024–2032). In 2023, Europe held a commanding 61.15% share of the bio-based chemicals market. The U.S. bio-based chemicals market is anticipated to expand considerably, with an estimated valuation of USD 40.9 billion by 2032, propelled by customers' increasing desire for bio-based products and evolving environmental protection legislation. The research is focused on improving these processes and increasing the production scale to meet market requirements while minimizing the impact on the environment [51].

9.6 Design for degradation: eco-friendly products

9.6.1 Ensuring products break down safely

Chemicals, the constituents of all matter, are the basis of our daily lives and are often engineered to be stable during use. Persistent chemicals accumulate in living organisms, disrupting biological processes and creating ecological dangers over time. Transient technology aims to synthesize stable chemicals that are degradable and nontoxic upon disposal. This approach addresses environmental issues by minimizing pollution and waste by avoiding products from staying in landfills or the environment. Materials designed for transient technology degrade safely into nontoxic components through mechanisms like light, water, or biological processes. The goal is to minimize the environmental impact while ensuring the material's reuse or recycling post-degradation [3].

9.6.2 Importance of biodegradability: addressing persistent environmental pollutants

The processes of biodegradation, hydrolysis, and photolysis are all used in the design of compounds that safely decompose after usage. Goods should break down effectively and leave no harmful residues behind. Because they frequently include chlorine atoms, which remove electrons from molecules and make it more difficult for bacteria to break them down, the majority of chemicals that survive in the environment are difficult to degrade. Knowledge bases, biodegradability models, and structural analysis are some of the techniques used to forecast which chemicals will persist and which will decompose properly. The development of more ecologically friendly products may be guided by the use of these techniques to forecast which chemicals will breakdown efficiently and which might persist [37].

9.6.3 Life cycle perspective: considering end-of-life impacts during product design

Industrialized sectors like paper, textile, and distilleries release harmful chemicals into the environment, posing significant risks to human health and ecosystems. These chemicals contain phenolic compounds, chlorinated chemicals, and endocrine-disrupting compounds. Traditional bioremediation techniques, relying on microbes, are ineffective for complex pollutants or take time to restore balance. Advanced techniques, like GEMs, are being used to improve bioremediation, but they have limitations due to their susceptibility to environmental factors like pH, temperature, and nutrient availability. In silico techniques are being used in bioremediation to predict pollutant degradation processes and assess their acute toxicity. These tools, based on QSAR models, machine learning algorithms, and molecular simulations, allow for better designed bioremediation strategies by simulating degradation pathways. While conventional bioremediation methods may not offer all the benefits, their combination with genetic engineering and computational tools is a significant breakthrough in improving efficiency and predictability in environmental cleanup operations [48].

9.6.4 Case study 10: biodegradable surfactants

9.6.4.1 Introduction

Surfactants are commonly used in cleaning agents. These agents decrease surface tension and enable the easy removal of dirt, grease, and other impurities. However, the environmental effects of traditional surfactants, especially those from petrochemical origins, are significant. Many conventional surfactants are not biodegradable in character, and this causes water pollution and damage to aquatic environments. Biodegradable surfactants that address these issues but still provide cleaning effectiveness needed in various applications will be developed. The new biodegradable surfactant formulations have entered the personal care and industrial cleaning field, and their environmental impact is measurable.

9.6.4.2 Innovative formulations

Biodegradable surfactants are formulated to degrade more rapidly in natural environments so that their long-term effects on the environment are at a minimum. Traditional surfactants, like LAS, remain in ecosystems for such a long time that they harm aquatic life and encourage pollution. Biodegradable alternatives are formulated from renewable resources, such as plant oils, and are designed to degrade rapidly when exposed to microorganisms in water and soil. Alkyl polyglucoside (APG) is a biodegradable surfactant derived from renewable feedstocks like glucose and fatty alcohols. It offers high cleaning power and biodegradability, with no toxic by-products produced during degradation, according to studies. Similarly, the surfactants derived from natural oils such as coconut oil or palm kernel oil have been found to be promising since they can degrade much better than synthetic ones [9, 43].

Other than plant-based surfactants, the area of research on biosurfactants produced by microorganisms is getting more attention. The formulated biosurfactants are biodegradable, as well as formulated using waste products, so they are sustainable. One such rhamnolipids produced by *Pseudomonas aeruginosa* has great biodegradability along with efficiency to be used as a cleansing agent [25].

9.6.4.3 Applications

Biodegradable surfactants are applied in the personal care industry and industrial cleaning industry seeking more environmentally friendly products.

In the personal care area, biodegradable surfactants are utilized in shampoos, body washes, and facial cleansers. Surfactants ensure these products clean the skin and hair effectively while being gentle to the environment. For example, the application of APGs and rhamnolipids in cosmetics and hygiene products has become increasingly popular due to their mildness and biodegradability, which is the "green" formulation trend in the beauty industry [33].

In industrial cleaning, biodegradable surfactants have been very critical in reducing the ecological footprint of cleaning agents applied in the food processing, automobile, and agricultural sectors. It is therefore important for biodegradable cleaning agents that no harmful chemical residues from such cleaning agents are permitted to accumulate in water systems. Surfactants derived from renewable resources, such as lauryl glucoside and other sugar-based surfactants, have only recently gained significant industrial application because of their biodegradability within days in wastewater treatment plants [47].

A good example of success is the application of biodegradable surfactants in cleaning up oil spills. Rhamnolipids and other biosurfactants have been used to minimize the environmental damage of oil spills by breaking down the oil more efficiently and quickly than traditional chemical dispersants. This application has been promising in minimizing long-term damage to marine ecosystems [7].

9.6.4.4 Environmental impact

The introduction of biodegradable surfactants will help reduce pollution of water bodies to a large extent. Most commonly used surfactants in conventional surfactants usually linger in the environment, making them toxic to aquatic species. In measuring the environmental impact of biodegradable surfactants, one needs to know the rate of degradation and its toxicity towards aquatic organisms. It has already been shown through various studies that surfactants such as APG and rhamnolipids degrade within a few days to weeks in natural environments and leave behind nontoxic by-products such as fatty acids and sugars, which easily get assimilated into the micro-organisms in that ecosystem. On the other hand, conventional surfactants degrade slowly, taking months or even years to break down and end up in aquatic food chains [43].

Biodegradable surfactants also play a role in alleviating pressure on wastewater treatment plants. As they degrade rapidly, they do not put pressure on the treatment process and therefore assist in achieving environmental standards. This is particularly pertinent because the pressure on industries to be cleaner and more environmentally friendly is increasing. This shift to biodegradable surfactants has been proven to lower the levels of hazardous chemical residues in treated water and reduce the impact on biodiversity. For example, it has been shown that the replacement of conventional surfactants with biodegradable ones in laundry detergents reduces aquatic toxicity by 40–50% [47].

9.6.4.5 Conclusion

The development and introduction of biodegradable surfactants are a great leap in reducing the environmental footprint of cleaning products. As such, with new formulations from renewable resources, they find excellent applications and contribute to environmental sustainability through quick degradation without harm. Personal care and industrial cleaning industries provide examples of success stories, because this type of surfactant holds great potential towards minimizing the pollution and toxic character of the water system. The increasing demand for "eco-friendly" products can see the role of biodegradable surfactants emerging at a forefront in reducing cleaning agent footprints on earth while offering a clean alternative both to the industries concerned and to the consumer.

9.7 Immediate assessment for pollution mitigation (real-time monitoring)

9.7.1 Advancing process monitoring

Analytical approaches require further advancement to provide real-time, in-process monitoring and control before the generation of hazardous chemicals. Conventional analytical chemistry necessitates substantial sample quantities, extensive solvent utilization, and significant energy consumption. The advancement of novel methodologies and precise mobile devices enables analyses to be conducted with minimal sample sizes at the sampling location and significantly reduced solvent usage. Consistent monitoring of the production process at every stage is essential to adhere to the principle of real-time analysis for pollution prevention, which calls for the improvement of analytical techniques to allow continuous monitoring of the chemical production process. The goal is to prevent the creation of hazardous substances, which would forestall the development of mistakes that might result in the production of toxic chemicals with negative effects on both humans and the environment. Reduced reliance on costly and energy-intensive waste management and cleaning operations is one goal of green chemistry, which seeks to eliminate pollution whenever possible by developing processes that produce few harmful by-products and waste [8, 31].

9.7.2 Analytical methods for real-time assessment

Analytical methods are used to track these reactions; analytical data, analyte characteristics, and sample preparation are the deciding factors in method selection. Researchers are working on methods to improve efficiency, accuracy, and reliability by doing away with sample preparation and waste management [32].

9.7.3 Chromatographic methodologies

9.7.3.1 High-performance liquid chromatography (HPLC)

High-performance liquid chromatography (HPLC) can cause contamination due to the mobile phase. Its main drawback is the high solvent usage, resulting in 500 L/year eluent, with over 50% containing poisonous and costly acetonitrile, a potentially harmful solvent. Various solutions have been proposed to reduce solvent usage, such as reducing column dimensions or using only water as a solvent. The solubility and polarity of organic compounds limit their application, but analytical scientists are developing new chromatographic techniques using environmentally friendly solvents like ethanol or ionic liquids [63].

9.7.3.2 Gas chromatography

It is the method of choice for chemical analysis when it comes to separation. The use of helium as a carrier gas in gas chromatography rather than organic solvents in HPLC makes it a more environmentally friendly method. Still, efforts to improve the analysis largely center on derivatization of the analyte and sample preparation processes [36].

9.7.3.3 High-temperature liquid chromatography

This technique is regarded as an effective method to reduce the volume of eluent necessary for the separation and analysis of the sample. Temperature is a crucial component that facilitates the differential migration phenomena; as temperature rises, viscosity decreases and diffusivity increases, enabling operation at elevated flow rates without compromising the resolution and accuracy of the analysis [54].

9.7.3.4 Ultra-high-performance liquid chromatography

Conventional HPLC devices are being gradually replaced by ultra-HPLC systems. They were put in place to reduce the increased back pressures caused by the small particle size. The runtime, solvent consumption, and waste output can all be reduced by shortening the column length [22].

9.7.3.5 Supercritical fluid chromatography (SFC)

Supercritical fluid chromatography (SFC) is used to investigate low concentrations of components and substances with a high molecular weight. Preliminary research suggests that the pharmaceutical, food, explosive, petroleum, polymer, and propellant industries will all benefit from the study. The main differences between gas and liquid chromatography and SFC are the eluent types and chromatographic conditions. SFC systems use carbon dioxide (CO_2), a gas at room temperature, as their mobile phase. To keep CO_2 in its fluidic state, these systems are subjected to extreme heat and pressure. Separating complex mixtures, such as chiral and natural chemical compounds, requires a highly pressurized stream of CO_2. In comparison to HPLC, this chromatography technology is faster, uses much less solvent, and is generally more cost-effective and ecologically friendly for separating different types of components [53].

9.7.4 Spectroscopic methods

Spectroscopy detection devices are often sustainable due to their minimal sample treatment requirements and low energy usage.

9.7.4.1 Mass spectrometry

Mass spectrometry, due to its high sensitivity, permits the utilization of minimal sample quantities. This technique is characterized by significant energy consumption and frequently necessitates appropriate sample preparation. Nevertheless, advancements such as nano electrospray techniques enable direct analysis of samples, and in certain instances, facilitate derivatization and detection without the need for reagents [59].

9.7.4.2 Infrared and near-infrared spectroscopy (IR/NIR)

Minimal sample needs, high sensitivity, and the ability to examine samples directly without prior treatment are just a few of the many advantages of infrared (IR) spectroscopy, a non-destructive technique. Because of their low cost, ease of use, and ability to provide real-time monitoring without residue, these analytical methods can be considered environmentally benign. No longer is it necessary to collect samples when there are so many devices available, some of which are portable, that allow for measurements to be taken in the field or even during the process itself with a variety of probes. Industry uses IR and near-IR (NIR) methods for continuous gas and particle emission monitoring, while the biomedical and pharmaceutical fields use IR and NIR methods for biomolecule and natural substance evaluation [58].

9.7.4.3 Raman spectroscopy

Raman spectroscopy does not require sample preparation due to its rapid, nondestructive, and noninvasive nature, which generates a molecular signature for each sample and enables observations in aquatic environments. This methodology is capable of assessing a diverse array of sample types, including medicines, materials, organisms, liquids, and bioprocesses, in both internal and external research environments. Polymorph identification, real-time process surveillance, and the identification of counterfeit or tainted pharmaceuticals are among the modern pharmaceutical applications [19].

9.7.4.4 Fluorescence spectroscopy

This method is used for both quantitative and qualitative investigation, and it is regarded as one of the most sensitive spectroscopic approaches to detect organic molecules. Industrial applications of fluorescence spectroscopy include the analysis of spectral data used in the evaluation of food, pharmaceuticals, wastewater, and natural samples, all of which provide crucial insights. Any solid, powder, film, or material solution can be analyzed using this technique. By using a fiber-optic sampler, depositing samples on silica quartz disks, or extracting them from a solution, many samples can be examined without preparation [1, 5].

9.7.5 Electrochemical techniques

Green chemistry's overarching goal is to do away with potentially harmful compounds used in analysis. Despite its widespread recognition as one of the most poisonous heavy metals, mercury has long served as the electrode material of choice in electrochemistry owing to its beneficial properties. Electrochemical measurements with alternative work electrodes have recently been the subject of investigation. These electrodes include bismuth-film electrodes for trace-metal analyses of high quality, carbon electrodes with a wide potential window, chemical inertness, affordability, and adaptability to a variety of sensing and detection applications, and surface-modified electrodes that can selectively detect contaminants while ignoring unwanted substances [61].

9.7.6 Pollution control through eco-chemistry case studies

9.7.6.1 Case studies in eco-chemistry: proactive approaches to environmental pollution control

From the case studies that were given, the following information regarding pollution control is provided:

9.7.6.1.1 Pharmaceutical industry

- According to estimates, the pharmaceutical business generates 25–100 kg of rubbish for every kilogram of active pharmaceutical ingredient (API) made. This wastefulness is well-known in the industry. Communities living in close proximity to manufacturing sites have health risks due to the high volume of trash, which worsens environmental contamination.
- The industry has achieved a remarkable decrease in hazardous waste by utilizing biocatalysts and other green chemistry principles. For instance, ibuprofen's production process has been fine-tuned to produce far less waste, going from 60 kilograms of waste for every kilogram of product to less than one kilogram. This has the dual benefit of reducing pollutants and making medicine production more sustainable.

- The introduction of in-line monitoring systems allows for real-time adjustments during production, which helps prevent the creation of off-specification products that would otherwise contribute to waste and pollution.

9.7.6.1.2 Plastic waste management

- The plastics industry is another area where eco-chemistry principles can significantly reduce pollution. The transition from linear to circular economy models is essential in managing plastic waste effectively. This involves designing products for reuse and recycling, which can help mitigate the environmental impact of plastic pollution.
- Implementing eco-friendly materials and processes in the plastics sector can lead to a reduction in the production of hazardous compounds, thereby decreasing the potential for environmental contamination.

These case studies illustrate how eco-chemistry can proactively address pollution by promoting sustainable practices and innovative solutions across various industries. The findings underscore the importance of prevention over remediation, highlighting that proactive measures are more effective and beneficial for the environment [42].

9.8 Inherently safer chemistry for accident prevention

9.8.1 Building safety into chemical processes

As a result of more government control and a focus on safety improvement, the chemical sector's safety record has been improved. The public's negative view of the industry, which is often associated with worries about safety and the environment, has had an impact on its safety record. It is crucial to manage the risks associated with combustion hazards and runaway reactions, the most common causes of mishaps in chemical processes. Significant damage and casualties can occur as a result of sudden pressure releases, such as tank ruptures, uncontrolled responses, or external fire pressure. Another subtle component could be static electricity. There are technical and managerial aspects to chemical process safety, and technical staff at any level can help with the overall management of safe facilities. As part of a process risk assessment, thermal process safety includes characterizing the energy potential of chemicals, reaction kinetics, and equipment concerns, as well as identifying and managing thermal hazards. Data on substances, estimation methods, thermodynamic calculations, and experimental results are all part of this process. Ongoing risk assessment and management throughout all phases of process development achieves inherent process safety and functional safety [64].

Figure 9.7: Safety measures for chemical process.

9.8.2 Chemical process safety

Rigorous process safety assessments are necessary before transitioning a process from the benchtop to a pilot plant. In exothermic reactions, external cooling is generally employed to dissipate heat from the reaction system. Inadequate cooling may result in heat generation surpassing the heat removal capability, causing elevated reaction temperatures that could instigate decomposition of the reaction mixture and precipitate runaway reactions or explosions (Figure 9.7). It is essential to assess reaction calorimetry and thermal instability hazards of chemical reactions to guarantee that a safe reaction temperature is kept far below the breakdown onset temperature. Prompt identification of exothermic reactions should initiate the evaluation of thermal hazard issues prior to conducting the reaction in research laboratories and/or scale-up facilities [18].

Chemical process safety can be attained by functional safety management and inherent safety during design with the help of process risk assessment and management. This includes cutting down on energy and material consumption, finding safer alternatives to potentially harmful substances, reducing exposure to potentially dangerous environments, and streamlining design processes to remove superfluous details. Reactive conversion, which involves the control of hazardous, combustible, and corrosive chemicals, is one way in which chemical technology is applied in product manufacturing. In industrial manufacturing, process technologies like fractional distillation and real-time monitoring are utilized. Environmental and safety technology are used to reduce risks in a proactive manner by creating safe process conditions. The process safety management system barriers are divided into three categories: plant, process, and personnel. Plant ensures rigorous equipment design, installation, inspection, and maintenance; process ensures management systems; and personnel ensures qualified personnel for safe operation [26, 30].

9.8.3 Inherently safer strategies

Inherently safer design (ISD) prioritizes the following four concepts as a process safety component: cut down, repurpose, moderate, and simplify. Machine malfunction or human error accounts for 60% of chemical accidents. To guarantee process safety, an ISD seeks to change process design in such a way as to eliminate risk. Still, it is unclear if the design is really risk-free [55].

There are two main methods for designing chemical process safety: managing hazards through engineering and management controls or eliminating them entirely. Process hazards can be minimized by engineering and management controls, reducing potential impact and preventing major injury or damage. The chemical process industries rely on systems; yet no system is foolproof, and failure can result in individual mishaps. Safer process design strategies include making process equipment smaller, using less dangerous substances or stages, keeping processing and storage environments moderate, and making process and plant design simpler. Miniaturizing chemical process industry (CPI) machinery enhances safety by decreasing the amount of potentially harmful materials emitted and the amount of energy that machinery contains. Minimizing the inventory of hazardous materials in a CPI plant is possible through many means that do not need significant changes to process technology. After the Bhopal, India, disaster in 1984, most CPI companies looked for ways to cut back on combustible and dangerous materials. They took the time to assess their current setup and procedures, finding ways to cut down on the number of dangerous materials they needed to operate [64, 65].

Implementation Strategies for ISD:

1.Minimize Equipment Size: Reduces hazardous materials release.

2.Substitute Hazardous Materials: Lowers risks.

3.Moderate Conditions: Minimizes potential hazards.

4.Simplify Design: Reduces errors and accidents.

Scheme 9.2: Implementation strategies for ISD [66].

Reducing inventory, making sure pipes are big enough to transmit the necessary quantity, and employing materials as gas instead of liquid are all ways to minimize hazardous material piping. When it comes to transporting hazardous goods, the Dow chemical exposure index is a useful tool for measuring intrinsic safety in proportion to prospective toxic exposure hazards. This helps provide a safe and efficient system. Within the hierarchy of controls, ISD is the foundational control that aims to eliminate or reduce risks. To improve efficiency, the hierarchy is divided into four sections: administrative, active-engineered, passive-engineered, and ISD. Methodical and explicit integration of ISD into an organization's process safety management (PSM) framework is necessary because ISD is an integral component of the framework [12, 66].

9.8.4 Prevention

Enclosed tanks and vessels have pressure relief valves installed to reduce the risk of explosions caused by overpressurization. By definition, prevention is "an action taken to diminish or eradicate the likelihood of specific adverse events occurring" and is generally considered more cost-effective than responding to and resolving the aftermath of disasters. The first four tiers of control are superior because they employ preventative efforts to lessen or eliminate risk via interventions in engineering, design, and elimination, soon after they take place

9.8.5 Mitigation

According to [67], mitigation is "the effort to reduce loss of life and property by diminishing the impact of disasters." The purpose of mitigation is to lessen the impact of a problem after it has already occurred, rather than to avoid it. To lessen the impact of an emergency and keep casualties to a minimum, it is wise to have a strategy ready in the event of a crisis.

9.8.6 Protection

Protecting, enclosing, or conserving an object from harm is similar to mitigation. Rather than preventing the disaster from happening, its purpose is to lessen the intensity of any harm that may come from it.

Automatic fire suppression systems in buildings, cathodic protection for subterranean storage tanks, and personal protective equipment are all examples of measures that reduce risk. Insurance, often known as risk transfer, serves as a safeguard for the things or entities that are insured.

9.8.7 Control

The management of risk through reducing the likelihood and impact of an exposure is what risk control is all about. Any action – process, regulation, technology, behavior, and so on – that modifies risk is defined as control in ISO Guide 73 (ANSI/ASSE Z690.1-2011). Controls might not reliably have the expected or desired effect on modification, as stated in the ISO standard. Loss control is a term used by certain experts, but risk management involves more than just reducing losses; it also regulates risk.

9.8.8 Risk reduction

Layers of controls or defenses including preventative, protective, mitigation, and control measures are usually necessary for an effective plan to reduce risk to an acceptable threshold. Interventions that aim to reduce risks are known as preventative and mitigation strategies.

Accidents, damages, and environmental pollution can all be avoided with a focus on safety. Alterations to production methods, the substitution of environmentally friendly consumables, process redesign and reengineering, and product reformulation to remove harmful chemicals are all part of primary prevention efforts. Process relief valves, alert systems, measurements of human and organizational control (such as defining work instructions, safety culture, and the usage of protective equipment), and so on can all be utilized in industrial processes to lessen the likelihood of accidents. Aside from accident feedback, which involves investigating accident reports, collecting and processing data, defining lessons learned, disseminating information, coming up with a solution and decision on prevention steps, and then putting them into action [28, 34] are required.

9.8.9 Detailed case study analysis of chemical safety procedures

Incident overview: The case involves the manufacturing of carbon disulfide, where a contractor's employee was found unconscious on a heap of furnace waste residues. Initially, it was assumed to be a natural death, but further investigation revealed it was due to asphyxia from poisonous gas exposure.

Chemical process: Carbon disulfide is produced by reacting sulfur with calcified charcoal at high temperatures in an electric furnace. The process generates hydrogen sulfide gas, which can become trapped in the residual mass at the bottom of the furnace. This gas is particularly dangerous as it can escape when the solidified mass is disturbed, especially in wet conditions.

9.8.9.1 Lack of safety procedures

- **Training deficiencies**: Staff, including the driver and cleaner, had not received training for handling hazardous jobs, which is crucial for ensuring safety during chemical operations.
- **Permit-to-work system**: There was no "permit-to-work" system in place for unloading hazardous chemicals, which is essential for managing risks associated with such operations.
- **Supervision and permission**: The production chemist took risks by unloading the tanker without proper permission from the department head, violating established safe operating procedures.
- **Emergency response failures**: After the incident, the injured worker was not given timely first aid. There was no medical officer or qualified first-aider present, and arrangements for immediate medical treatment were poorly executed. The injured worker was first taken to a nearby clinic instead of a hospital, delaying critical care.
- **Medical findings**: The injured worker suffered complete loss of vision in one eye and partial recovery in the other after surgery. This outcome underscores the dire consequences of inadequate safety measures and emergency response protocols.

Conclusion: This case study serves as a stark reminder of the importance of rigorous safety procedures, proper training, and effective emergency response in workplaces dealing with hazardous materials. The failures in these areas not only led to a tragic injury but also highlight systemic issues that need to be addressed to prevent future incidents [20].

9.8.9.2 Conclusion

Green chemistry is an emerging approach to transform hitherto traditional chemical processes that have posed environmental, economic, and societal challenges. The focus is now on sustainability, efficiency, and environmental stewardship so much that green chemistry heads the list of directions needed in the chemical industry if its future is to become a sustainable one. These results are a culmination of findings and lessons from case studies in green chemistry, effects in the global arena, and next steps toward continued evolution.

9.9 Reflecting on the role of green chemistry

Green chemistry lies at the interface of science and sustainability. It provides practical solutions to negative impacts from chemical production and consumption. These guiding principles are aimed at conserving resources, abolishing hazardous materials, and making the processes more efficient without losing quality or functionality in end products. The chemical industry does business according to the new paradigm called green chemistry, aligning itself to meet the needs of society regarding the environment as well as pushing innovation to further this cause in every aspect of the production.

9.9.1 Energy efficiency

Ambient temperature and pressure reactions have been game changers in energy conservation. Green chemistry saves energy by stopping the energy-intensive heating process and cooling, which were associated with industrial activity.

9.9.2 Process simplification

One of the simplified synthesis pathways that have changed chemical manufacturing in many ways is one-pot reactions. This pathway uses fewer steps to synthesize the product, resulting in less waste, fewer resources, and no intermediates or protective groups. Such benefits go beyond reducing waste to lower costs of production, decrease labor, and increase speeds of production.

9.9.3 Catalysis

Catalysis is at the heart of green chemistry: Reactions can take place under more mild conditions, reduce selectivity and waste, and realize cleaner, greener routes. Applications of homogeneous, heterogeneous, and enzymatic catalysts for low-emission processes and resource-efficient transformations include biodiesel producing solid acid catalysts and zeolite in petrochemical applications.

9.9.4 Global impacts

The principles and applications of green chemistry have a large-scale impact, translating to a global effect.

9.9.5 Environmental benefits

1. **Reduction in carbon emissions**
   Green chemistry reduces the levels of greenhouse gases since it lowers the energy required and eliminates hazardous by-products associated with manufacturing chemicals.
2. **Natural resource conserving**
   Resource-intensifying processes limit the exploitation of finite resources, thus conserving more for future exploitation. For example, using atom economy and renewable feedstocks are some of the ways green chemistry can minimize the use of nonrenewable material.
3. **Waste management**
   Green chemistry emphasizes waste minimization as directly relevant to the environmental burden of chemical waste disposal. It ensures a cleaner process with fewer toxic by-products by using recyclable catalysts and efficient reaction designs.

9.9.6 Societal well-being

Green chemistry leads to improvements in well-being due to the decreased exposure of people to hazardous chemicals, clean air and water, and a safe work environment for employees working in chemical industries. Therefore, such innovations help communities and ecosystems and promote public confidence in scientific and industrial practice.

9.10 Future research and implementations directions

Although green chemistry has been relatively successful, much is still to be learned and implemented. Continued research and thoughtful implementation may be necessary in order fully to realize potential.

9.10.1 Emerging trends

1. **Synthesis of new green catalysts**
   Research is now looking towards the design of catalysts that are not only efficient but more importantly benign to the environment. Bio-catalysts, metal-free organo-catalysts, and even recyclable heterogeneous catalysts will play a significant

role among others. Nanotechnology will lead toward highly active and efficient catalysts.

2. **AI in process optimization**
   AI and machine learning are becoming vital tools for optimizing chemical processes. By predicting reaction outcomes and identifying optimal conditions, AI accelerates the development of sustainable practices and reduces trial-and-error experimentation.
3. **Use of bio-based feedstocks**
   One of the leading trends of sustainable chemistry is the shift from petrochemical feedstocks to bio-based. Materials from agricultural and forestry residues are renewable, replacing the synthesis of fuels, polymers, and fine chemicals.

9.10.2 Policy and partnership

1. **Governments and industry partnerships**
   Green chemistry will be the driving force of policies and incentives that governments can employ to promote sustainable practices through tax benefits, research and development grants, as well as environmental compliance regulations.
2. **Cross-sectoral collaboration**
   Together with other types such as academia and industry, this collaboration accelerates the innovation process to provide a scalable solution when approaching pressing issues in the environmental sector.

9.10.3 Call for action

It is just the same as other forms of science: green chemistry relies on the cumulative action of scientists, organizations, and politicians.

1. **To academics**
   A curriculum with a focus on sustainability ensures that the next generations of science practitioners come qualified to prioritize responsibility for the environment.
2. **To industry professionals**
   Companies should make an effort to invest in green technology and sustainable manufacturing. Business strategies that focus on ESG provisions help advance corporate reputation while cultivating tangible environmental benefits.
3. **To Policymakers**
   Decision and regulatory makers must drive green chemistry through regulation, financial incentives, and international cooperation. On the policy side, renewable energy integration and resource efficiency and waste reduction promote global sustainability.

Green chemistry is a holistic approach towards the attainment of sustainable development. It reduces energy use, conserves resources, and minimizes waste. Its principles empower industries toward cleaner and more efficient processes that ultimately benefit the environment and mankind. Finally, inclusion of emerging technologies, a pro-policy environment, and a cross-sectorial approach should be critical enablers for the accelerating adoption of green chemistry. The journey for sustainability would be shared responsibility where Earth's resources would be conserved and ecosystems protected, ensuring a healthy future for the coming generations once the principles of green chemistry are embraced.

References

[1] Ahmad, M. H., Sahar, A., & Hitzmann, B. (2017). Fluorescence spectroscopy for the monitoring of food processes. *Advances in Biochemical Engineering/Biotechnology*, 161, 121–151.

[2] Anastas, N. D., & Warner, J. C. (2005). The incorporation of hazard reduction as a chemical design criterion in green chemistry. *Chemical Health & Safety*, 12(2), 9–13.

[3] Andrew, L. J., Lizundia, E., & MacLachlan, M. J. (2024) Designing for degradation: Transient devices enabled by (nano) cellulose. *Advanced Materials*, 2401560.

[4] Anastas, P. T., Kirchhoff, M. M., & Williamson, T. C. (2001). Catalysis as a foundational pillar of green chemistry. *Applied Catalysis A: General*, 221(1–2), 3–13.

[5] Anastas, P. T. (2018). The molecular design research network. *ToxicologicalSciences*, 161(2), 241–248.

[6] Atom economy https://www.scranton.edu/faculty/cannm/greenchemistry/english/organicmodule.shtml#:~:text=Atom%20Economy%20in%20Addition%20Reactions,atom%20economy%20of%20this%20reaction

[7] Banat, I. M., et al. (2014). Rhamnolipids: Biosurfactants with potential for environmental cleanup. *Critical Reviews in Environmental Science and Technology*. 44(9), 887–920.

[8] Banjare, M. K., Behera, K., Banjare, R. K., Tandon, M., Pandey, S., & Ghosh, K. K. (2024). Application of green chemistry for environmental remediation. In: Garg, S. & Bhatia, A. (ed.), *Green Chemistry Approaches to Environmental Sustainability: Status, Challenges and Prospective*. Amsterdam: Elsevier, pp. 67–91.

[9] Bérard, R., et al. (2013). Surfactants from renewable sources: Synthesis, properties, and applications. *Green Chemistry*. 15(4), 567–583.

[10] Betts, K., How industrial applications in green chemistry are changing our world, 2018/10/12, Available from: https://www.acs.org/content/dam/acsorg/membership/acs/benefits/extra-insights/green-chemistry-applications.pdf

[11] Brown, D. G., & Bostro, J. (2016). Analysis of past and present synthetic methodologies on medicinal chemistry: Where have all the new reactions gone?. *Journal of Medicine Chemistry*, 59, 4443–4458.

[12] Brown, K. R., Addo, A., Khan, F. I., & Amyotte, P. R. (2020). Inherently safer design principles in risk management. In: Amyotte, P. R. & Khan, F. I. (eds.), *Methods in Chemical Process Safety*. Cambridge, MA: Academic Press/Elsevier, Vol. 4, pp. 379–440.

[13] Cann, M. C., & Dickneider, T. A. (2004). Infusing the chemistry curriculum with green chemistry using real-world examples, web modules, and atom economy in organic chemistry courses. *Journal of Chemical Education*, 81(7), 977–980.

[14] Castañeda-Facio, A. O., & Rodríguez-Herrera, R. (2020). *Green Chemistry and Applications*. USA: CRC Press.

[15] Coish, P., Brooks, B. W., Gallagher, E. P., Mills, M., Kavanagh, T. J., Simcox, N., ... Anastas, P. T. (2018). The Molecular Design Research Network: An overview. *Toxicological Sciences*, 161 (2), 241–248. https://doi.org/10.1093/toxsci/kfx175

[16] Beach, E. S., & Paul, T. (2007). Green chemistry: The emergence of a transformative framework. *Green Chemistry Letters and Reviews*, 1(1), 9–24. https://doi.org/10.1080/17518250701882441

[17] Crawford, S. E., Hartung, T., Hollert, H., Mathes, B., van Ravenzwaay, B., & Steger-Hartmann, T. (2007). Green toxicology: A strategy for sustainable chemical and material development. *Green Chemistry Letters and Reviews*, 1, 9–24. doi:10.1080/17518250701882441

[18] Crowl, D. A., & Louvar, J. F. (2001). *Chemical Process Safety: Fundamentals with Applications*, (2nd ed.). Upper Saddle River, New Jersey, USA: Prentice Hall PTR, Pearson Education.

[19] Davis, K. L., Kemper, M. S., & Lewis, I. R. (2007). Raman spectroscopy for monitoring real-time processes in the pharmaceutical industry. In Šašić, S. (ed.), *Pharmaceutical Applications of Raman Spectroscopy*. New Jersey, United States: John Wiley & Sons, Inc., pp. 117–162. doi: 10.1002/9780470225882

[20] Deshmukh, L. M. (2005). *Industrial Safety Management*. New Delhi, India: Tata McGraw-Hill Education Pvt. Ltd.,.

[21] DeVito, S. C. (2016). On the design of safer chemicals: A path forward. *Green Chemistry*, 18(16), 4332–4347.

[22] Dong, M. W., & Zhang, K. (2014). Ultra-high-pressure liquid chromatography (UHPLC) in method development. *Trends in Analytical Chemistry TrAC*, 63, 21–30.

[23] Eissen, M., Mazur, R., Quebbemann, H. G., & Pennemann, K. H. (2004). Atom economy and yield of synthesis sequences. *Helvetica Chimica Acta*, 87(2), 524–535.

[24] European Commission. Directive 2008/98/EC on waste (Waste Framework Directive), 12/25/2018, Available from: http://ec.europa.eu/environment/waste/hazardous_index.Html

[25] Ghoreishi, S. M., et al. (2014). Biosurfactants as alternatives to conventional surfactants: Synthesis, properties, and applications. *Environmental Science and Pollution Research*. 21(9), 5163–5177.

[26] Hartmann, T., Studer, C., & Krug, H. F. (2017). Green toxicology: A strategy for sustainable chemical and material development. *Environmental Sciences Europe*, 29, 16. doi: 10.1186/s12302-017-0115-z

[27] Horn, J. (2000). The proton-pump inhibitors: Similarities and differences. *Clinical Therapeutics*, 22(3), 266–280.

[28] Kidam, K., Hussin, N. E., Hassan, O., Ahmad, A., Johari, A., & Hurme, M. (2014). Accident prevention approach throughout process design life cycle. *Process Safety and Environmental Protection*, 92(5), 412–422.

[29] Kidwai, M., & Mohan, R. (2005). Green chemistry: An innovative technology. *Foundations of Chemistry*, 7(3), 269–287.

[30] Kletz, T. A., & Amyotte, P. (2010). *Process Plants: A Handbook for Inherently Safer Design*. Boca Raton, Florida, USA: CRC Press.

[31] Koel, M., & Kaljurand, M. (2019). *Green Analytical Chemistry*, (2nd ed.). Cambridge, United Kingdom: Royal Society of Chemistry.

[32] Korany, M. A., Mahgoub, H., Haggag, R. S., Ragab, M. A. A., & Elmallah, O. A. (2017). Green chemistry: Analytical and chromatography. *Journal of Liquid Chromatography & Related Technologies*, 40, 839–852.

[33] Lehmann, J., et al. (2019). Green formulations in the cosmetics industry: Biodegradable surfactants as a sustainable choice. *Journal of Cleaner Production*, 234, 1130–1137.

[34] Lyon, B. K., & Popov, G. (2016). The art of assessing risk. *Professional Safety*, 61(03), 40–51.

[35] Matthews, S. J., & McCoy, C. (2003). Thalidomide: A review of approved and investigational uses. *Clinical Therapeutics*, 25(2), 342–395.

[36] McNair, H. M., Miller, J. M., & Snow, N. H. (2019). *Basic Gas Chromatography*, (3rd ed.). Hoboken, New Jersey, USA: John Wiley & Sons.

[37] Meng, X., Guo, Y., Wang, Y., Fan, S., Wang, K., & Han, W. (2022). A systematic review of photolysis and hydrolysis degradation modes, degradation mechanisms, and identification methods of pesticides. *Journal of Chemistry*, 2022(1), 9552466.

[38] Merck group, Corporate responsibility report 2017, 2018/10/24 Available from: http://reports.merckgroup.com/2017/cr-report/products/sustainable-products/sustainableproduct-design.html

[39] National Center for Biotechnology Information (2024). PubChem Compound Summary for CID 9568614, Esomeprazole. Retrieved December 9, 2024 from https://pubchem.ncbi.nlm.nih.gov/compound/Esomeprazole

[40] National Center for Biotechnology Information (2024). PubChem Compound Summary for CID 4594, Omeprazole. Retrieved December 9, 2024 from https://pubchem.ncbi.nlm.nih.gov/compound/Omeprazole

[41] Novartis, A. G., Materials & Waste, 2018/12/02, Available from https://www.novartis.com/our-company/corporate-responsibility/environmental-sustainability/materials-waste

[42] Poddar, A. K. (2023). Case studies in eco-chemistry: Proactive approaches to environmental pollution control. *European Chemical Bulletin*, 12(5), 5890–5901. https://doi.org/10.31838/ecb/2023.12.5.454.

[43] Schäfer, A., et al. (2005). Biodegradable surfactants: The environmental impact of modern surfactant technology. *Environmental Toxicology and Chemistry*. 24(5), 1021–1033.

[44] Sheldon, R. A. (2017). The E factor 25 years on: The rise of green chemistry and sustainability. *Green Chemistry*, 19, 18–43.

[45] Sheldon, R. A. (2017). The E factor 25 years on: The rise of green chemistry and sustainability. *Green Chemistry*, 19, 18–43.

[46] Sheldon, R. A. (1992). Organic synthesis—past, present and future. *Chemistry & Industry*, 23 (7 December), 903–906.

[47] Silva, M. I., et al. (2014). Sugar-based surfactants for industrial cleaning and environmental applications. *Environmental Engineering Science*. 31(8), 432–441.

[48] Singh, A. K., Bilal, M., Iqbal, H. M., & Raj, A. (2021). Trends in predictive biodegradation for sustainable mitigation of environmental pollutants: Recent progress and future outlook. *Science of the Total Environment*, 770, 144561.

[49] Song, Y., Wang, Y., & Geng, Z. (2004). Some exercises reflecting green chemistry concepts. *Journal of Chemical Education*, 81(5), 691–692.

[50] Sreeharsha, R. V., Dubey, N., & Mohan, S. V. (2023). Orienting biodiesel production towards sustainability and circularity by tailoring the feedstock and processes. *Journal of Cleaner Production*, 414, 137526.

[51] Sundaram, T., Rajendran, S., Gnanasekaran, L., Rachmadona, N., Jiang, J. J., Khoo, K. S., & Show, P. L. (2023). Bioengineering strategies of microalgae biomass for biofuel production: Recent advancement and insight. *Bioengineered*, 14(1), 2252228.

[52] Taipabu, M. I., Viswanathan, K., Wu, W., Hattu, N., & Atabani, A. E. (2022). A critical review of the hydrogen production from biomass-based feedstocks: Challenge, solution, and future prospect. *Process Safety and Environmental Protection*, 164, 384–407.

[53] Taylor, L. T. (2009). Supercritical fluid chromatography for the 21st century. *The Journal of Supercritical Fluids*, 47, 566–573.

[54] Teutenberg, T. (2010). *High-Temperature Liquid Chromatography: A User's Guide for Method Development*, Vol. 13. Cambridge, United Kingdom: Royal Society of Chemistry.

[55] Trevor, K. (2000). *By Accident . . . a Life Preventing Them in Industry*. London: PFV Publications, Peter Varey Associates.

[56] Trost, B. M. (1995). Atom economy-a challenge for organic synthesis: Homogeneous catalysis leads the way. *Angewandte Chemie International Edition in English*, 34(3), 259–281.

[57] Trost, B. M. (2002). On inventing reactions for atom economy. *Accounts of Chemical Research*, 35(9), 695–705.

[58] Vanarase, A. U., Alcalà, M., Jerez Rozo, J. I., Muzzio, F. J., & Romañach, R. J. (2010). Real-time monitoring of drug concentration in a continuous powder mixing process using NIR spectroscopy. *Chemical Engineering Science*, 65, 5728–5733.

[59] Venter, A., Nefliu, M., & Cooks, R. G. (2008). Ambient desorption ionization mass spectrometry. *Trends in Analytical Chemistry TrAC*, 27, 284–290.

[60] Vink, E. T., Glassner, D. A., Kolstad, J. J., Wooley, R. J., & O'Connor, R. P. (2007). The eco-profiles for current and near-future NatureWorks® polylactide (PLA) production. *Industrial Biotechnology*, 3(1), 58–81.

[61] Wang, J. (2002). Real-time electrochemical monitoring: Toward green analytical chemistry. *Accounts of Chemical Research*, 35(9), 811–816.

[62] Williams, P. T. (2005). *Waste Treatment and Disposal* (2nd ed.). Chichester, West Sussex, United Kingdom: John Wiley & Sons.

[63] Yabré, M., Ferey, L., Somé, I. T., & Gaudin, K. (2018). Greening reversed-phase liquid chromatography methods using alternative solvents for pharmaceutical analysis. *Molecules*, 23, 1065 (1–25).

[64] Yang, Q., Peres, S. C., Wang, Q., & Dastidar, A. G. (2022). Process safety from bench to pilot to plant. *ACS Chemical Health & Safety*, 29(2), 110–113.

[65] Yang, F. (2015). *Studies on Tandem Cyclization of Alkynones and Trifluoromethylation in Water at Room Temperature*. Lanzhou: Lanzhou University.

[66] Youcai, Z., & Sheng, H. (2017). Recycling technologies and pollution potential for contaminated construction and demolition waste in recycling processes. In: *Pollution Control and Resource Recovery: Industrial Construction and Demolition Wastes*. Amsterdam, The Netherlands: Elsevier (Butterworth-Heinemann), pp. 195–331.

[67] Federal Emergency Management Agency (FEMA). (2017). *Mitigation Framework Leadership Group: National Mitigation Framework*, (2nd ed.). U.S. Department of Homeland Security. https://www.fema.gov/media-library/assets/documents/32209

Priyanka Kumari*, Neeraj Kumari, and Noushi Zaidi

Chapter 10
Latest technologies and future perspectives in green materials

Abstract: Green materials and technology have advanced significantly as a result of the quick transition toward sustainability, with the goal of minimizing environmental effect while preserving industrial efficiency. In order to solve urgent global issues like waste accumulation, resource depletion, and climate change, this chapter examines the most recent advancements in bio-based polymers, recycled and upcycled materials, and hybrid biocomposites. Green solvents, low-energy composite fabrication, and new production methods like 3D printing are transforming a number of industries, including construction, biomedical, and packaging. Techno-economic analysis and life cycle assessment are also essential for assessing the financial and environmental feasibility of sustainable materials. The significance of international cooperation, legislative changes, and regulatory frameworks in hastening the adoption of green technologies is also emphasized in the conversation. The future of sustainable development could be drastically altered by new developments in self-healing and stimuli-responsive materials, despite obstacles such as scalability, cost competitiveness, and material performance restrictions. Green materials are positioned to spearhead the next phase of industrial change and guarantee a more robust and sustainable future by utilizing developments in synthetic biology, machine learning, and the concepts of the circular economy.

Keywords: Green materials, sustainability, self-healing, bio-based materials, life cycle assessment

10.1 Introduction

Presently, the whole world is changing rapidly and moving to achieve a sustainable environment by implementing green manufacturing process. The manufacturing processes are mainly focusing on green materials and green technology results to lessen the cost, resources, and energy with more efficiency as compared to traditional tech-

***Corresponding author: Priyanka Kumari,** Department of Chemistry, Shivaji College, University of Delhi, Delhi, India, e-mail: priyanka.dhaka1985@gmail.com
Neeraj Kumari, School of Basic and Applied Sciences, K. R. Mangalam University, Gurugram, India
Noushi Zaidi, Department of Chemistry, College of Science, King Faisal University, Al-Ahsa, Saudi Arabia

https://doi.org/10.1515/9783111618906-010

nology and materials. Green technology is that technology that does not have an adverse effect on the climate or environment using a sustainable source of energy. Green materials, also referred to as sustainable or eco-friendly materials, have surged to the forefront of industrial and academic research as the global community grapples with pressing environmental challenges, including climate change, resource depletion, and waste accumulation. These materials aim to minimize ecological impact across their entire life cycle—encompassing raw material procurement, manufacturing, use, and eventual disposal. Growing consumer awareness and legislative pressure are catalyzing a rapid shift toward greener production and consumption patterns, driving unprecedented innovation in material science.

The COVID-19 pandemic has affected the whole world in social, economic, and political ways. Therefore, there should be clear awareness about global challenges like global warming, ozone layer depletion, acid rain, etc. There are several protocols and agreements signed by the governments of various countries, however, no proper action has been taken. All the same, scientists are also working to find latest technologies for a sustainable environment for the coming generation, which generally help industries to find an alternate for waste disposal and use of bio-based eco-friendly materials. The green technology mainly covers all the areas like agriculture, material science, atmospheric science, environmental science, and hydrology. Green materials like starch, lignin, and cellulose mainly enhance the bioavailability and many other significant properties of the traditional products.

This chapter offers a comprehensive exploration of the latest technologies underpinning the field of green materials and discusses their future perspectives. From groundbreaking developments in bio-based polymers and upcycled materials to advanced fabrication techniques, this treatment underscores the versatility, complexity, and transformative potential of sustainable materials across various industry sectors. The chapter further delves into the crucial role of policy, regulation, and public–private partnerships in scaling up these innovations, while shedding light on ethical considerations and emerging global trends that will shape the trajectory of green materials for years to come.

10.2 Different types of green materials

Recently, green materials used in the synthesis of nanomaterials are gaining a lot of attention due to the increasing emphasis on sustainability, which aligns with the 12 principles of green chemistry. There are various materials like bio-based materials, recycled and upcycled materials, and biocomposites that are used in the synthesis of nanomaterials.

10.2.1 Bio-based polymers

Bio-based polymers have become a central focus in green materials research due to their potential to replace petrochemical-derived plastics. Derived from renewable resources such as corn starch, sugarcane, cellulose, algae, and other biomasses, these polymers aim to reduce reliance on fossil fuels and mitigate plastic pollution. These materials are intended to reduce the environmental footprint by replacing petroleum-based goods, lowering carbon emissions, and promoting the circular economy.

10.2.1.1 Polylactic acid (PLA)

Polylactic acid (PLA) is a biodegradable polymer formed through the fermentation of renewable resources like corn or sugar beet, producing lactic acid through polymerization process. PLA exhibits biocompatibility, transparency, and ease of operation, due to which it is more suitable for biomedical applications and food packaging. It is used as an alternative to conventional plastics like polyethylene terephthalate (PET), as it easily undergoes degradation and converts into nontoxic byproducts under industrial composting conditions. However, there are various limitations also, like low thermal resistance and brittleness. Therefore, research focuses on enhancing its thermal stability and mechanical strength through copolymerization or blending with other polymers [1].

10.2.1.2 Polyhydroxyalkanoates (PHAs)

Synthesized by bacterial fermentation of lipids or sugars, polyhydroxyalkanoates (PHAs) demonstrate excellent biodegradability in diverse environments, including soil and marine settings. By optimizing fermentation conditions, researchers can tailor PHA properties for applications ranging from single-use packaging to medical implants [2].

10.2.1.3 Cellulose derivatives

Cellulose derivatives are mainly naturally occurring, renewable biopolymers (e.g., cellulose acetate) found in the plant cell wall. They utilize lignocellulosic biomass, offering superior biodegradability, mechanical strength, and moderate chemical resistance. Due to their biodegradable and nontoxic nature, the derivative of cellulose aligns with the 12 principles of green chemistry. Therefore, researchers are doing efforts to reduce production costs and scale industrial processes due to their potential applications in textiles, filters, and biodegradable films. Cellulose acetate, a derivative of cellulose, formed by acetylation of cellulose, is used in textile industries and in cigarette filters on a large scale. Carboxymethyl cellulose, formed by carboxymethylation of cellulose, is widely used as a water-retaining agent in cosmetics and pharmaceutical industries [3].

10.2.2 Recycled and upcycled materials

While bio-based polymers have gained momentum, recycled and upcycled materials address the pressing need to manage post-consumer and post-industrial waste effectively. In order to reduce landfill waste and preserve natural resources, recycled materials are taken from post-consumer or industrial waste and used to make new products. Metals, glass, paper, and recycled plastics are typical examples. For example, textiles, packaging, and construction all employ recycled PET (rPET) from plastic bottles. Because of their great reprocessing energy efficiency, recycled metals like aluminum are widely employed in the automotive and aerospace industries. Recycling minimizes environmental contamination, in addition to consuming less energy than generating new materials. Recycling glass helps preserve raw resources like sand and limestone, while recycling paper saves trees and uses less water and energy. Recycled materials are essential for lowering the carbon footprint of many businesses by fostering a circular economy.

With upcycled materials, waste or abandoned objects are ingeniously molded into more valuable or useful items without being broken down into their component parts. This method concentrates on providing aesthetic or practical value while using less energy than recycling. Examples include reusing industrial pallets into furniture, turning denim scraps into handbags, and updating outdated furniture with contemporary décor. Upcycled materials, which combine sustainability and creativity, are popular in-home décor and fashion. Upcycling prolongs product life cycles, lowers trash production, and encourages resource efficiency, in contrast to standard recycling, which frequently degrades material quality. Upcycled materials support a more sustainable and environmentally friendly economy by promoting innovation. The following methods are generally used for the manufacture of recycled/upcycled materials:

10.2.2.1 Mechanical recycling

Post-consumer plastics such as PET and polyethylene (PE) are shredded, melted, and remolded to form new products. This approach is cost-effective but can compromise material properties if contaminants are present.

10.2.2.2 Chemical recycling

Advanced depolymerization methods break plastics down into their monomers, which can be repolymerized into high-quality resins akin to virgin plastics. This method holds promise for closed-loop recycling, although high energy costs remain a challenge [4].

10.2.2.3 Upcycling agricultural residues

Agricultural byproducts (e.g., corn husks and rice straw) can be converted into composite panels or specialty chemicals. Innovations in pyrolysis and enzymatic hydrolysis enable the transformation of low-value residues into high-performance materials with reduced environmental impacts.

10.2.3 Hybrid biocomposites and smart blends

An emerging frontier in green material synthesis involves hybrid biocomposite materials that integrate natural fibers, bio-based polymers, and even inorganic fillers to create synergies in mechanical, thermal, and degradability characteristics. These composites can be replacement of traditional synthetic materials, and used for packaging and construction purpose.

10.2.3.1 Natural fiber reinforcement

Fibers from bamboo, hemp, flax, and kenaf infuse composites with robust mechanical properties while maintaining lower densities than synthetic fiber alternatives. The composites infused with hemp fibers can be used for automobile panels due to their significant durability and environmentally friendly nature. The addition of calcium carbonate or clay to a biocomposite can enhance its thermal stability and processability without compromising its biodegradability [5].

10.2.3.2 Smart blends

Researchers are blending biopolymers with small amounts of traditional plastics or nanofillers to enhance barrier properties and extend shelf life. Such blends preserve the environmental benefits of bio-based constituents while meeting stringent industrial performance standards. When a biopolymer like PLA is blended with nanoclay or graphene, it improves barrier characteristics, which further extends the shelf life and is found to be more suitable in food packaging applications. When PLA is blended with nanofillers, the composite becomes stronger and more heat-resistant and can be used in electronic manufacture and for packaging purposes.

10.2.4 Green solvents

In the synthesis of nanomaterials, water-based extracts serve as alternatives to organic solvents. According to the principles of green chemistry, solvents should be less hazardous. With thc increasing demand for nanomaterials, it is essential to replace harmful reagents with natural alternatives derived from plants, algae, fungi, and bacteria. Various green solvents, such as water, supercritical CO_2, ionic liquids (ILs), and deep eutectic solvents (DESs), are being utilized as substitutes for conventional organic solvents.

10.2.4.1 Water as green solvent

Water is considered a universal green solvent due to its nontoxic nature, cost-effectiveness, and easy availability. Due to the polar nature of water, it can be used to facilitate various organic and inorganic reactions such as hydrolysis, acid–base reaction, and various other processes. It is an excellent medium for dissolving various polar compounds and facilitating ionic reactions due to its high dielectric constant. The hydrogen bond present in water molecules enhances the selectivity and reaction kinetics by the reduction of the requirement of hazardous substances. The high specific heat capacity and thermal stability of water mainly allows the reaction by controlling the heat and enhancing the energy efficiency of the reaction. In green chemistry, water is used as an alternative to various volatile solvents, hence reducing the environmental pollution and health risks. Green synthesis methods like the microwave-assisted method, which takes place in the presence of water further improves the reaction efficiency. Its compatibility with renewable resources makes it indispensable in the transition to greener industrial processes.

10.2.4.2 Supercritical CO_2

Supercritical carbon dioxide ($scCO_2$) is also known as supercritical fluid and is considered as green solvent due to its nontoxic, nonflammable, and recyclable nature. When CO_2 is heated and pressurized beyond its critical temperature (31.1 °C) and pressure (73.8 bar), it enters the supercritical state, thereby, exhibiting properties of both a gas and a liquid. It serves as an excellent alternative to traditional solvents used in polymer synthesis. Recently, it has played a significant role in the green synthesis of nanomaterials, particularly in the fabrication of semiconductors, contributing to the resolution of certain environmental challenges. In many chemical reactions, it serves as a medium during hydrogenation and oxidation of the processes, hence offering an alternative to many hazardous substances in a sustainable way. It is used as an alternate to perchloroethylene which, mainly acts as a dry-cleaning agent. It is also used during the extraction methods like extraction of essential oils from plants and decaffeination.

10.2.4.3 Ionic liquids

ILs are nonmolecular molecular solvents or salts composed of organic cations and inorganic cations/anions. These are found in the form of liquid at or below 100 °C. The solubility of ILs ranges from polar to nonpolar compounds. The ILs are environmentally friendly and used as an alternative to volatile organic solvents (volatile organic compounds, VOCs) due to their significant unique properties like high thermal stability, tunable polarity, and negligible vapor pressure. These solvents are mainly designed for some particular applications like organic molecule synthesis, catalyzation, extraction, etc. by controlling their cation–anion combinations. 1-Butyl-3-methylimidazolium hexafluorophosphate ([Bmim][PF6]), an IL, is used on a large scale in catalytic reactions, while choline chloride with urea forms a DES for metal extraction and cellulose processing. Additionally, ILs are used in the production of pharmaceuticals, biomass conversion, and CO_2 capture. Though problems like high prices and possible toxicity are still being researched, their recyclability further strengthens their green credentials.

10.2.4.4 Deep eutectic solvents

DESs are formed by the combination of hydrogen bond acceptor and hydrogen bond donor compounds in variable ratios. The eutectic mixture formed through hydrogen bond interaction has a lower melting point than the individual components. Secondly, these solvents are formed by the combination of solids, whose melting point is lower than 100 °C, which further combines results in the formation of solvents having melting points lower than the latter. The solvents are better than organic solvents due to their biodegradable nature, nontoxic nature, and ease of operation. By varying the ratio of hydrogen bond acceptor and hydrogen bond donor compounds, these solvents can be tailored to the needs of specific applications.

DESs exhibit superior extraction characteristics, with easily adjustable viscosity, high sustainability, and excellent solubility for both polar and nonpolar compounds. Notably, they remain in a liquid state even at temperatures below 0 °C. Originally, DESs were developed as extraction solvents for use in dispersive liquid–liquid microextraction, facilitating the extraction of organic compounds or metal ions from oil or aqueous samples.

10.3 Innovative processing and manufacturing techniques

Innovative methods for producing and processing green materials are revolutionizing industries through the use of sustainable resources, waste reduction, and energy conservation. These strategies, which include low-energy/solvent-free techniques, green composite fabrication, and additive manufacturing, aid in the transition to environmentally friendly production. An extension of these techniques, along with examples and their drawbacks, is provided below.

10.3.1 Additive manufacturing (3D printing)

Additive manufacturing, or 3D printing, is reshaping the fabrication of green materials by enabling precise, layer-by-layer deposition that reduces waste:

10.3.1.1 Extrusion-based techniques (FDM)

This technique is generally used for PLA and other biodegradable filaments. Researchers are experimenting with composite filaments reinforced with cellulose or graphene, aiming to boost mechanical performance. For example, PLA reinforced with cellulose fibers creates stronger and lighter components for automobiles and packaging industries. But the main limitation of this technique is that the manufacturing speed is slower than the conventional method and the recycling of printed parts is limited. In addition to these restrictions, continuous research aims to improve the mechanical characteristics and thermal resistors of these biocomposites. This development can further expand its applicability in various sectors and build a path for more stable and sustainable production methods [6].

10.3.1.2 Powder-based methods (SLS and binder jetting)

Applicable to powdered biomaterials or biodegradable polymer powders, these methods allow the creation of complex structures with reduced material usage. For instance, medical implants have made using powdered polycaprolactone (PCL). The requirement for specialized powders, which might not necessarily be biodegradable and the high initial equipment expenditures are the limitations of this method.

10.3.1.3 Bioprinting and direct ink writing

This method is perfect for biomedical applications, allowing the production of bioresorbable implants or scaffolds loaded with cells. For example, scaffolds for tissue engineering have been printed using direct ink writing and bioinks based on gelatine or alginate.

Limitations include difficulties in preserving their bioactivity after processing, high precision requirements, and limited scalability.

10.3.2 Green composite fabrication

Historically, the creation of composites has depended on energy-intensive production methods and resins produced from petroleum. Green composites, which incorporate renewable resources and low-energy fabrication techniques to lessen their environmental impact, have been developed in response to growing environmental concerns and the need for sustainable materials. For a number of industrial uses, such as the building, transportation, and renewable energy sectors, these composites provide a competitive substitute.

10.3.2.1 Bio-based resins

Bio-based resins are sustainable substitutes for synthetic polymers since they come from renewable resources. Soybean oil epoxies, cashew nutshell liquid (CNSL)-based resins, and lignin-based thermosets offer potential for reduced carbon footprint and improved biodegradability. These materials are utilized in construction panels, wind turbine blades, and automobile parts. Epoxy resins made from soybean oil provide improved biodegradability and lessen the dependency on fossil fuels. These resins offer a balance between performance and sustainability and have been included in adhesives, coatings, and automotive components. CNSL resins, for example, are used in adhesives and coatings that are intended to withstand corrosion. Sometimes, the variability in raw material quality, high production costs, and limited mechanical strength, in comparison to traditional resins, make these resins to be of limited use. Because of their mechanical strength and resistance to corrosion, these resins are frequently employed in coatings, adhesives, and composites. Nevertheless, because of their high production costs and inconsistent raw material qualities, they haven't been widely used. Lignin-based thermoset resins, a byproduct of the paper and biofuel industries, improve the sustainability of composite materials that can be produced. Although issues like mechanical strength restrictions and processing hurdles need to be resolved, these lignin-based resins exhibit potential for use in wind turbine blades, automotive parts, and building panels.

10.3.2.2 Low-energy molding

Techniques like vacuum-assisted resin transfer molding (VARTM) and cold pressing reduce thermal energy requirements and can be adapted for natural fibers, further cutting down on CO_2 emissions. Using natural fibers, VARTM has been used to create bio-composite wind turbine blades [5]. This technique ensures equal distribution while reducing waste, by pouring resin into a fiber-reinforced mold under vacuum pressure. For natural fiber-reinforced composites, like those found in biocomposite wind turbine blades, VARTM is especially appropriate [5]. This process greatly lowers the amount of thermal energy used and improves the sustainability of producing composites on a wide scale.

By enabling composite materials to be manufactured at lower temperatures than traditional high-temperature molding, cold pressing lowers CO_2 emissions and energy expenses. This method works best with natural fiber composites, such as those made of hemp, flax, and jute, which maintain their structural integrity at moderate temperatures and pressures. Green composites, which combine bio-based resins with low-energy manufacturing processes, are emerging as a practical option for businesses trying to lower their carbon footprint, without sacrificing material performance. To enable their wider use in commercial applications, ongoing research aims to enhance these sustainable materials' mechanical qualities, cost-effectiveness, and scalability.

10.3.3 Low-energy and solvent-free processing methods

By using less energy and using fewer or no hazardous solvents, advances in composite fabrication seek to lessen their impact on the environment. These techniques preserve the materials' mechanical and functional integrity while improving sustainability. Innovative processing methods that either do away with solvents or substitute them with eco-friendly substitutes are the focus of both industry and research. Key solvent-free or low-impact processing techniques utilized in the creation of green composites are listed below:

10.3.3.1 Reactive extrusion

Reactive extrusion is a very effective, solvent-free manufacturing technique that combines shape and polymerization in one phase. This approach reduces

(a) Emissions of VOCs by doing away with the requirement for organic solvents, reducing processing times through the direct extrusion conversion of monomers into polymers.
(b) Improves material qualities by means of in situ polymerization, which qualifies it for the creation of bio-based and biodegradable composites.
(c) Minimizes energy-intensive drying or solvent removal procedures and streamlines processing steps to lower production costs.

10.3.3.2 Supercritical fluid processing

An environmentally benign substitute for conventional solvent-based techniques is provided by supercritical fluid technology, especially $scCO_2$. This procedure uses $scCO_2$ as a recyclable, nontoxic medium for foaming, impregnation, and swelling polymers to reduce threats to the environment and human health by efficiently replacing dangerous organic solvents. This procedure also increases material performance and composite porosity, particularly in lightweight structural applications and permits exact control of processing parameters, resulting in additives and reinforcing chemicals dispersing uniformly across polymer matrices [7].

10.3.3.3 Aqueous-based coatings and laminates

Solvent-based coatings and laminates are gradually being replaced with aqueous-based formulations, particularly in the production of consumer items and packaging. These systems reduce VOC emissions significantly by using water as a dispersion medium rather than organic solvents derived from petroleum. They are less exposed to dangerous chemicals, making them safer for both producers and consumers and offer superior flexibility, adhesion, and barrier qualities for uses such as protective coatings, structural laminates, and food packaging films.

10.4 Characterization and performance optimization

10.4.1 Analytical techniques

Comprehensive characterization is crucial to ensuring that green composites meet the required mechanical, thermal, and chemical property benchmarks. Researchers can better understand the composition, structure, and long-term performance of materials by using a variety of analytical techniques.

10.4.1.1 Spectroscopic methods

The analysis of green composites' chemical makeup relies heavily on Fourier-transform infrared (FTIR) spectroscopy. Through the identification of distinctive functional groups, FTIR sheds light on sustainability to ensure the use of biodegradable components, the density of crosslinking, and the interactions between the reinforcing fibers and the polymer matrix. The peak shifts and intensity variations provide information on thermal and oxidative stability, molecule interactions, and structural homogeneity at the molecular level. Other spectroscopic techniques, such as Raman spectroscopy, are also employed.

10.4.1.2 Microscopy approaches

Researchers can investigate fiber–matrix interactions and microstructural characteristics of green composites using high-resolution images provided by microscopy techniques. Through surface morphology, fiber dispersion and void formation are all captured by scanning electron microscopy (SEM), which affects the mechanical performance. Transmission electron microscopy (TEM) provides in-depth information about nanostructures, emphasizing nanoscale interfacial bonding. In order to help create composites with exceptional durability, atomic force microscopy (AFM) maps mechanical characteristics such as modulus fluctuation and surface roughness [8].

10.4.1.3 Thermal analysis

The proper processing behavior and thermal stability are essential for green composites. Important methods for thermal analysis include phase transitions, such as glass transition (T_g) and melting point (T_m), which are determined using differential scanning calorimetry, which is essential for process optimization. Thermogravimetric analysis assesses weight loss as a function of temperature, offering information on material lifespan and breakdown mechanisms. Viscoelastic characteristics are measured under a range of mechanical and thermal circumstances using dynamic mechanical analysis, which aids in the long-term prediction of composite performance.

10.4.2 Multiscale modeling and simulation

Advanced computational approaches provide predictive capabilities for optimizing green composite formulations and their mechanical behavior under different environmental conditions.

10.4.2.1 Atomistic modeling

Molecular dynamics simulations offer insights into polymer chain interactions, hydrogen bonding, and crystallinity levels in bio-based polymers [9]. This method forecasts the strength of the hydrogen connection between reinforcements and biopolymers by assessing phase behavior and crystallinity, which affects biodegradability and mechanical strength. The atomistic modeling also evaluates the transmission of gases and moisture, which is important for uses in medical implants and food packaging [9].

10.4.2.2 Continuum mechanics

Finite element methods can simulate macro-scale properties like tensile strength, impact resistance, and flexural modulus, guiding structural design in automotive and aerospace sectors. There are various other applications of this method such as evaluating lightweight possibilities and crash resistance, forecasting fatigue behavior under cyclic loads in the aerospace industry, and assessing impact resistance and flexural strength for the creation of sustainable infrastructure. Researchers can improve structural performance and fine-tune composite formulations while preserving environmental sustainability by combining simulation results with experimental data.

10.4.3 Accelerated aging and reliability testing

Given the nascent stage of many green materials, **accelerated aging tests** help estimate long-term performance:

10.4.3.1 Environmental chambers

These chambers subject samples to high humidity, salt spray, and temperature cycling to replicate decades of service life in weeks. Temperature swings and high humidity are used to evaluate hydrothermal aging. Salt spray testing is used in coastal and marine applications to assess corrosion resistance. To replicate the harsh climatic conditions that structural materials encounter, freeze–thaw cycles are used. Before a material is used in the actual world, these tests aid in determining its lifespan and identifying failure mechanisms.

10.4.3.2 UV weathering

UV weathering uses artificial UV light to simulate solar radiation, critical for outdoor applications like construction materials or packaging [10]. Long-term exposure to sunlight can deteriorate the mechanical qualities of green composites used outdoors. UV weathering tests use artificial UV radiation sources to simulate decades of exposure to sunshine in a brief period of time. This approach aids in evaluating discoloration and photodegradation, which are essential for branding and aesthetics, and for studying embrittlement and surface cracking, which compromise mechanical integrity and oxidative stress-induced changes in chemical composition.

10.4.3.3 Biomechanical stress tests

These stress tests are relevant for biodegradable medical implants, ensuring they retain structural integrity during tissue regeneration and degrade safely thereafter. These tests mainly consist of degradation studies conducted in vitro and test for cytotoxicity and biocompatibility. These experiments mimic physiological circumstances to track changes in pH, weight loss, and mechanical strength over time, assessing if biodegradable implants can promote tissue regeneration without failing too soon. This is known as load-bearing analysis and makes sure that medical uses don't cause any unfavorable biological reactions.

10.5 Emerging research in self-healing and responsive green materials

Self-healing sustainable materials are becoming more and more popular because of their ability to increase material lifespan, cut waste, and boost performance in a variety of industries, such as electronics, healthcare, and construction. In order to provide high-performance, low-maintenance materials that are also ecologically benign, recent developments have concentrated on encapsulating systems, dynamic covalent bonding, and stimuli-responsive materials.

10.5.1 Microvascular and encapsulation methods

Self-healing devices use inbuilt processes to fix damage on their own, extending the material life and lowering maintenance expenses. These methods, which have uses in coatings, composites, and structural materials, are inspired by biological healing processes.

10.5.1.1 Microencapsulation

Microencapsulation involves encapsulating healing agents (e.g., epoxy and bio-based resin) within microcapsules that rupture when cracks form, releasing a healing agent to bond fractured surfaces. The process of microencapsulation needs coating of therapeutic substances in polymeric microcapsules scattered across a material matrix. The microcapsules burst as a fracture spreads, allowing the injured area to cure itself by releasing the healing agent (such as liquid polymers, epoxy resins, or bio-based resins). This method has proven successful in enhancing durability and resilience to environmental deterioration in polymer composites, concrete, and coatings [10].

10.5.1.2 Vascular networks

These networks embed a network of microtubules within materials, mimicking biological circulatory systems. When damage occurs, the healing agent is pumped to the site of failure, enabling repeated healing events. Vascular self-healing materials, which draw inspiration from biological circulatory systems, have an implanted network of capillaries or hollow microtubules that are filled with a healing agent. Multiple healing cycles are made possible by the material's ability to detect defects and pump an agent to the damaged area. These technologies are especially helpful in high-stress settings where material integrity may be compromised by repeated damage, such as in aerospace and automotive applications.

10.5.2 Dynamic covalent chemistry

Dynamic covalent bonds (e.g., imine and Diels-Alder reactions) allow for reversible cross-linking. Under the right circumstances, materials can self-repair because dynamic covalent connections allow reversible cross-linking within polymer networks. These systems are useful for flexible electronics, biomedical devices, and recyclable polymers because they can respond to environmental stimuli like heat, light, or mechanical stress.

10.5.2.1 Reversible polymer networks

These networks adapt to mechanical stress or thermal stimuli, self-repairing microcracks without external intervention. When exposed to external stimuli, polymers with dynamic covalent bonds—such as imine, Diels-Alder, disulfide, or boronic ester linkages—can experience bond rearrangement. This characteristic makes it possible for materials to self-heal and change their shape, preserving structural integrity in the face of mechanical stress or temperature changes, without the need for outside assistance. These polymers hold great promise for 3D-printed bioresorbable materials, flexible sensors, and electronic skins.

10.5.2.2 pH- or light-responsive bonds

These bonds are particularly useful in biomedical applications, where local changes (e.g., inflammation) trigger the healing process [11]. In reaction to changes in the environment, materials with pH-sensitive or photoreactive functional groups can selectively initiate healing processes. For instance, in biomedical applications, pH-responsive materials might help regenerate tissue by releasing healing chemicals when exposed to acidity caused by inflammation. Similar to this, light-responsive materials provide precise control over the healing process by being activated by UV or visible light [11].

10.5.3 Stimuli-responsive and adaptive green materials

Responsive materials alter their properties (e.g., shape, color, and permeability) upon encountering specific triggers such as temperature, moisture, or electric fields. Wearables, self-cleaning surfaces, and controlled drug release systems are just a few of the applications made possible by smart materials that can react to environmental stimuli and give better performance. When exposed to temperature, moisture, electric fields, or other environmental stimuli, these materials change in shape, color, permeability, or mechanical characteristics.

10.5.3.1 Shape-memory biopolymers

These polymers return to a predetermined form under heat or light, offering advanced capabilities in wearable devices or implantable systems. When exposed to certain triggers, including heat or light, shape-memory polymers (SMPs) are made to revert to a predetermined shape. Biodegradable SMPs made from renewable resources, including PLA or composites based on chitosan, have potential applications in responsive packaging, smart textiles, and medical implants. The ability to program these polymers to alter their shape in response to outside stimuli enhances their usefulness in engineering and biomedical applications.

10.5.3.2 Smart coatings

These coatings switch from hydrophobic to hydrophilic states for self-cleaning surfaces or controlled release of active compounds in agriculture. Environmentally responsive features, such as the ability to transition between hydrophobic and hydrophilic states, are incorporated into smart coatings. Self-cleaning surfaces, antifouling coatings, and controlled-release systems for use in pharmaceutical or agricultural applications are made possible by this characteristic. These coatings improve sustainability while offering superior performance in harsh conditions by utilizing bio-based and biodegradable components.

10.6 Environmental and economic assessments

For sustainable material development and waste management techniques to be feasible and yield long-term benefits, both economic and environmental assessments are necessary. The two main approaches utilized for this are life cycle assessment (LCA) and technoeconomic analysis (TEA). Additionally, standards and labelling aid in directing industries and consumers toward sustainable decisions.

10.6.1 Life cycle assessment (LCA)

LCAs quantify the cumulative environmental impact of a product, from resource extraction to end-of-life disposal. A product's environmental effects are systematically assessed over the course of its life cycle using a process known as life cycle assessment, or LCA. It aids in the identification of important areas for waste reduction, emissions management, and sustainable sourcing improvement for businesses and policymakers.

10.6.1.1 Cradle-to-grave

Using a cradle-to-grave methodology, cycle assessment looks at a material or product's whole life cycle, including extraction of raw materials, where sustainable sourcing and extraction practices are assessed by determining energy consumption, waste production, and emissions during the manufacturing process. Logistics' carbon footprint can be evaluated using this methodology by examining the amount of energy used. Industries can reduce environmental impacts and optimize production methods by examining each step.

10.6.1.2 Hotspot identification

Critical environmental hotspots, or phases of a product's life cycle that contribute disproportionately to energy consumption, pollution, or greenhouse gas emissions, can also be identified using LCA. For instance, the manufacture of plastic is a major hotspot due to the substantial emissions produced during the polymerization process. Using cleaner production techniques, increasing energy efficiency, or moving to bio-based alternatives are a few examples of mitigation techniques.Industries can make well-informed decisions about process changes and sustainable alternatives by concentrating on these hotspots [8].

10.6.2 Techno-economic analysis (TEA)

TEA evaluates a technology's financial feasibility while taking sustainability into account, in contrast to LCA, which concentrates on environmental impact. When assessing whether sustainable solutions can compete on a commercial basis, TEA is essential.

10.6.2.1 Production cost modeling

In this step, capital and operating expenses (CAPEX and OPEX) are calculated, and they include the cost and accessibility of feedstocks (e.g., petroleum-based vs. bio-based polymers), energy consumption (includes expenses for pollution, water use, and electricity), costs of labor, and upkeep for competent personnel and equipment maintenance.

Industries can discover financial obstacles and areas for cost reduction with the help of a clear cost model.

10.6.2.2 Market assessment

TEA evaluates a technology's financial feasibility while taking sustainability into account, in contrast to LCA, which concentrates on environmental impact. When assessing whether sustainable solutions can compete on a commercial basis, TEA is essential. To ascertain the possible profitability of sustainable products, TEA additionally incorporates a market analysis [12].

10.6.3 Standardization and labeling

Standards and certifications are used by industry and consumers to confirm the sustainability of materials. By guaranteeing that goods sold as "eco-friendly" adhere to stringent environmental standards, these labels also avoid greenwashing.

10.6.3.1 Biodegradability certifications

Composability and biodegradability standards are established by following certification programs:

(a) Plastics are guaranteed to break down in industrial composting systems by ASTM D6400.

(b) The European standard EN 13432 for compostable packaging ensures full biodegradation after six months, and disintegration within 12 weeks.

(c) OK Compost (TÜV Austria) confirms verifying that materials can be composted at home and in an industrial setting.

These certifications encourage appropriate disposal and assist businesses in adhering to international sustainability regulations.

10.6.3.2 Eco-labels

Genuinely sustainable products are distinguished from deceptive "greenwashed" claims by following eco-labels:

(a) Products that fulfill stringent performance and environmental requirements are certified by the Green Seal (USA).

(b) The oldest eco-label in Europe, the Blue Angel (Germany), is given to goods that use few resources and produce few emissions.

(c) USDA Certified Bio-based Product verifies that a product has biological elements that have been certified as renewable.

Businesses and consumers can make environmentally conscious decisions by adhering to standardized labeling [13].

10.7 Sector-specific applications

In order to reduce environmental effects and maintain economic viability, it is imperative that different industries embrace sustainable materials and procedures. To cut waste, carbon footprint, and resource efficiency, various industries have started using recyclable, biodegradable, and bio-based materials.

10.7.1 Packaging and consumer goods

The packaging sector remains a dominant consumer of plastics. One of the biggest users of plastics is still the packaging sector, which makes it a vital area for sustainable innovation. Compostable containers and bio-based films provide practical answers to the escalating issues of microplastic pollution and landfill overflow. Active packaging, which uses oxygen-scavenging or antimicrobial chemicals to prolong product shelf life and minimize food waste, is one of the major developments in this field [14]. Furthermore, reusable systems are becoming more popular, especially in the food and beverage sector, where models of refillable packaging greatly reduce single-use waste and encourage the circular economy. By lowering reliance on new plastic materials, these sustainable alternatives not only help the environment but also save money over time [14].

10.7.2 Automotive and aerospace

Weight reduction is paramount for enhancing fuel efficiency and lowering emissions. Because lighter materials directly improve fuel efficiency and lower greenhouse gas emissions, weight reduction is a top priority in the automobile and aerospace industries. Natural fiber composites, including panels reinforced with hemp or flax, are a promising invention that is being used more and more in dashboards, structural elements, and car interiors. These materials preserve performance and durability while providing a renewable substitute for conventional synthetic composites. Additionally, to further reduce total mass and fuel consumption, bio-based thermoplastics are being used in nonstructural or semi-structural parts of automobiles and airplanes [15]. The industry's objective of being carbon-neutral is in line with the incorporation of these sustainable resources, which also helps to reduce the environmental impact.

10.7.3 Construction and infrastructure

Construction, responsible for a significant fraction of global CO_2 emissions, is progressively adopting green materials. The construction sector, which contributes significantly to CO_2 emissions worldwide, is gradually moving toward sustainable building methods and green materials. Among these innovations include the usage of mycelium-based and hempcrete blocks, which provide superior insulating qualities, while lowering a building's carbon impact. Because these materials biodegrade, an eco-friendly end-of-life situation is guaranteed. Furthermore, by substituting aluminosilicate precursors like fly ash and slag for traditional Portland cement, geopolymer binders offer a sustainable substitute. This invention dramatically lowers CO_2 emissions, in addition to energy consumption [16]. The use of local bio-based or recycled feedstocks, such as wood-based polymers and recycled aggregates, to create load-bearing walls with little material waste is another ground-breaking advancement in the industry.

10.7.4 Biomedical and healthcare

By providing bio-based and biodegradable substitutes for medicinal uses, sustainable materials have likewise completely transformed the biomedical and healthcare sectors. Tissue engineering scaffolds, which use polymers like PLA, PCL, and gelatin-methacryloyl to encourage cellular growth while gradually decaying safely, are among the most important advancements in this discipline. By removing the need for surgical removal, these biodegradable scaffolds enhance patient outcomes. Biodegradable polymer coatings on magnesium-based implants offer a revolutionary approach in orthopedics by lowering the need for follow-up surgery to remove implants. Additionally, chitosan-based films, which have antibacterial qualities and enable controlled drug release, have been developed as a result of improvements in wound dressings. These films improve wound healing and infection prevention [17]. These developments make healthcare more sustainable by improving patient care and lowering medical waste.

The transition to recyclable, bio-based, and biodegradable materials is having a major positive impact on the environment and the economy across these several industries. Adopting sustainable alternatives aids businesses in cutting waste, emissions, and increasing productivity, all of which assist in creating a more resilient and sustainable future.

10.8 Policy and regulatory frameworks

Government regulations and policies have a significant impact on the shift to circular economies and sustainable materials. In order to promote innovation and guarantee the broad acceptance of greener options, legislative measures, financial incentives, and international partnerships are essential. Policymakers help firms to transition to environmentally friendly practices while preserving economic viability by enforcing strict rules, providing financial incentives, and encouraging producer responsibility.

10.8.1 Government regulations and incentives

One of the main forces influencing the landscape of material innovation is government action. To reduce plastic waste and encourage sustainable alternatives, several nations have passed legislation. The prohibition on single-use plastics, which has been put into effect in places like the European Union and several parts of Asia, is among the most significant policies. These limitations have greatly increased the need for recyclable and biodegradable materials, as have tariffs on single-use plastics. Governments are also using tax credits and subsidies to encourage research and development in advanced recycling technologies, bio-based polymers, and green composites. By lowering entry barriers for producers of sustainable materials, this kind of funding promotes innovation and the marketing of environmentally friendly goods.

10.8.2 Extended producer responsibility

Manufacturers are held accountable for the end-of-life care of their products under extended producer responsibility rules, which make sure businesses actively participate in recycling and trash reduction initiatives. Product take-back programs are a crucial component of this system, requiring companies to consider recycling when designing their goods. Companies are encouraged by these programs to reduce packaging waste and design goods that are simple to disassemble and reuse. Furthermore, deposit-return programs have shown great success in encouraging the retrieval of post-consumer products. These systems aid in preserving a consistent supply of high-quality recyclable materials that may be processed using mechanical or chemical recycling techniques by providing financial incentives for customers to return unwanted goods [13].

10.8.3 International collaborations and funding

Collaboration across national boundaries is necessary to address global environmental concerns. To assist large-scale pilot programs and the development of green technologies, governments, academic institutions, and industry leaders come together through international efforts like Horizon Europe, Mission Innovation, and the Global Environment Facility (GEF). Through open-access publications and databases, these partnerships ensure that scientific breakthroughs in sustainable materials are publicly available and promote knowledge exchange. These initiatives hasten the industry-wide adoption of environmentally friendly solutions by promoting openness and international collaboration. International collaborations also aid in the distribution of financial risk by combining resources and knowledge to assist high-potential technologies that might otherwise have trouble obtaining finance. The commercialization of next-generation sustainable materials is accelerated by this collaborative strategy, which reduces the financial risks connected to exploratory inventions.

The regulatory environment is constantly changing to support the adoption of sustainable materials through a combination of strict laws, financial incentives, corporate responsibility, and international collaborations. By encouraging innovation in green technologies, these frameworks not only promote environmental responsibility but also boost economic growth. The future of material science is set to become more efficient, sustainable, and internationally networked as industries adjust to new laws.

10.9 Challenges, opportunities, and ethical considerations

Green materials have made great strides, but a number of obstacles prevent their broad use, and public opinion and ethical issues will determine how they develop in the future. Scaling up biorefineries and recycling plants is sometimes difficult by uneven feedstock quality, which lowers process efficiency. Technical and market restrictions are among the main obstacles. Furthermore, certain bio-based polymers are not as heat-tolerant or moisture-resistant as conventional plastics, which restricts their use in high-performance industries [12]. Cost competitiveness is another significant barrier because shifting oil prices can make petrochemical plastics more affordable, requiring economies of scale or regulatory changes to make bio-based substitutes feasible. Beyond these technical difficulties, resource equality and the procurement of raw materials present ethical issues. Food prices and local ecosystems may be impacted by the displacement of agricultural land caused by the large-scale production of feedstocks for bioplastics, such as corn and sugarcane. Additionally, to avoid exploitation and preserve sustainable production, it is crucial to guarantee equitable remuneration to farmers and communities participating in biomass supply chains [18]. In addition to these difficulties, market adoption is significantly influenced by consumer education and public perception. While greenwashing, in which businesses overstate sustainability promises, erodes consumer trust and hinders adoption, misconceptions regarding biodegradability, compostability, and recyclability can cause confusion. Initiatives for community involvement, clear labeling, and instructional programs in schools are necessary to remove these obstacles and educate customers about the real advantages and drawbacks of green products. In order to shape a more environmentally conscious materials sector, it will be essential to address these issues while seizing the chance for sustainable innovation [19-20].

10.10 Future outlook

The future of green materials is being shaped by advancements in technology, circular economy principles, large-scale adoption, and shifting market dynamics. Technological convergence is expected to drive significant breakthroughs, with synthetic biology enabling enzyme engineering to create customized enzymes that enhance polymerization and depolymerization, thereby supporting near-zero waste cycles. Additionally, machine learning algorithms are transforming material development by predicting optimal polymer compositions and manufacturing parameters, reducing both costs and development time [9]. These innovations complement circular economy integration, where green materials are crucial in eliminating waste and maximizing resource efficiency. Closed-loop systems are becoming more prevalent, with manufacturers and retailers reclaiming used products for refurbishment, remanufacturing, or recycling, thus minimizing virgin resource extraction. Moreover, design for disassembly promotes modular architectures that simplify repairs and component replacement, extending product life cycles and reducing overall waste. However, scaling and global adoption remain key challenges. The establishment of centralized collection, sorting, and reprocessing facilities is crucial for maintaining stable feedstock quality, but disparities in infrastructure pose challenges for emerging economies, necessitating international partnerships and technology transfers to bridge the gap. Meanwhile, market and consumer dynamics continue to evolve, with increasing demand for sustainable products. Achieving price parity with petroleum-based alternatives is critical for mainstream adoption, as cost remains a decisive factor. Companies also leverage sustainability trends for brand differentiation, aligning their products with environmentally conscious consumer preferences to strengthen their market position [13]. As these factors converge, the global shift toward sustainable materials will accelerate, fostering long-term environmental and economic benefits [21].

10.11 Conclusion

Green materials are no longer a niche research topic; they represent a robust and rapidly evolving domain integral to sustainable development strategies worldwide. Advances in bio-based polymers, recycled materials, and self-healing systems have expanded the functional capabilities of eco-friendly alternatives, bridging performance gaps that once hindered their broader adoption. Meanwhile, innovations in processing methods, such as additive manufacturing and solvent-free techniques, reinforce the competitiveness of green materials against conventional counterparts.

However, significant challenges persist. Achieving industrial-scale production, navigating complex regulatory landscapes, and overcoming performance trade-offs demand concerted efforts from scientists, engineers, policymakers, and industries. Ethical considerations—related to land use, resource distribution, and social equity—must also be addressed to ensure that green material solutions are truly sustainable and beneficial to all stakeholders.

Looking ahead, ongoing multidisciplinary collaborations will harness computational tools, synthetic biology, and emerging manufacturing paradigms to push the boundaries of what green materials can achieve. Coupled with supportive regulatory frameworks and educated consumer bases, these materials hold immense promise for transforming industries—from packaging and automotive to construction and healthcare—into pillars of a circular, equitable, and environmentally resilient future.

References

[1] Smith, J. (2021). Bio-based polylactic acid: Properties, applications, and future outlook. *Journal of Materials Science*, 56(4), 1321–1335. https://doi.org/10.1007/s10853-020-05384-4

[2] Anderson, P., & Lee, D. (2020). Polyhydroxyalkanoates for sustainable packaging. *Polymer Reviews*, 60(2), 156–172. https://doi.org/10.1080/15583724.2020.1737460

[3] Zhang, R., Gupta, V., & Santos, L. (2022b). Advances in cellulose-based biopolymers: Processing, properties, and applications. *Carbohydrate Polymers*, 287(1), 119350. https://doi.org/10.1016/j.carbpol.2022.119350

[4] Li, M., & Patel, K. (2019). Advances in upcycling agricultural residues for high-value composite production. *Composites Part B: Engineering*, 162(1), 456–468. https://doi.org/10.1016/j.compositesb.2019.03.011

[5] Rodriguez, A., & Akash, B. (2018). Sustainable resin systems in green composites. International *Journal of Composite Materials*, 7(1), 12–22. https://doi.org/10.5923/j.cmaterials.20180701.02

[6] Mendez, R., Zhou, J., & Carter, W. (2022). Bioprinting strategies for eco-friendly medical devices. *Biofabrication*, 14(3), 034002. https://doi.org/10.1088/1758-5090/ac7b3a

[7] Khan, F. R., Fernandez, M., & Oliveira, J. (2020). Supercritical fluid processing of biopolymers: A sustainable approach. *Green Chemistry*, 22(5), 1578–1592. https://doi.org/10.1039/C9GC04214A

[8] Garcia, E., & Thompson, D. (2020). Life-cycle assessment of bio-based polymers. *Green Chemistry Letters and Reviews*, 13(2), 142–159. https://doi.org/10.1080/17518253.2020.1731227

[9] Huang, Y., & Choi, W. (2019). Multi-scale modeling of biopolymer-based composites. *Materials & Design*, 182(2), 108012. https://doi.org/10.1016/j.matdes.2019.108012
[10] Zhang, L., O'Connor, M., & Park, J. (2021). Self-healing mechanisms in biobased polymer systems. *Progress in Polymer Science*, 117(6), 101396. https://doi.org/10.1016/j.progpolymsci.2021.101396
[11] Park, S. H., Kim, J., & Liu, Y. (2020). Dynamic covalent polymer networks: A pathway to self-healing biobased materials. *Macromolecules*, 53(22), 10184–10196. https://doi.org/10.1021/acs.macromol.0c01682
[12] Kim, H., & Rasmussen, K. (2021). Techno-economic analysis of recycled bioplastics. *Journal of Industrial Ecology*, 25(2), 382–396. https://doi.org/10.1111/jiec.13010
[13] Williams, R. (2019). Standards and certifications for green materials. *Environmental Research Letters*, 14(12), 125007. https://doi.org/10.1088/1748-9326/ab5621
[14] Hernandez, A., Miller, S., & Tan, R. (2021). Bio-based plastics in food packaging: Safety and sustainability considerations. *Food Packaging and Shelf Life*, 29(3), 100744. https://doi.org/10.1016/j.fpsl.2021.100744
[15] Castro, M. T., Huo, X., & Li, Y. (2023). Advanced bio-based composites in automotive and aerospace applications. *Composite Structures*, 307(2), 116433. https://doi.org/10.1016/j.compstruct.2023.116433
[16] Lopez, F., Simmons, B., & Wang, H. (2023). Additive manufacturing in sustainable construction. *Sustainable Structures*, 5(1), 15–27. https://doi.org/10.1016/j.susstruc.2023.100015
[17] Tran, Y., & Wu, G. (2022). Chitosan-based wound dressings with antibacterial properties. *Journal of Biomedical Materials Research Part B: Applied Biomaterials*, 110(5), 1821–1833. https://doi.org/10.1002/jbm.b.34975
[18] Peters, R., & Garcia, G. (2022). Land use and socioeconomic implications of bio-based polymers. *Biomass and Bioenergy*, 163(1), 106492. https://doi.org/10.1016/j.biombioe.2022.106492
[19] Bao, C., Serrano-Lotina, A., Niu, M., Portela, R., Li, Y., Lim, K. H., Liu, P., Wang, W. J., Bañares, M. A., & Wang, Q. (2023). Microwave-associated chemistry in environmental catalysis for air pollution remediation: A review. *Chemical Engineering Journal*, 466, 142902. https://doi.org/10.1016/j.cej.2023.142902
[20] (ASTM D6400). (2019). Standard specification for labeling of plastics designed to be aerobically composted in municipal or industrial facilities, ASTM International. https://doi.org/10.1520/D6400-19
[21] Zhang, F., O'Malley, M., & Chang, R. (2022a). Circular economy approaches to cellulose-based bioplastics. *Resources, Conservation & Recycling*, 178(1), 106059. https://doi.org/10.1016/j.resconrec.2022.106059

Pooja*, Nandini Pabreja, Aman Maurya, Laishram Saya, and Ajay Kumar Mishra

Chapter 11
Greener approach for next generation materials: Biofuel, biorefinery

Abstract: The basic need of industries is energy but the source of energy is the main concern. Not only must the source of energy be renewable but the process of producing the source should also be greener. The utilization of next-generation biofuel and biorefinery is vital for the long-term sustainability of the commercial, chemical, agriculture, industrial and associated energy sectors because of the growing demands of fossil fuels and in particular, industries need to respond to the challenges of sustainability and over usage of nonrenewable resources. In order to address the sustainable strategy, there is a need to switch over to renewable resources and manufacture of environmentally compatible materials. This transformation is possible by the use of accessible, widely available, and cheap source of carbon, which is biomass, and biorefineries are employed for converting biomass into valuable products. Biorefineries are setup to achieve circular bioeconomy by promoting resource restoration and its regeneration. This chapter focuses on the carbon-neutral, green, and sustainable approach of next-generation biorefineries. Their source of biowaste, pretreatment of biowaste using green methods, set up, operation, production of biofuels like biodiesel, biomethane, sustainable activation fuel (SAF), biogas, and biohydrogen by utilization of second-generation biowaste and processes like controlled pyrolysis, hydrothermal liquification or gasification, incineration or carbonization, aerobic fermentation, anaerobic digestion, microwave activation, and enzymatic reactions using phytochemicals have been discussed. It also highlights the importance and usage of commercially feasible first-generation renewable feedstock and commercially competitive third-generation feedstock like microalgae and macroalgae, a comparison of all three generations of renewable feedstock, technical developments in the industrial sectors based on green energy, issues, and challenges and future perspectives of the biofuels.

Keywords: Sustainable activation fuel, biomass, pyrolysis, hydrothermal liquification, phytochemicals

***Corresponding author: Pooja**, Department of Chemistry, Sri Venkateswara College, University of Delhi, Delhi, India, e-mail: pooja.chem123@gmail.com

Nandini Pabreja, Aman Maurya, Laishram Saya, Department of Chemistry, Sri Venkateswara College, University of Delhi, Delhi, India

Ajay Kumar Mishra, Department of Chemistry, University of the Western, Cape Town, South Africa

https://doi.org/10.1515/9783111618906-011

11.1 Introduction

Increasing population, fast urbanization, economic development, and geopolitical instability has led to the increasing need for fossil fuels and has led to energy insecurity in the recent years. Not only do the exhaustible and nonrenewable resources contribute to this insecurity, but they also have adverse environmental effects such as global warming, biodiversity loss, and greenhouse effect [1]. The environmental effects of relying solely on petroleum (crude oil) as an energy source have been rising, alongside concerns over sustainability and economic implications. The combustion of crude oil results in a rise in the quantity of hazardous greenhouse gas (GHG) emissions in the environment [2]. Currently, fossil fuels remain the dominant source of energy globally; however they are considered unsustainable, are directly linked to air and water contamination [3], and are forecast to be depleted in the future. Fossil fuel combustion is widely acknowledged as a primary cause of climate change, especially global warming, due to the release of carbon dioxide (CO_2) emissions. Implementing proactive measures and investigating alternative, more sustainable energy sources is crucial to decrease reliance on fossil fuels and minimize CO_2 emissions [4]. Consequently, it is essential to implement sustainable and green energy alternatives derived from renewable sources such as biodiesel, biomethane, biogas, bioethanol and biohydrogen [5]. Biomass, an organic material generated by living organisms, such as plants, is the source of "biofuels," a type of combustible fuel that can be produced multiple times and is a source of green energy [6]. It has lower nitrogen and sulfur content as compared to fossil fuels [7], making them a more environmentally friendly option. In consequence, growing worries about the environment and the increasing crude oil costs have driven research into alternative energy sources. Unlike petroleum-based products, green fuels (biofuels) are biodegradable and have a more environmentally friendly profile. Fossil fuels have a lengthy record of carbon capture, but they are not classified as biofuels because they contain carbon that is no longer part of the carbon cycle. Biofuels encompass a range of products, including solid biomass, liquid fuels, biogases, and biomass conversion products, among several others. Methane, methanol, bio-oil, biodiesel, ethanol, biobutanol, and other liquid and gaseous biofuels can be generated from biomass, and with the utilization of these biofuels, future energy security and sustainability can be attained. Biofuels reduce the extent of pollution load and environmental impact by lowering not just air pollutants during combustion but also the emission of GHGs [8, 9]. The procedure or the methodology involved in the generation of biofuels also needs to be safer and less hazardous for the environment and human health. So, the biofuels made from biomass resources using green and environmentally friendly methods are gaining attention from researchers and scientists worldwide [10].

Recent advancements in integrated biorefineries have greatly enhanced the production of a wide range of bioproducts from waste lignocellulosic components. These address a broad range of applications and comprise biofuels, platform chemicals, resins, bioplastics, additives, and other bio-based materials [11, 12]. Future supply chains for green and sustainable chemical products have been established through the use of low environmental impact technologies and the incorporation of green chemistry into biorefineries. The initial step in these future biorefineries should be the safe extraction of surface chemicals, followed by processing using renewable feedstock, and the final step of gasification to generate syngas [13, 14]. Biorefinery is based on the concept of the conventional grassroots crude oil refinery and petrochemical complex. Biorefineries open up multiple choices for the production of various types and forms of fuels, platform chemicals, chemical intermediates for processing, and heat and power from a wide range of biomass materials. By the combination of mechanical, chemical, thermochemical, and biochemical conversion techniques with microbial fermentations, the advanced biorefinery combines a number of upstream, mid-stream, and downstream biomass processing phases to create renewable biofuels and chemicals that can use a variety of agroforestry and marine feedstocks [15].

Despite the rapid growth of the biofuel industry, there is an absence of consensus about sustainability according to technical, economical, and environmental measures. It becomes a necessity to examine the trends and drivers in biofuels, from a point of producing sustainable next-generation biofuels [16]. Scientific and technological investigating procedures, including efficient pretreatments of biomass hydrolysis, successfully integrated into the biorefinery scenario requires enhanced understanding of the colonies of microbes at each step of the methanization process, and the development of advanced techniques to recover and collect volatile fatty acids (VFAs) produced during acidogenesis. In addition, it needs further efficiency in the conversion VFAs produced during acidogenesis into important molecules, especially when involving engineered strains, ongoing studies on the thermochemical conversion of produced gases, and optimized methods for the determination of biochemical methane potential and process performance measurements. At the final stage, for the evaluation of environmental consequences, sustainability and cost of processing, along with an in depth research on process scale-up, a life cycle assessment (LCA), and technoeconomic analysis are needed [17].

In this chapter, production of next-generation biofuels has been discussed through green methods and a comparison of biofuels based on different generations is made out. Common ways for conversion of biomass through thermochemical processes (gasification, combustion, or pyrolysis) and biochemical processes (pretreatment, hydrolysis, physical, chemical, or biological) or a combination have been focused [18]. Also, the potential and future perspectives of biorefineries are discussed, with proper design, including methods like extraction, production (gaseous and liquid biofuels), cogeneration, and production [19]. Commercially competitive microalgae biorefineries have also been discussed.

11.1.1 Biofuel

A biofuel is derived from a source of energy that originates from the fixation of biological carbon dioxide (CO_2). Biofuels are fuels that are generated from live plants, animals, or by-products that are less than 20 years old. Biofuels are a renewable source that stores energy generated from the sun. Plants are commonly regarded as renewable energy sources due to their ability to be regrown. Unlike petroleum-based products, green fuels (biofuels) are biodegradable and have a more environmentally friendly profile. Fossil fuels have a lengthy record of carbon capture, but they are not classified as biofuels because they contain carbon that is no longer part of the carbon cycle. Biofuels encompass a range of products, including solid biomass, liquid fuels, biogases, and biomass conversion products, among several others. These fuels, also commonly referred to as agrofuels, originate from agricultural products. Biomass and biowastes have the potential to serve as alternative options. Extensively used biomass biofuels are the strongest type, thanks to a surge in interest in renewable energy sources, including microbiologically produced biofuels, bioalcohols, methane, and hydrogen. The fermentation of starch and sugars by microorganisms results in the production of bioethanol, biobutanol, biomethanol, biodiesel, and various other liquid biofuels, in contrast to the trans esterification of vegetable oils, which yields biodiesel. Prior to the discovery of a more economical source of fossil oil, microbial fermentation was utilized in multiple applications, including solvents, greases, cleansers, and crucial compounds, to support the expanding chemical industry. Bioethanol and biodiesel are the two most commonly used biofuels in transportation, whereas biogas has a relatively limited number of large-scale applications in several European regions. Biobutanol has gained recognition as a potential liquid biofuel, particularly due to its attributes, including high energy levels, low volatility, and the ability to absorb moisture. Simultaneously, it can be used either straight or blended with gasoline, necessitating no engine alterations. These types of fuels are categorized as second- or third-generation biofuels and are characterized by the conversion technology that is currently being researched and improved, with projects being tested at the pilot or demonstration stages. The predominant process for producing biodiesel is through extracting oils from vegetable oil feedstock, resulting in the formation of fatty acid methyl esters (FAME) and fatty acid ethyl esters. Anaerobic digestion (AD) or fermentation of biomass results in the production of hydrogen, methane, and CO_2.

Biomass is converted into various forms of energy (Figure 11.1), including fuel and heat, through different conversion methods, which involve processing it into liquids or gases, breaking it down, and direct use. Biomass is converted into a liquid or gaseous state through a deconstruction process, which can take place at either low or high temperatures. Low-temperature processes like gasification turn biomass into hydrogen gas, a clean source of energy, and pyrolysis transforms it into both biochar and bio-oil, which can be used either as a fuel or as soil conditioners. Bio-oil is produced through the hydrothermal liquefaction process, which utilizes water under highly pressurized

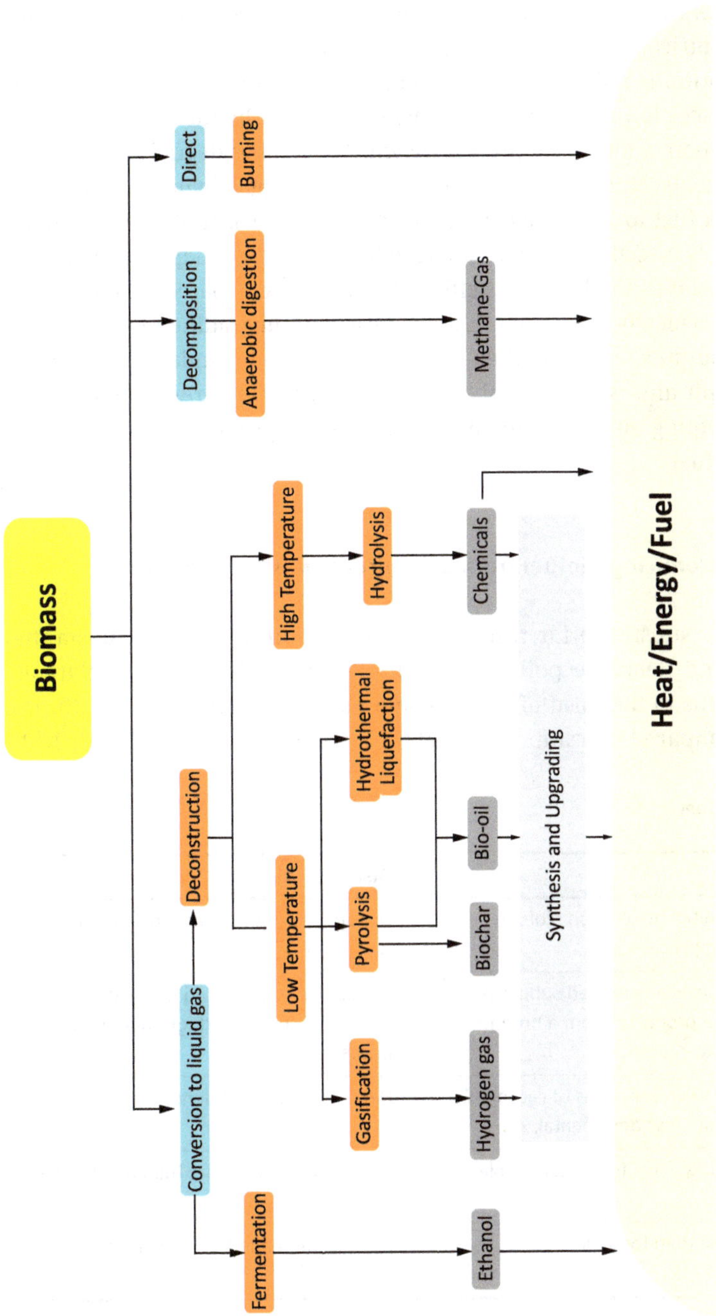

Figure 11.1: Flowchart showing the conversion of biomass to biofuels through various processes.

and heated conditions. At high temperatures, biomass undergoes hydrolysis, resulting in the breakdown of intricate organic materials into chemicals, which several industries can utilize. Fermentation is a distinct conversion process, where microorganisms degrade organic substances to produce ethanol, a widely utilized biofuel. AD is the main method of decomposition, a biological process by which microbes degrade organic materials in environments lacking oxygen, resulting in methane as a byproduct. Methane can then be used as a fuel to generate electricity or provide heat. In the direct utilization method, biomass is directly burned to produce heat or energy, skipping intermediary steps, which makes it a simple yet less-efficient technique compared to other methods. These pathways typically culminate in a synthesis and upgrading phase to boost the quality of their outputs, ensuring they satisfy the required standards for various applications. The main aim is to make the most of biomass as a renewable energy source, thereby promoting environmentally friendly energy production and reducing dependence on fossil fuels.

11.1.1.1 Benefits of producing biofuel from agricultural waste materials

Biofuels, serving as a substitute for traditional energy sources, offer a substantial amount of energy and decrease pollution levels [20], Biofuels produce relatively fewer carbon emissions in the form of CO_2, CO, and tetrahydrocannabinol (THC). The benefits of biofuel compared to traditional fossil fuels are listed in Table 11.1 [11, 21].

Table 11.1: Benefits of biofuel.

Aspects	Advantages	Disadvantages
Energy security	Feasibility of initial-generation biofuel manufacturing processes.	Disruptions to the food supply chain.
Energy availability	Fossil diesel fuel is a restricted supply, but biofuels can be produced from a broad spectrum of raw materials.	Primarily manufactured from edible oil, this can lead to food shortages and higher prices.
GHG emission	Substantially fewer emissions of carbon (CO_2, CO, and THC) are less detrimental.	Decrease in fuel efficiency.
Cost	This product originates from sustainable materials.	Currently priced higher than fossil diesel fuel.
Usage	Less flammable than fossil diesel,	Not recommended for use in cold conditions

11.1.1.2 The drawbacks of biofuel

Several drawbacks exist when using biofuels as a transportation fuel, particularly in four key areas: cost, nitrogen oxide emission, viscosity, and corrosion resistance. Conversely, reclaiming land through exploitative methods and growing annual crops such as maize would lead to heightened soil erosion and a decrease in soil fertility. Other disadvantages are listed in Table 11.1.

11.1.1.3 Method of production of biofuels from agricultural waste materials

(a) **Biofuel production through AD**
AD technology is employed to decompose organic matter in an oxygen-free environment and generate biogas. High levels of various minerals, including phosphate and ammonium, are found in the remaining wastewater from this digestive process. Excessive wastewater applied to agricultural land could lead to concerns about pollution or financial difficulties if further treatment is necessary prior to discharge, though it is possible to use nutrient-rich wastewater to cultivate photosynthetic microalgae. Combining conventional AD techniques with microalgae cultivation may help lower the cost of effluent treatment [22].

AD efficiently converts low-energy biomass, including algae rich in lipids, carbohydrates, and proteins, into methane (CH_4), for energy or renewable fuels. Algae's lack of lignin makes them more biodegradable than lignocellulosic biomass. Pretreatments like energy-intensive lipid extraction enhance this process by breaking down cell walls. Byproducts include liquid digestate, useful for fertilizer and algae cultivation, and solids suitable for composting [23]. Global food and energy shortages threaten sustainable development, highlighting the need for renewable energy like biofuels. Biofuels, with lower GHG emissions than fossil fuels, offer a sustainable alternative but face criticism for using food crops as feedstocks. Second-generation biofuels address this by utilizing nonedible sources like crop residues [23, 24].

AD is a key technology for converting organic waste into biofuels such as methane. While first-generation biofuels relied on edible crops, AD enables the efficient use of agricultural waste, making it a more sustainable waste-to-energy solution [25].

(b) **Biofuel production via hydrothermal carbonization (HTC)**
HTC, or wet torrefaction, has recently become the preferred method for treating wet waste. The HTC process is an exothermic reaction that reduces the oxygen and hydrogen content of biomass through hydrolysis [26]. HTC is a process of conversion of wet biomass into biofuel at a low temperature range (180–350 °C) and autogenous pressure, i.e., the pressure generated by the reactor itself during the carbonization activity [27], whereas microwave-assisted HTC operates at even lower temperatures, that is, 160–200 °C [28]. HTC can be efficiently performed via decoupled temperature and

pressure hydrothermal (DTPH) reaction at 200 °C and 20 MPa pressure. HTC has been shown to be a reliable technology for converting biomass into a variety of bio-products, including solid fuel or hydrocar, bio-oil, soil conditioner to improve soil fertility and maximize crop yields, carbon catalyst that can be used to make fine chemicals, affordable carbon adsorbents used for water purification or as a chemical barrier for different heavy metals or CO_2 sorption, and, finally, carbon material for effective fuel cells [29]. Three key phases make up the carbonization reaction: (a) dehydration of the carbohydrate to produce hydroxymethyl furfural; (b) polymerization of these furfurals to create polyfurans; and (c) further carbonization through intermolecular dehydration [30].

(c) **Biofuel production through pyrolysis**
Six new conversion approaches have recently been added to the two main procedure categories for biomass transformation: thermochemical and biochemical conversion. While only pyrolysis can provide a number of high-value products, including chemicals, charcoal, bio-oil, and biogas [31], second-generation biofuels are mostly made from municipal, industrial, and agricultural waste, including lignocellulosic biomass, plastic waste, food waste, and medical waste. This has the future perspective to produce profitable items and spark the economic, social, and environmental interest. Second-generation biofuels have faced a number of challenges, most notably high transportation and pretreatment costs, which have prevented their research, promotion, and commercialization [32]. Assuming the many advantages, waste oil pyrolysis will probably have a significant impact on the energy market in the near future [33]. By heating organic compounds to high temperatures in an inert atmosphere or without oxygen, the pyrolysis method converts them into the desired products. The distribution of products from pyrolysis varies, depending on the feedstocks and heating rates; these products can be used in industries such as chemicals (biogas, bio-oil, and char) or as an energy source. Research has demonstrated that traditional pyrolysis yields both desired products, like bio-oil, and less-desirable products, like char. This is problematic because these by-products encourage the production of coke and clog the filter's pores. The pyrolysis process can be made more efficient by adding an appropriate catalyst, which will increase the yield of bio-oil and decrease tar products [34].

(d) **Biofuel gasification**
An optimal method for the conversion of biowaste into heat and electricity through engines, turbines, and boilers is known as gasification [35]. This process includes the utilization of low-density biomass like lignocellulosic biomass and produces heat and electricity. This process includes co-gasification studies, reactor development, and enhancement in tar removal process. The primary byproduct of gasification is gas, which is converted into fuel gas and used to create one of the highly flammable gases, that is, hydrogen gas. Biochar and bio-oil are additional byproducts that are obtained in low yields [36, 35]. In contrast to pyrolysis, gasification can take place at high tem-

peratures, between 800 °C and 1,200 °C, in the presence of air. Low hydrogen content is the result of a gasification reaction that uses air and steam as gasifying agents [37, 38].

(e) **Biodiesel production through microwave-assisted transesterification**
The term "biodiesel" describes FAME, which are frequently made using alcohol and extracted oils and fats from biomass and other waste materials [30]. The procedure, commonly referred to as transesterification, can be carried out with or without a catalyst, such as an enzyme, metal oxide, acid, or alkali [39]. Figure 11.2 provides a schematic overview of microwave-assisted transesterification (MATr). MW power, time, agitation speed, and alcohol-to-oil or catalyst ratios, in relation to the biodiesel yield, are among the characteristics under investigation. The yield is said to be more significantly impacted by MW power and reaction time than by any of the other factors. Due to its significant advantages over traditional heating, including quicker chemical reaction rates, shorter reaction times, and thus lower energy usage, MATr has been increasingly popular in recent years [40, 41].

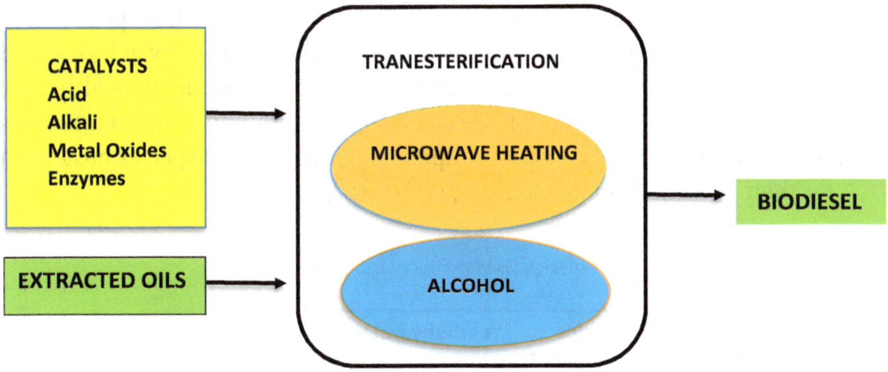

Figure 11.2: Schematic diagram of MATr.

11.1.1.3.1 First generation

First-generation biofuels are produced from food crops, including sugar cane [42], beet, corn, and soy. Up to now, bioethanol and biodiesel have been the most extensively researched of the various first-generation biofuels. Fermentation of corn and sugarcane for bioethanol production is currently in favor [43]. Transesterification of oilseed feedstocks is also a popular method for producing biodiesel. The abundance of raw materials like sugar and oil is linked to high productivity in these crops. The sustainability of this approach is being questioned due to the significant land area required for farming, and concerns have been raised about the potential threats it poses to the food industry. In addition, the net carbon benefit of bioethanol derived from corn is often less positive than that achieved by utilizing fossil fuels [44]

11.1.1.3.2 Second generation

Second-generation biofuels are typically produced from nonfood crops. The majority of these consist of lignocellulosic material derived from crop residue, forestry waste, agro-processing by-products, energy crops, etc. [45] Biofuels derived from raw materials other than crops, such as microbes and microalgae, are referred to as third-generation biofuels. According to the current state of scientific understanding and expected technological advancements, these are viewed as a viable energy source that is free from the drawbacks inherent in the first- and second-generation biofuels. Facilities designed to demonstrate and test emerging biomass processing technologies, as well as pilot plants, are currently under development; nonetheless, they have yet to be brought into full-scale commercial operation.

11.1.1.3.3 Third generation

Algae are autotrophs that belong to the bottom classes of plants [46]. There are more than 30,000 species inhabiting diverse areas such as ponds, lakes, hot springs, tree barks, snow, etc. and have simple reproductive structures. From the physiological and biochemical view, algae are more or less similar to higher plants, as they also have chlorophyll and almost the same carbohydrate and protein end products. Besides that, they also conserve varying amount of lipids. Most of the algal species have been analyzed for their chemical composition [47] (Table 11.2).

Table 11.2: Composition of some algae species (% in dry biomass) [48].

Strain	Proteins	Carbohydrates	Lipids
Tetraselmis maculata	52	15	3
Chlorella zofingiensis	11.2	11.5	56.7
Chlorella FC2 IITG	10.4	15	3
Anabaena cylindrica	43–56	25–30	4–7
Botryococcus braunii	40	2	33
Chlamydomonas reinhardtii	48	17	21
Chlorella pyrenoidosa	57	26	2
Chlorella vulgaris	41–58	12–17	10–22
Dunaliella salina	57	32	6
Dunaliella bioculata	49	4	8
Dunaliella tertiolecta	29	14	11
Euglena gracilis	39–61	14–18	14–20
Porphyridium cruentum	28–39	40–57	9–14
Prymnesium parvum	28–45	25–33	22–39
Scenedesmus dimorphus	8–18	21–52	16–40

Table 11.2 (continued)

Strain	Proteins	Carbohydrates	Lipids
Scenedesmus quadricauda	47	–	1.9
Spirogyra sp.	6–20	33–64	11–21
Spirulina maxima	60–71	13–16	6–7
Spirulina platensis	42–63	8–14	4–11
Pseudochoricystis ellipsoidea	10.2	34	38
Chlorogloeopsis fritschii	41.8	37.8	8.2

11.2 Green approach to biofuels and biorefineries

11.2.1 Methods

11.2.1.1 Algae-based biorefinery

Microalgae have the potential to be used as feedstocks for the production of biofuels [49]. These are single-celled photosynthetic organisms found in various environments. Algal strains exhibit adaptability across a range of temperatures, pH levels, and nutrient availability. Additionally, they can be cultivated on nonarable land and do not necessarily need access to freshwater. Achieving high productivity is relatively straightforward in warm, tropical, and subtropical areas throughout the year [48]. The growth rates of these plants are 30 times greater than those of other plant species.

The constant availability of biomass throughout the year enables a continuous biofuel production process, comparable to the operation of petroleum refineries. Microalgae play a significant role in sequestering atmospheric carbon due to their high photosynthetic efficiency, exceeding that of other plants. Many microalgae are capable of producing oil at rates exceeding even those of the most productive oil crops. These biofuels meet the requirements of being both economically viable and environmentally sustainable [50]

A biorefinery system that is both sustainable and economically viable should make the most of all biomass components, including proteins, carbohydrates, and lipids. An algal-based biorefinery achieves the production of a broad spectrum of biofuels and chemicals through the integration of various advanced technologies for biomass processing. The chosen technologies should have a minimal environmental impact in order to ensure the feasibility of the biorefinery's operation. The sustainability of a biorefinery is influenced by social, economic, and environmental factors.

The availability of feedstock plays a crucial role in operating a biorefinery. Microalgae, which are accessible year-round, offer promising feedstock for fractionation into valuable compounds within a biorefinery. Algal production on a commercial scale is typically achieved by growing the algae in either open raceways or in transparent containers, referred to as photo bioreactors. A combination of initial closed-system cultivation, followed by open-system algae growth, is showing promise in terms of output. There is a strong correlation between algal culture and both water bioremediation and the mitigation of carbon dioxide. Industrial wastewater, desalination plant wastewater, and flue gas emissions can be used for algal cultivation.

Functional biomass components can be processed to produce both energy products, such as biodiesel, methane, ethanol, and hydrogen, and nonenergy products, including nutraceuticals, fertilizers, animal feed, bioactive compounds, other bulk chemicals, and polymers.

The accumulation of bioproducts by algae, which include proteins, lipids, carbohydrates, starch, and vitamins, can be significantly influenced by the growth conditions and the specific type of algal species. Generally, green algae accumulate lipids and starch, diatoms store only lipids, whereas golden algae build up lipids and carbohydrates as their main reserve material. The amount of lipid not only varies but its composition does as well, depending on the type of algal strain used. Microalgae lipids can be selectively fractionated, and this process can be converted into biodiesel and two key essential fatty acids: DHA and EPA, or be utilized in the chemical industry. In addition to lipids, algae synthesize a variety of compounds, including proteins, isoprenoids, and polysaccharides. Proteins are used for food and animal feed, including essential nutrients such as PUFA and vitamins, whereas carbohydrates are primarily converted into industrial chemicals and energy sources. Along with algal biomass, downstream processing can also yield a range of polysaccharides, enzymes, pigments, antibiotics, antioxidants, metabolites, and minerals. A variety of liquid fuels, including biodiesel, bioethanol, biogasoline, biomethanol, biobutanol, and biohydrogen, can be produced. Oxygen generated during the photosynthetic growth of algae is also recoverable. Conversion of algal biomass can be achieved through various biochemical and physical pathways to produce multiple biofuels and byproducts. These processes can be of biochemical, thermochemical, or chemical origin.

11.3 Future perspectives

Achieving a value chain from the biomass feedstock for the bio-based goods in the final market is the main goal of developing a sustainable biorefinery. Several conversion techniques are skillfully coupled to improve biomass consumption and produce a sustainable output [37]. Numerous biomass feedstocks, such as lignocelluloses, microalgae, and different kinds of organic waste, are used in the biorefinery concept. The bioeconomy's present constraints on the production of various high-value products will be met by the biorefinery idea and process integration.

11.4 Conclusion

By focusing on the optimal use of biomass and reusing and recycling waste residues, a circular bioeconomy can be adopted to increase the production of biofuels in a sustainable manner. Biofuels may generate an infinite quantity of energy because they are made from biomass, a renewable resource. In comparison to fossil fuels, they may also benefit society and the environment if properly handled. Prior to biofuel technology being widely used worldwide, developing and impoverished nations would see both positive and bad effects from the increase in biofuel production.

Abbreviations

AD	Anaerobic digestion
CH_4	Methane
CO	Carbon monoxide
CO_2	Carbon dioxide
DHA	Docosahexaenoic acid
EPA	Eicosapentaenoic acid
FAME	Fatty acid methyl esters
HTC	Hydrothermal carbonization
PUFA	Polyunsaturated fatty acid
MATr	Microwave-assisted transesterification
THC	Tetrahydrocannabinol
VFAs	Volatile fatty acids

References

[1] Duran, S. K., Kumar, P., & Sandhu, S. S. (2021). A review on microalgae strains, cultivation, harvesting, biodiesel conversion and engine implementation. *Biofuels*, 12, 91–102. https://doi.org/10.1080/17597269.2018.1457314

[2] Zabermawi, N. M., Alsulaimany, F. A. S., El-Saadony, M. T., & El-Tarabily, K. A. (2022). New eco-friendly trends to produce biofuel and bioenergy from microorganisms: An updated review. *Saudi Journal of Biological Sciences*, S1319562X22001000. https://doi.org/10.1016/j.sjbs.2022.02.024

[3] Alaswad, A., Dassisti, M., Prescott, T., & Olabi, A. G. (2015). Technologies and developments of third generation biofuel production. *Renewable and Sustainable Energy Reviews*, 51, 1446–1460. https://doi.org/10.1016/j.rser.2015.07.058

[4] He, S., Barati, B., Hu, X., & Wang, S. (2023). Carbon migration of microalgae from cultivation towards biofuel production by hydrothermal technology: A review. *Fuel Processing Technology*, 240, 107563. https://doi.org/10.1016/j.fuproc.2022.107563

[5] Abdullah, S. H. Y. S., Hanapi, N. H. M., Azid, A., Umar, R., Juahir, H., Khatoon, H., & Endut, A. (2017). A review of biomass-derived heterogeneous catalyst for a sustainable biodiesel production. *Renewable and Sustainable Energy Reviews*, 70, 1040–1051. https://doi.org/10.1016/j.rser.2016.12.008

[6] Malode, S. J., Prabhu, K. K., Mascarenhas, R. J., Shetti, N. P., & Aminabhavi, T. M. (2021). Recent advances and viability in biofuel production. *Energy Conversion and Management: X*, 10, 100070. https://doi.org/10.1016/j.ecmx.2020.100070
[7] Tashtoush, G. M., Al-Widyan, M. I., & Albatayneh, A. M. (2007). Factorial analysis of diesel engine performance using different types of biofuels. *Journal of Environmental Management*, 84, 401–411. https://doi.org/10.1016/j.jenvman.2006.06.017
[8] Anukam, A., Mohammadi, A., Naqvi, M., & Granström, K. (2019). A review of the chemistry of anaerobic digestion: Methods of accelerating and optimizing process efficiency. *Processes*, 7, 504. https://doi.org/10.3390/pr7080504
[9] Prasad, S., Singh, A., Korres, N. E., Rathore, D., Sevda, S., & Pant, D. (2020). Sustainable utilization of crop residues for energy generation: A life cycle assessment (LCA) perspective. *Bioresource Technology*, 303, 122964. https://doi.org/10.1016/j.biortech.2020.122964
[10] Khan, S., Naushad, M., Iqbal, J., Bathula, C., & Sharma, G. (2022). Production and harvesting of microalgae and an efficient operational approach to biofuel production for a sustainable environment. *Fuel*, 311, 122543. https://doi.org/10.1016/j.fuel.2021.122543
[11] Nisar, A., Hashum, K., Bashir, M., Mubeen, N., Younus, S., Mehmood, S., Haq, F., & Haroon, M. (2023). Advancing sustainable biofuel production from agricultural residues: A comprehensive mini-review. *SCE*, 116–129. https://doi.org/10.37256/sce.5120243763
[12] Dheeran, P., & Reddy, L. (2018). Biorefining of lignocelluloses: An opportunity for sustainable biofuel production. In: Kumar, S., & Sani, R. K. (eds.), *Biorefining of Biomass to Biofuels, Biofuel and Biorefinery Technologies*. Cham: Springer International Publishing. pp. 1–23. https://doi.org/10.1007/978-3-319-67678-4_1
[13] H. Clark, J., E. I. Deswarte, F., & J. Farmer, T. (2009). The integration of green chemistry into future biorefineries. *Biofuels Bioprod Bioref*, 3, 72–90. https://doi.org/10.1002/bbb.119
[14] Taylor, G. (2008). Biofuels and the biorefinery concept. *Energy Policy*, 36, 4406–4409. https://doi.org/10.1016/j.enpol.2008.09.069
[15] Shukla, M., & Kumar, S. (2018). Algal growth in photosynthetic algal microbial fuel cell and its subsequent utilization for biofuels. *Renewable and Sustainable Energy Reviews*, 82, 402–414. https://doi.org/10.1016/j.rser.2017.09.067
[16] Zinoviev, S., Müller-Langer, F., Das, P., Bertero, N., Fornasiero, P., Kaltschmitt, M., Centi, G., & Miertus, S. (2010). Next-generation biofuels: Survey of emerging technologies and sustainability issues. *ChemSusChem*, 3, 1106–1133. https://doi.org/10.1002/cssc.201000052
[17] Koubaa, M. (2024). Integrated biorefinery for a next-generation methanization process focusing on volatile fatty acid valorization: A critical review. *Molecules*, 29, 2477. https://doi.org/10.3390/molecules29112477
[18] Naik, S. N., Goud, V. V., Rout, P. K., & Dalai, A. K. (2010). Production of first and second generation biofuels: A comprehensive review. *Renewable and Sustainable Energy Reviews*, 14, 578–597. https://doi.org/10.1016/j.rser.2009.10.003
[19] Moncada B., J., Aristizábal M., V., & Cardona A., C. A. (2016). Design strategies for sustainable biorefineries. *Biochemical Engineering Journal*, 116, 122–134. https://doi.org/10.1016/j.bej.2016.06.009
[20] Datta, A., Hossain, A., & Roy, S. (2019). An overview on biofuels and their advantages and disadvantages. *Asian Journal of Chemistry*, 31, 1851–1858. https://doi.org/10.14233/ajchem.2019.22098
[21] Xue, J., Grift, T. E., & Hansen, A. C. (2011). Effect of biodiesel on engine performances and emissions. *Renewable and Sustainable Energy Reviews*, 15, 1098–1116. https://doi.org/10.1016/j.rser.2010.11.016
[22] Cai, T., Park, S. Y., Racharaks, R., & Li, Y. (2013). Cultivation of Nannochloropsis salina using anaerobic digestion effluent as a nutrient source for biofuel production. *Applied Energy*, 108, 486–492. https://doi.org/10.1016/j.apenergy.2013.03.056

[23] Ayala-Parra, P., Liu, Y., Field, J. A., & Sierra-Alvarez, R. (2017). Nutrient recovery and biogas generation from the anaerobic digestion of waste biomass from algal biofuel production. *Renewable Energy*, 108, 410–416. https://doi.org/10.1016/j.renene.2017.02.085
[24] Abomohra, A. E.-F., Shang, H., El-Sheekh, M., Eladel, H., Ebaid, R., Wang, S., & Wang, Q. (2019). Night illumination using monochromatic light-emitting diodes for enhanced microalgal growth and biodiesel production. *Bioresource Technology*, 288, 121514. https://doi.org/10.1016/j.biortech.2019.121514
[25] Ward, A. J., Lewis, D. M., & Green, F. B. (2014). Anaerobic digestion of algae biomass: A review. *Algal Research*, 5, 204–214. https://doi.org/10.1016/j.algal.2014.02.001
[26] Funke, A., & Ziegler, F. (2010). Hydrothermal carbonization of biomass: A summary and discussion of chemical mechanisms for process engineering. *Biofuels Bioprod Bioref*, 4, 160–177. https://doi.org/10.1002/bbb.198
[27] Lu, X., Jordan, B., & Berge, N. D. (2012). Thermal conversion of municipal solid waste via hydrothermal carbonization: Comparison of carbonization products to products from current waste management techniques. *Waste Management*, 32, 1353–1365. https://doi.org/10.1016/j.wasman.2012.02.012
[28] Deng, C., Lin, R., Kang, X., Wu, B., Ning, X., Wall, D., & Murphy, J. D. (2022). Co-production of hydrochar, levulinic acid and value-added chemicals by microwave-assisted hydrothermal carbonization of seaweed. *Chemical Engineering Journal*, 441, 135915. https://doi.org/10.1016/j.cej.2022.135915
[29] Oliveira, I., Blöhse, D., & Ramke, H.-G. (2013). Hydrothermal carbonization of agricultural residues. *Bioresource Technology*, 142, 138–146. https://doi.org/10.1016/j.biortech.2013.04.125
[30] Liu, F., Yu, R., & Guo, M. (2017). Hydrothermal carbonization of forestry residues: Influence of reaction temperature on holocellulose-derived hydrochar properties. *Journal of Materials Science*, 52, 1736–1746. https://doi.org/10.1007/s10853-016-0465-8
[31] Salama, E.-S., Hwang, J.-H., El-Dalatony, M. M., Kurade, M. B., Kabra, A. N., Abou-Shanab, R. A. I., Kim, K.-H., Yang, I.-S., Govindwar, S. P., Kim, S., & Jeon, B.-H. (2018). Enhancement of microalgal growth and biocomponent-based transformations for improved biofuel recovery: A review. *Bioresource Technology*, 258, 365–375. https://doi.org/10.1016/j.biortech.2018.02.006
[32] Ong, H. C., Chen, W.-H., Singh, Y., Gan, Y. Y., Chen, C.-Y., & Show, P. L. (2020). A state-of-the-art review on thermochemical conversion of biomass for biofuel production: A TG-FTIR approach. *Energy Conversion and Management*, 209, 112634. https://doi.org/10.1016/j.enconman.2020.112634
[33] Su, G., Ong, H. C., Mofijur, M., Mahlia, T. M. I., & Ok, Y. S. (2022). Pyrolysis of waste oils for the production of biofuels: A critical review. *Journal of Hazardous Materials*, 424, 127396. https://doi.org/10.1016/j.jhazmat.2021.127396
[34] Zaidi, A. A., Khan, A., AlMohamadi, H., Anjum, M. W., Ali, I., Naqvi, S. R., Kokuryo, S., Miyake, K., & Nishiyama, N. (2023). Catalytic pyrolysis of rice husk over defect-rich beta zeolites for biofuel production. *Fuel*, 348, 128624. https://doi.org/10.1016/j.fuel.2023.128624
[35] Yang, Y., Liew, R. K., Tamothran, A. M., Foong, S. Y., Yek, P. N. Y., Chia, P. W., Van Tran, T., Peng, W., & Lam, S. S. (2021). Gasification of refuse-derived fuel from municipal solid waste for energy production: A review. *Environmental Chemistry Letters*, 19, 2127–2140. https://doi.org/10.1007/s10311-020-01177-5
[36] Lee, J., Kwon, E. E., & Park, Y.-K. (2020). Recent advances in the catalytic pyrolysis of microalgae. *Catalysis Today*, 355, 263–271. https://doi.org/10.1016/j.cattod.2019.03.010
[37] Goswami, L., Kayalvizhi, R., Dikshit, P. K., Sherpa, K. C., Roy, S., Kushwaha, A., Kim, B. S., Banerjee, R., Jacob, S., & Rajak, R. C. (2022). A critical review on prospects of bio-refinery products from second and third generation biomasses. *Chemical Engineering Journal*, 448, 137677. https://doi.org/10.1016/j.cej.2022.137677

[38] Robbins, M. P., Evans, G., Valentine, J., Donnison, I. S., & Allison, G. G. (2012). New opportunities for the exploitation of energy crops by thermochemical conversion in Northern Europe and the UK. *Progress in Energy and Combustion Science*, 38, 138–155. https://doi.org/10.1016/j.pecs.2011.08.001

[39] Chang, K.-L., Lin, Y.-C., Jhang, S.-R., Cheng, W. L., Chen, S.-C., & Mao, S.-Y. (2017). Rapid jatropha-castor biodiesel production with microwave heating and a heterogeneous base Catalyst Nano-Ca (OH)2/Fe3O4. *Catalysts*, 7, 203. https://doi.org/10.3390/catal7070203

[40] Arpia, A. A., Chen, W.-H., Lam, S. S., Rousset, P., & De Luna, M. D. G. (2021). Sustainable biofuel and bioenergy production from biomass waste residues using microwave-assisted heating: A comprehensive review. *Chemical Engineering Journal*, 403, 126233. https://doi.org/10.1016/j.cej.2020.126233

[41] Fatimah, I., Rubiyanto, D., Taushiyah, A., Najah, F. B., Azmi, U., & Sim, Y.-L. (2019). Use of ZrO2 supported on bamboo leaf ash as a heterogeneous catalyst in microwave-assisted biodiesel conversion. *Sustainable Chemistry and Pharmacy*, 12, 100129. https://doi.org/10.1016/j.scp.2019.100129

[42] Goldemberg, J., & Guardabassi, P. (2010). The potential for first-generation ethanol production from sugarcane. *Biofuels Bioprod Bioref*, 4, 17–24. https://doi.org/10.1002/bbb.186

[43] Sarris, D., & Papanikolaou, S. (2016). Biotechnological production of ethanol: Biochemistry, processes and technologies. *Engineering in Life Sciences*, 16, 307–329. https://doi.org/10.1002/elsc.201400199

[44] Searchinger, T., Heimlich, R., Houghton, R. A., Dong, F., Elobeid, A., Fabiosa, J., Tokgoz, S., Hayes, D., & Yu, T.-H. (2008). Use of U.S. croplands for biofuels increases greenhouse gases through emissions from land-use change. *Science*, 319, 1238–1240. https://doi.org/10.1126/science.1151861

[45] Vohra, M., Manwar, J., Manmode, R., Padgilwar, S., & Patil, S. (2014). Bioethanol production: Feedstock and current technologies. *Journal of Environmental Chemical Engineering*, 2, 573–584. https://doi.org/10.1016/j.jece.2013.10.013

[46] Ramos-Suárez, J. L., Arroyo, N. C., & González-Fernández, C. (2015). The role of anaerobic digestion in algal biorefineries: Clean energy production, organic waste treatment, and nutrient loop closure. In: Singh, B., Bauddh, K., & Bux, F. (eds.), *Algae and Environmental Sustainability*. New Delhi: Springer India. pp. 53–76. https://doi.org/10.1007/978-81-322-2641-3_5

[47] Babich, I. V., Van Der Hulst, M., Lefferts, L., Moulijn, J. A., O'Connor, P., & Seshan, K. (2011). Catalytic pyrolysis of microalgae to high-quality liquid bio-fuels. *Biomass and Bioenergy*, 35, 3199–3207. https://doi.org/10.1016/j.biombioe.2011.04.043

[48] Béchet, Q., Shilton, A., & Guieysse, B. (2016). Maximizing productivity and reducing environmental impacts of full-scale algal production through optimization of open pond depth and hydraulic retention time. *Environ. Sci. Technol*, 50, 4102–4110. https://doi.org/10.1021/acs.est.5b05412

[49] Mata, T. M., Martins, A. A., & Caetano, N. S. (2010). Microalgae for biodiesel production and other applications: A review. *Renewable and Sustainable Energy Reviews*, 14, 217–232. https://doi.org/10.1016/j.rser.2009.07.020

[50] Schenk, P. M., Thomas-Hall, S. R., Stephens, E., Marx, U. C., Mussgnug, J. H., Posten, C., Kruse, O., & Hankamer, B. (2008). Second generation biofuels: High-efficiency microalgae for biodiesel production. *Bioenergy Research*, 1, 20–43. https://doi.org/10.1007/s12155-008-9008-8

Hema Kundra*, Ambika Tundwal, Archana Dagar, Himani,
and Chandra Mohan

Chapter 12
Utilization of fly ash for a sustainable environment: Innovations in waste management, construction, and renewable energy applications

Abstract: Fly ash, a by-product of coal combustion has always been associated with environmental pollutants due to its large-scale production and improper disposal. However, its large potential for use in different industries can drastically enhance environmental sustainment. The objective of this chapter was to research how effectively fly ash can be repurposed to reduce waste and ensure a healthy environment in sectors like waste management, construction, and renewable energy. The study addresses the utilization of fly ash in sustainable construction material production, such as concrete and bricks, whereby the waste is reduced along with the consumption of natural resources. The ability of fly ash to be used in developing geopolymers and greenhouse gas emissions in construction is also considered. The chapter thus reflects the application of fly ash in renewable energy systems, for example by synthesizing it to zeolite for the treatment of wastewater and integrating it into solar technologies. By propelling these applications, fly ash can be an actor in the circular economy, with major environmental challenges addressed. The chapter will go further to discuss the challenges and recommendations for improving fly ash utilization toward minimum environmental impact as well as support for sustainable development initiatives.

Keywords: Green nanomaterials, Natural resource, Eco-friendly, Waste Management, Biosynthesis

12.1 Introduction

Increasing burdens of waste and depletion of resources have catalyzed interest in sustainable, green technologies. Among the promising approaches under these is the use

***Corresponding author: Hema Kundra**, Guru Tegh Bhadur Institute of Technology, Guru Gobind Singh Indraprastha University, Delhi, India, e-mail: chemicalphysics5824@gmail.com
Ambika Tundwal, Archana Dagar, Himani, Guru Tegh Bhadur Institute of Technology, Guru Gobind Singh Indraprastha University, Delhi, India
Chandra Mohan, School of Basic and Applied Sciences, K. R. Mangalam University, Gurugram, India

https://doi.org/10.1515/9783111618906-012

of waste streams to synthesize nanomaterials (NMs). These "green NMs" are produced using environmentally friendly methods and renewable resources and reduce toxic by-products that enable circular use of materials. Nanotechnology tackles problems in the pre-treatment, hydrolysis, and output of biofuels [17]. The smaller NMs, with dimensions ranging from 1 to 100 nm, have high selectivity and porosity, active sites, and high interfacial surface area [12]. NMs are generally associated with good crystallinity, chemical stability, better absorption capacity, and higher catalytic activity [2]. NMs are used very effectively in many different scientific and technological areas, such as water purification, mass transfer, cancer therapy, biological sensing, drug delivery, and production of biofuels [3]. NMs are classified into different types: carbonaceous materials, enzymes or biocatalysts, metals, metal oxides, and nonmetal oxides [4].

12.1.1 Definition and scope

Nanoscale materials are those substances that possess at least one dimension less than about 100 nm. This extremely small size provides a large ratio of surface area to volume and enhances the number of active sites for various reactions and processes. These nanoparticles (NPs) also show the ability to present different morphologies that have widened their applications due to different fields [5, 18]. In addition, nanostructured materials react much faster with other molecules compared to large particles [6]. NMs have already made a worthy commercial impact and the awareness of NMs will increase due to their unique optical scale properties impacting several fields, such as bioenergy, electronics, mechatronics, medicine, pharmaceutical, ionic liquids, polymer, and many more.

12.1.2 Green chemistry principles in nanomaterial synthesis

Green-synthesized NPs have huge uses as they are biocompatible and allow controlled substance release [7]. The biomedical applications of these NPs in drug delivery systems are of high importance [8], such as in the treatment of cancer by Devanesan et al. [5] and Yusefi et al. [19] and diagnostic imaging. However, they find different applications in agriculture, which play key roles in crop security and nutrient supply, and hence their importance in plant vigor augmentation. The applications tend to facilitate increased sustainability and reduce more negatives such as adverse environmental impacts to farming practices. Green NPs are also employed in environmental remediation, which include purification of water and soil, antimicrobial agents, and control of air pollution [10]. The removal of contaminants and enhancement of the remediation process help in making the environment greener and cleaner. Moreover, green NPs have the potential to be used in photovoltaics and energy storage. This will result in enhancing the efficiency and performance of solar cells and storage devices and thus provide the means to develop sustainable solutions for energy. Besides the fields mentioned above, such NPs are versatile, with a chance to provide better formulations and safer products in cosmetics and food industries [11]. They also work as sensors, where substances can be detected and traced precisely. Additionally, they are used in textiles and electronics for the improvement of functional properties and offer scope for novel applications [16]. Green NPs have greater versatility and innovation potential than others, ranging from nanoelectronics to nano absorbents and catalysis.

12.1.3 The role of green nanotechnology in sustainable development

Green nanotechnology plays a vital role in advancing sustainable development by focusing on eco-friendly NMs and processes that minimize environmental impact. It encompasses the use of nanoscale innovations to create sustainable solutions in energy, waste management, and water purification. For example, green NMs like nanoscale catalysts improve energy efficiency and reduce emissions in industrial processes. Additionally, nanotechnology-based water filtration systems can remove pollutants effectively, providing clean drinking water. By promoting recycling, reducing toxic materials, and enhancing renewable energy technologies, green nanotechnology supports the transition to a sustainable, eco-conscious society.

This book chapter elucidates how green-synthesized NPs are finding applications in medicine, agriculture, environmental remediation, energy, and several other fields to revolutionize industry, ushering its way to a greener future. The eco-friendliness, biocompatibility, and cost-effectiveness of green synthesis methods draw the interest of researchers and enterprises seeking greener alternatives; however, imperative-related issues such as those concerning reproducibility, stability, and control over size and shape are critical to make them more practical in their applications [19].

12.2 Sources of waste for green nanomaterial synthesis

Green NM synthesis can utilize various waste sources, transforming waste into valuable resources. This approach not only reduces environmental pollution but also supports the circular economy by repurposing waste. Here's an overview of different waste sources and the associated challenges and opportunities in their use:

12.2.1 Industrial waste (e.g., metal-rich sludge, fly ash, and e-waste)

Recently, waste materials like metal-rich sludge, fly ash, and e-waste have emerged into focus for their potential use as valuable sources for green NMs. Such industrial waste is full of treasure in terms of valuable metals, minerals, and elements that can be converted to useful NMs toward environmental sustainability and diminution of pollution. Reutilization of wastes thus not only reduces the pile-up of waste in landfills but also saves natural resources, thus fitting well within the circular economy philosophy.

Fly ash consists of combustion materials for coal in power plants and holds significant amounts of silicon dioxide (SiO_2), alumina (Al_2O_3), and iron oxides (Fe_2O_3). NMs produced from fly ash can be used as an effective adsorbent in heavy metals and organic pollutants in water treatment plants. In addition, fly ash NPs are used in strengthening construction materials as concrete with lessened environmental impacts to building operations.

Another promising source of NMs is e-waste containing precious metals, including gold, silver, platinum, and palladium [14]. These metals are excellent candidates for the development of nanocatalysts for energy generation, pollution control, and green chemistry, recovering lost resources in the recycling process of e-waste, and reducing their impacts on the environment by serving potential applications in the fields of advanced electronics, sensors, and energy storage devices.

Metal oxides are rich in essential metals extracted from mining, steel production, and wastewater treatment. The former can be reduced into NPs for catalysis applications and battery technology. Recycling this industrial waste can be seen as a way of mitigating pollution, conserving finite natural resources and reducing consumption of primary-mined minerals. In return, the production of sustainable NMs promotes recycling waste to valuable materials. It therefore minimizes pollution.

12.2.2 Agricultural waste (e.g., biomass and crop residues)

Agricultural waste, including biomass, crop residues, and fruit peels, are rich sources of renewable raw materials for the preparation of green NMs. These wastes are large and in general underutilized, but rich in valuable organic compounds such as cellulose, lignin, hemicellulose, polysaccharides, and others, which can be processed into NMs with diverse applications. In this way, the method, apart from providing alternative, sustainable options in place of the materials traditionally used, promotes environmental preservation and development in the countryside.

In contrast, the NM most representative of agricultural waste is nanocellulose, obtained from the cellulose in biomass such as wood, straw, rice husks, or crop residues. Nanocellulose is reportedly recognized for its high mechanical strength, biodegradability, and biocompatibility, thereby presenting suitable properties for a wide range of applications. Nanocellulose is applied in the packaging industry to produce biodegradable films. An ecological friendly alternative to conventional plastic that will reduce plastic waste and the associated environmental harm, these produced bioplastics have a lower environmental footprint than those based on petroleum. The trend also enjoys dominance in the application related to water purification technologies, where nanocellulose-based filters have been proved efficient in the removal of heavy metals, oils, and other polluters from water, making water systems clean.

Other useful wastes that come from agriculture are those from lignin. Lignin is a complex natural polymer attached to the plant cell walls. Nano-lignin can be produced from the lignin and has a prospect for usage in the development of science in the field of bio-based composites for construction and packaging applications. Nanolignin has proved to exhibit antioxidant and antimicrobial properties that may be useful in many biomedical applications, including wound healing and drug delivery systems. Transformation of such lignocellulosic biomass into carbon-based NMs or biochar is highly promising for remediation of future environments to clean soils and waters.

Crop residues, like corn stover, wheat straw, and rice husks, are also combusted post-harvest. These crop residues can now be processed into NMs with less combustion of crop residues. Biochars as nano adsorbents of toxic metals and organic pollutants scavenge from water and soil and can clean the environment, contributing to sustainable agriculture.

This would develop new industries and new employment opportunities in rural economies. It produces a new industry and new jobs, besides reusing agricultural waste; using agricultural waste would further help to reduce the environmental impacts of traditional farming, specifically open-field burning and effects of overusing synthetic fertilizers. It serves as a practical solution to waste reduction, but it also plays a role in making specific industries sustainable, from packaging to health and environmental remediation. In theory, it fits really well with the philosophy of a circular economy, wherein agro waste residues are transformed into valuable, usable products that spur economic growth while ensuring sustainability in the environment.

12.2.3 Municipal and biomedical waste

Municipal as well as biomedical waste are gross sources of waste materials and often very difficult to deal with. The waste consists of a rich source of organic and biomolecular components that, if processed correctly, may eventually become functional materials for an enormous range of applications. The use of municipal and biomedical waste for the production of NMs is compatible with the objectives of sustainable development, which consists of counteracting the amassing waste, reducing environmental pollution, and upholding the tenets of a circular economy.

Municipal waste, especially organic matter, can be deemed as primary sources of carbon for carbon-based NMs such as graphene and carbon nanotubes [13]. Graphene is one of the most promising NMs because of its high electrical conductivity, mechanical strength, and thermal properties, which make it very versatile. Such properties make it ideal for energy storage devices like supercapacitors and batteries as well as water filtration systems. Graphene-based materials produced from municipal waste save landfills and bypass the use of raw graphite, whose mining is degrading the environment. Municipal waste recycling into carbon NMs can be a strong enabler for achieving the solution of urban waste management in the form of transforming waste into precious high products, hence contributing to sustainability [1].

The other critical area where green nanotechnology will create a revolution is biomedical waste. Biomedical waste includes syringes, gloves, bandages, and even expired pharmaceuticals. Well-managed biomedical waste can be processed to extract valuable biomolecules like proteins, enzymes, and other bioactive compounds that may be used in the synthesis of biocompatible NMs for many medical applications. For example, the NPs derived from the shell of crustaceans or synthesized from waste surgical residue in a hospital can be applied in drug delivery because it is biodegradable; it can be used for encapsulation and drug release. Silver NPs produced from medical waste are famously antimicrobial. These find numerous applications in wound healing and in devices used in a clinical setup, besides infection control.

These can be further processed to form NPs that are biocompatible and suitable for the targeted delivery of drugs or application in medical imaging. Pharmaceutical waste can be used in synthesizing liposomes and polymeric NMs; this provides a sustainable means of medical therapies with minimal environmental discharges. The derived NMs are biodegradable, hence there is less concern about the pollution caused by synthetic nondegradable materials over long periods.

Not only is the repurposing of municipal and biomedical waste into green NMs profoundly helpful in terms of public health; it reduces health risk exposures by reducing the negative environmental consequences resulting from improper waste disposal, which include contaminants entering water bodies or hazardous chemicals being released from biomedical wastes poorly disposed. With all these advanced recycling technologies being applied to all these materials, they can easily be transformed into valuable products without posing dangers to human health, thus ensuring a safer and more sustainable future for both the urban and the healthcare systems.

Generally, municipal and biomedical waste converted into NMs can be used as a route to the solution of critical issues regarding the management of waste and will help ensure adequate sustainable sources of raw materials for nanotechnology. Such a process will transform old material into new material to produce high-value resources, thereby rendering NMs that are likely to improve their application in areas from energy storage to medicine. Consequently, they form part of the green production landscape of NMs.

12.2.4 Challenges and opportunities in waste source segregation and recovery

One of the challenges in the use of wastes in the synthesis of green NM is the efficient segregation and recovery of materials from waste. Industrial, agricultural, municipal, and biomedical wastes often are not segregated; hence, a long process of extracting specific materials without contamination is required. This means efficiency in terms of recovery depends on developing effective and relatively low-cost methods for sorting and processing. It is also a great challenge when it comes to safety regulations and health risks associated with handling biomedical as well as industrial waste. But the breakthroughs in processing technologies, combined with incentives for waste segregation, overcome the obstacles. Awareness creation, imposition of harsh waste management policies, and investment in recovery technologies unlock the full potential that waste holds as a resource in green NM production.

12.3 Sustainable synthesis methods of green nanomaterials

The sustainable synthesis of green NMs aims to reduce environmental and health risks by using eco-friendly processes and minimizing toxic chemicals. Here's an exploration of various sustainable synthesis methods, categorized by approach:

12.3.1 Biological methods: microbial and plant-mediated NP synthesis

The use of processes for natural methods means that biological synthesis methods of NMs are mainly constituted by nanosynthesis of NMs that naturally come about with the assistance of microorganisms or plants, thereby providing a more natural and economic approach. Of microbial synthesis, there are bacteria, fungi and algae as sources and serve to create NPs as byproducts of their metabolic activities. Some bacteria reduce the metal ions, resulting in the formation of metallic NPs used in biomedical and environmental applications. This plant-mediated synthesis method utilizes extracts that are rich in a variety of biomolecules, including enzymes, proteins, or antioxidants that serve as natural reducing agents in the formation of NPs. A process such as this is not only environment-friendly but also scalable. Biological sources are abundant and renewable.

12.3.2 Physical and chemical methods

1. Green NMs synthesized by physical and chemical methods incorporate novel, sustainable methodologies with minimal toxic reagent usage or even total omission.
2. Environment-friendly solvents: Here, solvents are water or bio-based, instead of the hazardous chemicals used for making NPs. These often have a low toxic potential, are biodegradable, and pose lesser threats to the environment.
3. Microwave-assisted synthesis: This synthesis method uses microwave energy to heat the reactants extremely fast. It significantly reduces time and energy usage compared to classic methods. It generally produces nanoscale particles uniformly and generally gives uniform size and shape. It also reduces the by-products that form, which ensures cleaner production processes.
4. Ionic liquids: Ionic liquids are recyclable, nonvolatile solvents that are thermally stable and have an ability to dissolve various substances under weak conditions. Ionic liquids, without the formation of harmful vapors during the synthesis of NMs, provide a route for the sustainable formation of metallic, oxide, and polymeric NPs.

12.3.3 Hybrid techniques and innovations for scaling green synthesis

Hybrid strategies, which involve integration of biological, physical, and chemical methods, are designed to enable large-scale synthesis of green NMs. For instance, the association of microbial synthesis with microwave-assisted processing may enhance yield and uniformity of NPs while decreasing resource input. Other recent developments in this field are continuous flow reactors, where NMs are synthesized and separated in real-time. This will make the process scalable. Besides, automated integration with machine learning will guarantee that a reaction condition leading to optimized NMs is effective and repeatable to produce green NMs. Hybrid approaches that adopt several sustainable technologies point the way toward industrial-scale NM production that is both “industrially” demanding and environmentally sustainable.

12.4 Characterization techniques for green nano materials

Characterizing green NMs is essential for understanding their properties, ensuring quality, and assessing their potential applications. Here's an overview of key techniques in NM characterization:

12.4.1 Surface and structural analysis: TEM, SEM, and XRD

1. Transmission electron microscopy (TEM): The TEM offers a high-resolution picture of the structures of NMs at the atomic level. Researchers can study a particle's morphology, size, and shape. This is excellent for assessing uniformity of NPs and detecting structural defects, which form important applications in catalysis, drug delivery, and environmental remediation.
2. SEM: SEM captures high-resolution images of the surface morphology and texture of NMs, which may be used to identify surface features and particle aggregation. It is also useful in providing visualization of the interaction of NPs with another substance, possibly polymer matrices in composite materials.
3. X-ray diffraction (XRD): The XRD technique provides information about the crystalline structure of NMs and gives some lattice parameters, crystal size, and phase purity. Information about the crystalline and phase composition will be useful in predicting the material's stability and reactivity as well as the application requirements. In these, some typical examples are energy storage and structural reinforcement.

12.4.2 Optical properties: UV–Vis spectroscopy and FTIR

1. UV–visible spectroscopy: UV–vis spectroscopy examines optical properties of nanoscale materials, such as absorbance and bandgap, which have been found to be very important for applications, including photocatalytic and solar. From the absorbance pattern of light, it can predict the size and shape of NPs because both dimensions are related to their electronic behavior as well as the optical properties.
2. FTIR: FTIR detects specific molecular vibrations on the surface of NMs for functional group identification. FTIR is useful for assessing surface chemistry, organic capping agents, or biocompatible coatings that affect stability, dispersibility, and interaction with biological systems.

12.4.3 Zeta potential and particle size analysis

1. Zeta potential: The zeta potential is the measure used to find out the surface charge of the NP that states its stability in suspension. A high positive or negative value of the zeta potential usually shows that it has excellent stability due to electrostatic repulsion that prevents aggregation between the particles, particularly when the system needs to be held at a stable dispersion, like in drug delivery and treatment of wastewater.
2. Particle size analysis: This method, among others, analyzes the distribution of particle sizes for NMs and estimates the polydispersity index of an NM. Considering a number of applications, the particle size is strongly correlated with consistency and functionality in real applications because of catalytic efficiency, bioavailability, and reactivity properties.

12.4.4 Importance of standardizing characterization for waste-derived nanomaterials

Standardization of characterization techniques of waste-derived NMs becomes an important step in controlling the quality and reproducibility of the NMs with safe use. Because of the heterogeneous nature of waste, standardized procedures ensure that the NMs derived from the wastes, on their part, meet established criteria regarding purity, size, and integrity of structure. This will also provide a basis for interstudy comparison as well as satisfying regulation and instilling confidence in their application in industries and sectors. Standardized characterization can also aid in determining any possible contaminant or inhomogeneity due to sources of origin of the wastes utilized and further ensure the safe and reliable utilization of NMs derived from wastes in sustainable applications.

12.5 Applications of green nanomaterials from waste

Green NMs, synthesized from waste, hold immense potential across various fields due to their eco-friendly nature and low production costs. Here's a breakdown of their applications across environmental, energy, and medical sectors:

12.5.1 Environmental applications

12.5.1.1 Heavy metal removal and water purification

Waste-based green NMs represent such promising examples as iron oxide or biochar NPs, capable of producing highly efficient removal of contaminant heavy metals, such as lead, mercury, and arsenic, from water. Large surface area and favorable surface chemistry make them effective for adsorption of pollutants and very useful for the treatment of wastewater and purification of water. Use of NMsis a sustainable method to deal with water quality issue while decreasing reliance on chemical-based water treatments.

12.5.1.2 Air pollutant mitigation

Green NMs, like nano-TiO_2, produced also from industrial wastes, can photocatalyze the degradation of harmful air pollutants such as NO_x and VOCs. Such materials, in the presence of sunlight, can degrade the material into smaller, harmless byproducts through producing reactive species. Thus, there is an impetus toward clean air in cities. This is a cheap method that is effective for pollution control and is, therefore, very favorable for use in urban areas where the concern lies in the quality of air.

12.5.1.3 Soil remediation

Green NMs, extracted from agricultural or industrial waste, prove useful for soil remediation by encapsulating inside toxic compounds and heavy metals, thus not letting them into the food chain. Biomass waste-based NPs, especially biochar, possess wonderful adsorptive property and thus reduce soil contaminants. For example, such applications would be of high importance in areas prone to industrial activities or mining since it undoubtedly affects the agriculture sector and hence challenges the health of human beings.

12.5.2 Energy applications

12.5.2.1 Green nanomaterials in energy storage and conversion (e.g., supercapacitors and solar cells)

Due to the formation of green NM from waste, they have large surface area along with high conductivity, which is used for the synthesis of carbon nanotubes and graphene in the applications of energy storage devices. Some of these include batteries and supercapacitors. They have improved the capacity for energy storage while enabling a faster charge/discharge rate. It is applied both to supply and demand areas for renewable energy. Their applications lead to improved efficiency due to better light absorption, including solar cells. By their use in sunlight-based energy, their efficiency is cost-effective.

12.5.2.2 Catalytic applications toward biofuels and hydrogen production

Waste-derived nanocatalysts, which include metal oxides recovered from electronic waste, find applications in biofuel production and hydrogen generation. Catalysts increase the efficiency at which a reaction happens, meaning biomass can be converted into biofuels and water split for the generation of hydrogen as an energy resource. In this aspect, green NMs facilitate reduction of dependence on fossil fuels and promote clean energy transition.

12.5.3 Medical and pharmaceutical applications

12.5.3.1 Drug delivery systems

It has been extensively investigated that green NMs out of agro-waste-natural polymers can be used for drug delivery. Biodegradable encapsulation of drugs through NPs reduces the undesirable effects, thereby improving the efficiency of therapy, as they may be directly targeting tissues or cells. In this regard, the current trend of medical products that have biocompatibility and eco-friendliness is tuned in with the green route of drug delivery.

12.5.3.2 Antimicrobial applications

Waste materials can be used to synthesize antimicrobial NPs, such as silver or zinc oxide, which are applied against bacterial and fungal infections. These green NMs are inserted into coatings, dressings, and medical devices to thwart infection in various hospital environments. Usage of waste-derived NPs contributes to a new and more effective and cost-friendly sustainability in antimicrobial solutions against the rising threat of resistance to antibiotics.

12.5.3.3 Biocompatible scaffolds for tissue engineering

Waste-derived green NMs, such as nanocellulose from agricultural biomass, may also be used as biocompatible scaffolds for tissue engineering. These materials provide support to cells and the generation of tissues can aid in the development of bioengineered tissues and organs. Biocompatible and biodegradable NMs are perfect for sustainable regenerative medical applications, in accordance with the goals of green and sustainable healthcare.

12.6 Case studies and success stories

Green nanotechnology has already shown considerable potential for practical applications, with several successful case studies and implementations highlighting its value in industry, sustainability, and economic impact. Here's an exploration of key areas where green NMs derived from waste have demonstrated measurable success:

12.6.1 Successful industrial implementations

In different industries, sectors are discovering the possibility of waste-derived green NMs in their processes. For instance, here is a case from India that has been successful in using a biochar-based NM, extracted from agro-waste crop residues, for the efficient removal of heavy metals and organic pollutants in a wastewater treatment plant. It increased the quality of water as well as reduced the costs of treatment, so it became a cost-effective, environment-friendly solution compared to other traditional methods, for large-scale water purification. Apart from that, agricultural waste can be converted into a source of NM production. This reduces the impact of waste disposal and water contamination impact that might have arisen from its direct application or improper disposal.

Companies in the electronics sector are using silver NPs extracted from e-waste to manufacture coatings used for cleaning antivirus microbes off smartphones and health equipment. While it reduces toxic waste generated from electric products, it also facilitates production at lower cost. A top construction company has also adopted silica NPs, which are obtained from rice husk ash, an agricultural by-product in concrete mixtures. These NPs increase the strength, toughness, and lifetime of concrete and make it more resistant to environmental wear and tear. It has proven not only to be cost-effective but also to have reduced the company's carbon footprint considerably by using agricultural waste, thereby underlining the ways in which sustainable materials can be integrated into industries with immense demand. These examples demonstrate the significant range of waste-based green NMs that lead to sustainability but are still pioneering industrial innovation.

12.6.2 Comparative case studies of conventional versus waste-derived nanomaterials

Comparing traditional NMs with those fabricated from waste, it has been illustrated that there is an important potential for superior sustainability in green NMs. For example, comparing traditional TiO_2 NPs against TiO_2 NPs derived from industrial sludge, there is no apparent loss in photocatalytic activity for pollutant degradation in the air. However, the green NMs provided key environmental benefits; they were prepared from industrial waste, and the manufacturing of virgin raw materials was minimized along with the amount of waste generated. In addition, the production cost for NPs prepared from waste was less due to the low-cost feedstocks applied to prepare waste-derived TiO_2 materials, which made them both economical as well as more environmentally benign compared to the conventional TiO_2.

Another study compared the CNTs produced directly from the derivatives of fossil fuels with those produced from wastes of biomass. Biomass-derived CNTs acquired identical conductivity and strength characteristics, which are especially of extreme importance for applications in supercapacitors and batteries. Besides, the manufacturing of biomass-derived CNTs consumed far fewer resources: CNTs were prepared from readily available biomass wastes, whereas the synthesis involving fossil fuel derivatives needed to be refined into the desired carbon-based chemical. The synthesis process, in addition, caused less pollution and used smaller amounts of carbon. This chapter therefore pointed out that carbon nanotubes derived from waste could be effectively used as alternatives to the conventional fossil fuel-based material, without sacrificing performance capability, and are therefore an economically viable and sustainable option in the long run. In this regard, the development of waste-derived NMs serves to realize the demand for high-performance materials with minimal environmental impact.

12.6.3 Economic and environmental impact analysis

For instance, waste-derived NMs have provided significant economic and environmental implications concerning industrially seeking alternative sustainable materials compared to the more traditional ones. The synthesis of NMs using waste from different industries presents a dual advantage since it reduces drastically the use of raw materials, while solving problems related to the disposal of waste in the widest economy of the circular economy. An example is the inclusion of NPs from fly ash, a residue from coal combustion, in building materials, like concrete. Its adoption has proved cost-effective and environmentally helpful. Using NPs of fly ash in construction could reduce more than 30% of the raw material's costs against traditionally used materials, which often resort to virgin resources. In addition, this method reduces emissions from greenhouse gases by about 15%, as it minimizes the need for energy-intensive processes and reduces the carbon footprint associated with making traditional building materials. This example manifests the profound economic savings and environmental benefits that waste-derived NMs can confer on every industry.

Environmental impact and economic benefits are also offered by green NMs. In contrast to the traditional chemical method, more eco-friendly chemical processes have made waste-derived NM not dependent on toxic chemicals needed in the production process, such as the rice husk and corn stover, which are agricultural waste, converted into NPs and useful for soil remediation and water treatment purposes. This process reduces the demand for harmful chemical treatments and minimizes waste generation, resulting in an overall lower environmental footprint. Additionally, reusing waste materials to produce NMs saves companies money on landfill and incineration processes. In this regard, solid waste diverted from landfills results in environmental savings as well helps in resource conservation by companies. This would work to help in achieving the sustainability goals and supports shifting into a circular economy, which minimizes waste while valuably reusing worthwhile resources in a productive and environmentally responsible way.

12.7 Challenges and limitations in waste-derived green nanomaterials

While waste-derived green NMs offer promising solutions for sustainability, there are several challenges and limitations that must be addressed for broader adoption. Here's a breakdown of some primary hurdles:

12.7.1 Hurdles of regulation

The regulatory framework for green NMs from waste streams is yet very young. Most countries do not have specific regulations for NMs in terms of synthesis, handling, and disposal of NMs specifically from waste streams. Such gaps in regulations make it vague about what is safe, what environmental impacts to accept, and how these need to be regulated for quality, thus slowing down approval and limiting wider scope adoption. Waste-based NMs are also under much tougher scrutiny as compared to the traditionally sourced NMs due to apprehension in the regulatory agencies that the toxins from waste sources may leach out and are thus difficult to mass commercialize. Further, regulatory standards must be harmonized, especially in green NMs, to ensure innovation along with safety and efficacy.

12.7.2 Scale-up and commercialization barriers

The most significant challenge in scaling up production from laboratory scale to industrial scale is that large-scale waste-derived green NMs are very challenging. Due to the requirement of specific equipment for processing, continuous input for extracting and processing waste into NM form poses tough challenges. In addition, quality and performance of NMs synthesized from such diverse waste sources can be challenging to maintain due to varying composition of waste that may cause significant problems with reproducibility and consistency of the final products. Also, commercial feasibility is compromised by very high technology and infrastructure investments at the outset, along with the research and development required to adapt the optimized processes for any type of waste. In fact, cost effective and scalable synthesis methods would be necessary innovation to resolve these issues, and in-house industry partnership would be necessary to set up production.

12.7.3 Ethical and environmental concerns

Although green NMs are marketed and touted as the most sustainable, ethical and ecological issues still surround them. For instance, NMs recovered from the environment require energetic processing, which can be balanced if not sustainably managed. Some by-products of green NMs could end up in toxic classes or involve solvents with ecological risks. This has an ethical concern with the origin; while agricultural and industrial waste is are acceptable, biomedical waste raise questions on health safety risk as well as the risk of contamination.

More importantly, their effects when out of commission are yet to be fully appreciated. NPs can behave erratically in soil or water, where they remain for thousands of years and alter the very fabric of ecosystems. Such environmental effects require long-term observation periods, and tools of responsible disposal or recycling should be set up. Such balances between the commercial rewards of green nanotechnology and these ethical and environmental concerns are hence important for responsible innovation in this field.

12.8 Future directions and opportunities

The potential for waste-derived green NMs to contribute to sustainable development is vast, with advancements in technology and collaborative efforts promising to overcome current challenges. Here's a look at key opportunities that could shape the future of this field:

12.8.1 Advancement in artificial intelligence for waste-to-nanomaterial process optimization

Waste-no-NM production holds significant potential for AI applications in efficiency and optimization terms. Algorithms and predictive analytics from the developed machine learning techniques can determine what are termed optimum conditions for synthesizing NMs, such as the ideal temperature, pressure, and reaction time. AI-driven models predict more complex relationships between process variables and NM properties, allowing for a better yield with consistent product quality. It will also enable subtle patterns and correlations to be captured, which tend to go unnoticed with traditional methods in large datasets, thereby providing more refined and controlled NM synthesis.

Another way AI would help in the rationalization of the source is by recognizing new materials found in waste to be applied in NM production. It can also determine the composition of wastes and predict which will most likely produce the most effective NMs, with less reliance on traditional raw resources. AI's success in reducing experimentation through a trial-and-error approach will help speed up development of new techniques in synthesizing NMs, and hence waste-derived NM production becomes scalable and economical. Ultimately though, it is all of those and AI that are going to further drive the conditions of optimum production as well as predict outcomes toward more sustainable and efficient waste-to-NM process shifts [15].

12.8.2 Collaborative models for industrial waste recovery and nanomaterial production

A successful partnership between industrial and research establishments, as well as waste management firms, would unlock the entire potential of industrial waste to bring green NMs into being. Strategic partnering would help industries ensure a stable supply of invaluable waste like e-waste from the electronics sector or fly ash from power plants, reusable as NMs. It encourages a circular economy as it reuses the virgin resources and reduces the input to landfills. Waste management companies can sort and pre-treat the materials and research institutions can further purify and advance the synthesis methods to ensure NMs are sustainable and efficient.

To upscale such waste-to-NM processes, industrial symbiosis networks can be envisaged wherein the wastes of one sector could be utilized by another. The government could take initiatives in that direction by providing incentives, a regulatory framework for NMs, and standards or regulations for green NMs. This will not only ensure the quality and safety of the products but also encourage broader industry participation. Finally, these extrapolated models hold immense potential to mitigate some of the adverse environmental impacts, contribute positively to industrial sustainability, and promote large-scale commercialization of waste-derived green NMs.

12.8.3 Recommendation of policies for encouraging green nanomaterials

Supportive policies are essential in speeding up the research and development and deployment of green NMs emanating from waste. For instance, governments can encourage companies to invest in nanotechnology research and development by offering tax breaks, grants, or providing subsidies. Policies that may be introduced to foster the adoption of green NMs in various sectors such as construction, electronics, and healthcare could therefore make the demand stable enough, thereby encouraging companies to opt for greener alternatives. Besides, standards and accreditation of green NMs shall result in ensuring quality and safety for both consumers and producers. Rules that make regulation of waste recovery easier and those that remove obstacles facing the recycling of industrial waste increase the access rate to the material and further open up potential uses for the material.

Improvements in artificial intelligence, collaborative industrial models, and supportive policies can be leveraged to transform green nanotechnology into the future. AI can be used to enhance the processes for converting waste into NMs, thereby ensuring greater efficiency and scalability. Collaborative industrial models between industries, research institutions, and waste management companies ensure a steady supply of waste materials and promote resource efficiency. Incentivizing government policies that promote investment and usage of green nanotechnology, in addition to regulations allowing the recovery of waste, would overcome the challenges.

12.9 Conclusion

The journey of transforming waste into high-value NMs has proven to be a significant advancement in the realm of sustainable technology. Waste-to-NM technologies, especially when leveraging green nanotechnology, present a transformative opportunity not only to minimize environmental pollution but also to create valuable resources from what would otherwise be discarded. Here's an overview of the broader perspectives on this evolving field:

12.9.1 Waste-to-nanomaterial technologies: A high-value solution

High-value solutions can be observed in waste-to-NM technologies, as such technologies directly target two of the most pertinent global challenges: waste management and resource scarcity. Waste from industries, agriculture, and even healthcare can be converted into usable NMs for water treatment, energy storage, and health care applications. Extracting valuable material from waste contributes to circularity, where there is less wastage and the longevity of materials is increased. We upcycle waste from waste sources into useful products, generating a new material source that might be cheaper as well as more sustainable and environmentally friendly. This will curb dependence on virgin material significantly, minimize the environmental footprint, and spur innovation in material science.

12.9.2 Green nanotechnology and its vision for a circular economy

Green nanotechnology emphasizes work toward a circular economy, where better reutilization, refurbishment, and recycling of the same products, materials, and resources take place. Waste-derived green NMs in their diverse applications range from water treatment to energy conversion and are sensible steps in sustainability because the waste is not directed to the environment, and nonrenewable resources are also necessary only minimally. The green nanotechnology is also about creating materials in a circular economy that is ecologically friendly and cost-effective, with resource efficiency as the goal. This is because green nanotechnology helps construct a different economic model that is resilient, sustainable, and self-sustaining by encouraging the recycling of waste and environmentally benign synthesis methods, reducing impacts on the environment and promoting growth in key industries.

12.9.3 Key influences shaping the future of green materials

The following will shape the future of green materials:

1. Technological advances: The advancement of AI, machine learning, and advanced material synthesis techniques could improve the process from waste to NMs, thereby making it more efficient and scalable.
2. Policy support: The support of governments and rules by regulatory authorities will be crucial in the achievement of policy support, incentives, and the laying down of safety standards for promoting the implementation of green nanotechnology.
3. Collaborative networks: Interindustry, research institutions, and waste management organizations will increasingly accelerate the production of NMs from waste sources. The industrial symbiosis models would act as the anchors in developing sustainable low-cost sources of NMs.
4. Public awareness and consumer demand: This is necessary because, with awareness of sustainability, more consumers and industries would demand green products, including those produced from the waste-derived NMs.

12.9.4 Broader socioeconomic benefits

Socioeconomic benefits of green nanotechnology from waste: More importantly, this nanotechnology, through adoption, will give other significant socioeconomic benefits.

Economic growth: NMs from waste open new markets and opportunities in employment fields as diverse as waste management, material science, and nanotechnology. There are also the benefits to the user industries in lower costs and increased efficiency.

1. Environmental protection: Green nanotechnology reduces the environmental impact of industrial processes and consumer products by providing sustainable alternatives to traditional materials, thus countering the impacts of ecological degradation through waste disposal.
2. Social equity: Sustainable measures can help boost social equity since they permit access by local communities to new industries and emerging technologies dealing with waste recovery. It has created job opportunities, hence improving their dwelling conditions, especially in areas experiencing severe problems in managing waste.

12.10 Future thoughts

An amalgam of waste conversion into valuable NMs through green nanotechnology stands on the cusp of changing sustainability in numerous sectors. With the convergence of technological advancements, collaboration efforts, and supportive policies, the potential offered by this technology for transformation in a more sustainable circular economy cuts across varied socioeconomic benefits speaking to economic growth, environmental conservation, and others.

References

[1] Ahmad, R., et al. (2018). Immobilization of Aspergillus niger cellulase on multiwall carbon nanotubes for cellulose hydrolysis. *Bioresource Technology*, 252, 72–75.

[2] Barzinjy, A. A., Hamad, S. M., Aydın, S., Ahmed, M. H., & Hussain, F. H. (2020). Green and eco-friendly synthesis of Nickel oxide nanoparticles and its photocatalytic activity for methyl orange degradation. *Journal of Materials Science: Materials in Electronics*, 31, 11303–11316.

[3] Contreras, J. E., Rodriguez, E. A., & Taha-Tijerina, J. (2017). Nanotechnology applications for electrical transformers – A review. *Electric Power Systems Research*, 143, 573–584.

[4] Debnath, B., Majumdar, M., Bhowmik, M., Bhowmik, K. L., Debnath, A., & Roy, D. N. (2020). The effective adsorption of tetracycline onto zirconia nanoparticles synthesized by novel microbial green technology. *Journal of Environmental Management*, 261, 110235.

[5] Devanesan, S., Jayamala, M., AlSalhi, M. S., Umamaheshwari, S., & Ranjitsingh, A. J. A. (2021). Antimicrobial and anticancer properties of Carica papaya leaves derived di-methyl flubendazole mediated silver nanoparticles. *Journal of Infection and Public Health* , 14, 577–587.

[6] Fahimmunisha, B. A., Ishwarya, R., AlSalhi, M. S., Devanesan, S., Govindarajan, M., & Vaseeharan, B. (2020). Green fabrication, characterization and antibacterial potential of zinc oxide nanoparticles using Aloe socotrina leaf extract: A novel drug delivery approach. *Journal of Drug Delivery Science and Technology* , 55, 101465.

[7] Fang, C., Ma, Z., Chen, L., Li, H., Jiang, C., & Zhang, W. (2019). Biosynthesis of gold nanoparticles, characterization and their loading with zonisamide as a novel drug delivery system for the treatment of acute spinal cord injury. *Journal of Photochemistry and Photobiology B: Biology* , 190, 72–75.

[8] Gao, Y., Xu, D., Ren, D., Zeng, K., & Wu, X. (2020). Green synthesis of zinc oxide nanoparticles using Citrus sinensis peel extract and application to strawberry preservation: A comparison study. *LWT - Food Science and Technology*, 126, 109297.

[9] Jain, R., Mendiratta, S., Kumar, L., & Srivastava, A. (2021). Green synthesis of iron nanoparticles using Artocarpus heterophyllus peel extract and their application as a heterogeneous Fenton-like catalyst for the degradation of Fuchsin Basic dye. *Current Research in Green and Sustainable Chemistry* , 4, 100086.

[10] Kalaba, M. H., Moghannem, S. A., El-Hawary, A. S., Radwan, A. A., Sharaf, M. H., & Shaban, A. S. (2021). Green synthesized ZnO nanoparticles mediated by Streptomyces plicatus: Characterizations, antimicrobial and nematicidal activities and cytogenetic effects. *Plants*, 10, 1760.

[11] Kumar, S., Boro, J. C., Ray, D., Mukherjee, A., & Dutta, J. (2019). Bionanocomposite films of agar incorporated with ZnO nanoparticles as an active packaging material for shelf life extension of green grape. *Heliyon*, 5, e01867.

[12] Kumari, A., et al. (2021). Recent advances in carbon nanotubes-based biocatalysts and their applications. *Advances in Colloid and Interface Science*, 297, 102542–102545.

[13] Mubarak, N. M., et al. (2014). Immobilization of cellulase enzyme on functionalized multiwall carbon nanotubes. *Journal of Molecular Catalysis B: Enzymatic*, 107, 124–131.

[14] Owoseni-Fagbenro, K. A., Saifullah, S., Imran, M., Perveen, S., Rao, K., Fasina, T. M., Olasupo, I. A., Adams, L. A., Ali, I., & Shah, M. R. (2019). Egg proteins stabilized green silver nanoparticles as delivery system for hesperidin enhanced bactericidal potential against resistant S. aureus. *Journal of Drug Delivery Science and Technology*, 50, 347–354.

[15] Rashwan, A. K., Karim, N., Xu, Y., Hanafy, N. A. N., Li, B., Mehanni, A. H. E., Taha, E. M., & Chen, W. (2022). An updated and comprehensive review on the potential health effects of curcumin-encapsulated micro/nanoparticles. *Critical Reviews in Food Science and Nutrition*, 63, 9731–9751.

[16] Sangeetha, M., et al. (2022). Vibration, acoustic and emission characteristics of the chlorella vulgaris microalgae oil in compression ignition engine to mitigate environmental pollution. *Chemosphere*, 239, 133475–133479.

[17] Sekoai, P. T., Ouma, C. N. M., Du Preez, S. P., Modisha, P., Engelbrecht, N., Bessarabov, D. G., & Ghimire, A. (2019). Application of nanoparticles in biofuels: An overview. *Fuel*, 237, 380–397.

[18] Uddin, S., Safdar, L. B., Iqbal, J., Yaseen, T., Laila, S., Anwar, S., Abbasi, B. A., Saif, M. S., & Quraishi, U. M. (2021). Green synthesis of nickel oxide nanoparticles using leaf extract of Berberis balochistanica: Characterization, and diverse biological applications. *Microscopy Research and Technique*, 84, 2004–2016.

[19] Yusefi, M., Shameli, K., Yee, O. S., Teow, S. Y., Hedayatnasab, Z., Jahangirian, H., Webster, T. J., & Kuča, K. (2021). Green synthesis of Fe3O4 nanoparticles stabilized by a Garcinia mangostana fruit peel extract for hyperthermia and anticancer activities. *International Journal of Nanomedicine*, 16, 2515.

Rishi Raj*, Shivakshi, Anil Kumar, and Gaurav Khandelwal

Chapter 13
Environmental risk assessment of green material: A circular economy approach

Abstract: The chapter discusses the potential environmental risks that green materials can pose, despite their benefits in reducing waste and conserving resources. Green materials, while essential for sustainable development, need to be assessed for their toxicity, biodegradability, and ability to be recycled or reused effectively.

The chapter focuses on understanding these risks within the framework of a circular economy. A circular economy is a system that promotes reducing waste, recycling, and reusing materials to keep resources in use for as long as possible. The chapter also examines how green materials can fit into this system and whether they are truly sustainable in the long term.

Real-world examples are provided, showing how green materials have been successfully integrated into circular economy models. However, it also addresses the challenges in accurately assessing the environmental risks of these materials. The chapter highlights the limitations of current assessment methods and suggests improvements to make the process more reliable.

In addition to these challenges, the chapter explores strategies to reduce environmental risks. These include innovations in material design, better policies, and industry best practices to create safer, more sustainable green materials. Looking ahead, the chapter discusses emerging trends and technologies that could further improve green materials and their role in a circular economy. It also points out future research areas that can help strengthen the connection between green chemistry, circular economy, and environmental risk assessment.

By offering a clear overview of the benefits and potential risks of green materials, this chapter aims to guide future development, research, and policy decisions to support a more sustainable future.

Keywords: Green materials, environmental risk, circular economy, sustainability, biodegradability

***Corresponding author: Rishi Raj**, Amity Institute of Applied Sciences, Amity University, Noida, e-mail: r.rishi.0807@gmail.com
Shivakshi, Amity Institute of Applied Sciences, Amity University, Noida, India, e-mail: sharmashivakshi224@gmail.com
Anil Kumar, Department of Applied Sciences, Bharati Vidyapeeth's College of Engineering, New Delhi, India
Gaurav Khandelwal, Bio-inspired MEMS and Biomedical Devices — Engineering and Technology Institute Groningen, University of Groningen, Netherlands

https://doi.org/10.1515/9783111618906-013

13.1 Introduction

13.1.1 Definition and importance of green materials

Green materials, also referred to as sustainable or eco-friendly materials, are developed from renewable resources or designed to minimize their environmental impact throughout their life cycle, encompassing production, use, and disposal phases. These materials are expected to have minimal environmental impacts yet be fully functional, and hence are needed in the current pursuit of sustainability. Green materials can substitute conventional materials, which consume a lot of resources and energy and emit carbon in their production. Using green materials, industries can fight cases of environmental degradation, embrace the use of circular economy, and help in sustainable economic development. Besides ecological values, green materials have socioeconomic impact, such as promoting new developments in technologies, in industries including green construction where bamboo and recycled steel are used instead of timber and metals as they provide high strength and are environment-friendly [68, 71].

Green materials are important in many industries, for example, in packaging, electronics, and many others where use of sustainable materials contributes to the conservation of virgin resources and reduces pollution [72]. For instance, biopolymers in packaging like polylactic acid (PLA) replace petrochemical-based plastics and lower down the concept of waste. The progress and embedding of green materials within related industries not only serve overall environmental aims and purposes but they also facilitate socioeconomic objectives, which promote sustainable structure of the economy toward production and consumption [60].

13.1.2 Principles of circular economy

The circular economy (CE) is an economic model designed to decouple growth from resource use by promoting the continuous circulation of resources [61]. In contrast to the linear economy, which is structured as a "take–make–dispose" process, the CE emphasizes recycling, reusing, and remanufacturing, aiming to retain resources within the production cycle for as long as possible. This model supports green material as they are designed for reuse, recyclability, and biodegradability in closed-loop systems that promote the optimization of resources to minimize wastage (Figure 13.1). The CE [43] now has certain principles such as designing waste out, using materials more often, and reestablishing natural systems; the products should either be safely released back into the ecosystem or give value to the next user, which can help in realizing sustainable development objectives [13].

In practice, circular design is implemented through product strategies that allow disassembly and repurposing. For example, using biodegradable materials that can decompose without polluting ecosystems embodies the principle of designing out waste. The CE framework not only aligns with environmental objectives but also provides economic benefits by reducing the need for raw materials and fostering resilience against resource scarcity [59]. Thus, the transition towards a CE is a strategic imperative for sustainable development, and green materials play a pivotal role in realizing this vision by ensuring that resources are maximally utilized within the economy before being reintegrated or safely decomposed [32, 67].

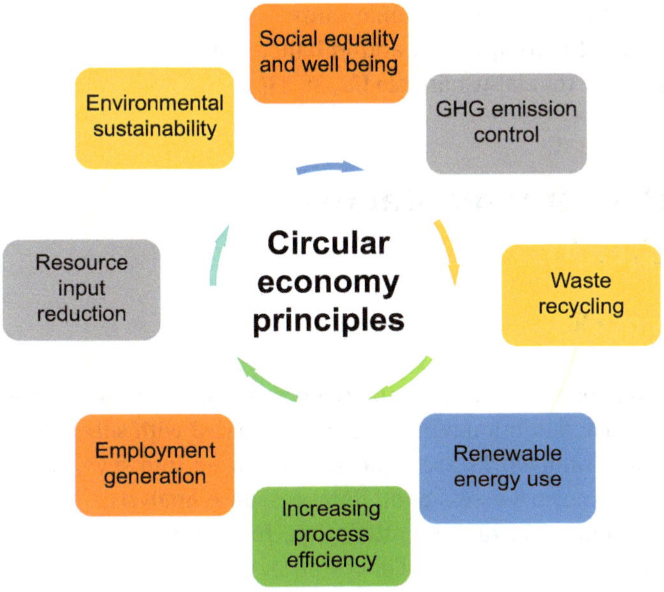

Figure 13.1: Principles of circular economy for waste management [35].

13.1.3 The role of environmental risk assessment in sustainability

Environmental risk assessment (ERA) is a critical tool used to evaluate the potential adverse effects that substances, activities, or processes may have on the environment, particularly in the context of green materials. While green materials are developed with sustainability in mind [17], it is essential to ensure that they do not inadvertently introduce new environmental hazards. ERA helps in an organized evaluation of different forms of risk such as toxicity, biodegradability, and long-term environmental impact; hence ERA assists in getting a true perception of whether indeed these materials are environmentally friendly or whether they are merely exporting the risks elsewhere. For instance, while biodegradable polymer can degrade into harmless components under one environmental setting, they can also produce microplastics or toxic leachates under another environmental setting implying the need for a life cycle assessment (LCA) [9, 64].

In this way, using ERA, decision-makers, and industry managers can understand which green materials are useful and what threats they can pose to ecology when implementing decoupling. This means that all the materials bearing the green label are safe for use while at the same time advance the cause of sustainability. Furthermore, policies of ERA continue to integrate product development with legislative and regulatory requirement, ensuring enhanced industry responsibility and public confidence in sustainable products. Through ERA the promotion of green building materials within a CE framework means that ERA initiates the support of sustainable transitions whilst offering assurance of environmental compatibility and social acceptance to prevent any negative ecological consequences and enhance the development of a steady, minimal-impact economy that is sustainable [22, 31, 33].

13.2 Overview of environmental risk assessment (ERA)

13.2.1 Core concepts of ERA

ERA is a fundamental methodology in environmental science, enabling the systematic evaluation of potential ecological and human health risks associated with substances, processes, and activities. ERA frameworks are typically structured into several key stages: hazard identification, exposure assessment, dose–response analysis, and risk characterization. The initial phase, hazard identification, focuses on determining whether a specific agent poses a potential risk. For instance, toxic substances, emissions, or industrial processes are assessed for their potential to cause ecological disruption or health issues. This stage is followed by exposure assessment, which quantifies the frequency and intensity of environmental receptor contact, such as exposure of soil, water bodies, or species to contaminants [54].

In the dose–response analysis, ERA considers the levels of exposure and the chances of adverse result occurrences and use empirical data to establish the risk severity. It is at this stage that assessments are made to convert evidence into information that is useful, especially for regulatory agencies and policymakers who depend on risk likelihood calculations. Finally, risk characterization is the integration of the data obtained from the three previous steps to provide an enhanced understanding of the risk to be faced to develop appropriate plans for risk management in case of environmental risk [3]. Due to its broad-spectrum approach ERA is an effective instrument in strategic environmental management, providing a scientific background for norms and requirements that have to be met to reduce adverse effects on the environment [21, 66].

13.2.2 Significance of life cycle assessment (LCA) in green material evaluation

LCA is an integral part of the ERA process, especially when evaluating the sustainability of green materials. LCA provides a comprehensive, systematic approach to assess environmental impacts across all stages of a material's life cycle, including extraction, manufacturing, use, and disposal. Integrating LCA into ERA enhances the analytical depth by incorporating a full life-cycle perspective, ensuring that environmental risks are not shifted across different phases, a phenomenon known as "burden shifting." For instance, materials considered eco-friendly in their production phase might generate harmful by-products during degradation, especially if not properly evaluated. Through LCA, industries and regulators can gain insights into cumulative environmental impacts, facilitating informed decision-making and supporting CE principles.

In this context, LCA enables stakeholders to design and select green materials that meet sustainability standards without unintended ecological consequences. The framework allows for the evaluation of metrics such as greenhouse gas emissions, energy consumption, and pollution levels throughout a product's life. By identifying hotspots in resource consumption and waste generation, LCA helps industries design more sustainable materials and products aligned with environmental policies and sustainable business practices [47].

13.2.3 Challenges in assessing environmental hazards

Despite the critical importance of ERA, several challenges complicate its application, especially for new and emerging materials. One significant challenge is data variability, as environmental conditions vary extensively across regions and ecosystems, making exposure and dose–response analyses less predictable and thus reducing the reliability of ERA models. Uncertainty in environmental data, influenced by complex variables such as climate, geographic location, and species diversity, can introduce inaccuracies into risk assessments [37]. Additionally, many ERA models were developed for traditional pollutants and are less effective at evaluating complex, modern materials like biodegradable polymers or nanomaterials [8].

Another significant problem is the absence of universally accepted procedures for evaluating green materials, many of which are characterized by specific degradation characteristics unacknowledged by existing models. Some of the materials, for instance, biodegradable ones may decompose into microplastics or other secondary pollutants posing long-term ecological risks if not well evaluated. Bridging these gaps requires the cross-cutting of interdisciplinary cooperation like infusing real-time data analysis on assessment models and defining new frameworks for new materials. In addition, to support sustainability and green materials, there is a need to adapt existing guidelines and make up for new ones so that the risk assessment paradigms grow in parallel with the material science field [69].

By incorporating LCA and leveraging emerging technologies within ERA frameworks, environmental assessments of green materials can be enhanced to ensure they meet both ecological and regulatory standards. This combined approach offers a robust, multidimensional framework for evaluating green materials, preventing ecological risks, and supporting sustainable innovation.

13.3 Toxicity and environmental impacts

13.3.1 Toxicity assessment techniques

Toxicity assessment plays a crucial role in identifying the potential hazards associated with green materials, both to ecosystems and human health. Various methodologies – such as in vitro, in vivo, and computational techniques – are employed in this analysis to provide a comprehensive understanding of toxicity levels. In vitro assays, like the MTT and neutral red uptake assays enable initial screenings to evaluate cellular viability and cytotoxicity upon exposure to potential toxicants. Such assays offer a less invasive means to measure the biochemical impact of substances on cells, offering preliminary insights that can inform more extensive evaluations [48].

Continued application of LCA and integration of the emerging technologies in the assessment of green materials and their compliance in ecosystems to regulatory compliances can be improved through integration of ERA toolkits. This combined approach provides a comprehensive, versatile approach to assess green materials and mitigate ecological threats and enable sustainability [19].

13.3.2 Ecotoxicity and biodegradation pathways

Ecotoxicity assessment is essential to understand the impacts of green materials within aquatic and terrestrial ecosystems [5, 25]. Using indicator species like *Daphnia magna* and zebrafish, researchers can evaluate both acute and chronic effects under controlled exposure conditions. These organisms help gauge the relationship between exposure levels and health, behavioral, and reproductive outcomes, shedding light on the potential ecosystem-wide impacts of green materials [6, 63].

Biodegradation pathways are equally vital, as they determine how materials break down in natural environments. Ideally, biodegradable substances should decompose completely into nontoxic by-products like carbon dioxide and water. However, partial degradation can produce persistent organic pollutants (POPs) or microplastics, which may accumulate and harm ecological systems over time. Following

OECD guidelines, biodegradability testing methodologies assess degradation rates and completeness, ensuring that materials degrade safely under various environmental conditions, minimizing ecological harm [49].

13.3.3 Quantitative tools for long-term impact analysis

To assess the long-term environmental impact of green materials, quantitative tools are indispensable. One prominent model is USEtox, a UNEP-SETAC model [38] used in LCAs that provides characterization factors to evaluate human toxicity and freshwater ecotoxicity. USEtox integrates exposure and effect factors to estimate potential risks across diverse scenarios and product life stages, thereby supporting comprehensive environmental assessments. This model aids in determining potential hazards and in guiding the sustainable adoption of green materials [14].

Further, toxicity of new chemicals can be predicted by the QSAR models or the quantitative structure–activity relationship. Since QSAR models can predict toxicological properties, their evaluation reduces risk at the research stage of new materials. In parallel with these models, probabilistic risk assessment frameworks can accommodate variability and uncertainties of data and make predictions on long-term effects more accurate [53]. In total, the integration of toxicity assessment techniques, ecotoxicity evaluations, and sophisticated modeling tools guarantees that green materials are sustainable with the least harm to the environment [15].

13.4 Circular economy and resource efficiency

13.4.1 Designing for recyclability and reuse

Incorporating recyclability and reuse at the design stage is pivotal for a CE. This approach emphasizes creating products and materials that maintain functionality and structural integrity over several life cycles, thus reducing environmental impact and conserving natural resources. By designing for ease of disassembly and material recovery, products can seamlessly reenter production cycles. There is, for instance, the advantage of efficient extraction and reuse of components in such designs as the modular design, which is plenty in the electronics and construction industries. New nanotechnologies have led to the creation of green material such as biopolymers, PLA to name but a few, which under certain conditions break down into harmless products, thereby alleviating the pressure on the landfill and emission of greenhouse gases as couples to traditional materials [30].

Such approaches are in tandem with extended producer responsibility (EPR), which holds manufacturers responsible for the management of products that they introduce into the market. It also helps in sustainable business archetypes where recyclable design reduces costs of disposing wastes with the environment [55]. At the production level, the use of environmentally safe materials forms a closed cycle of reusable material, linking the green environment with economic returns.

13.4.2 Measuring resource efficiency of green materials

Resource efficiency is a crucial measure of sustainability within a CE, focusing on optimizing resource input to maximize outputs while minimizing environmental impact[34]. Tools like material flow analysis (MFA) and LCA are widely used to assess resource flows and the environmental footprint of materials across all stages of production, use, and disposal [44]. For instance, these tools help quantify energy usage, waste generation, and recycling potential, which are essential for understanding the full scope of a material's sustainability.

Metrics like the energy return on investment (EROI) and circularity indicators provide valuable insights, quantifying the percentage of recycled content in products versus virgin material usage [12]. Such data not only supports sustainable manufacturing practices but also facilitates the development of market value-based indicators that integrate resource efficiency with economic viability. These indicators allow businesses and policymakers to assess the environmental and economic benefits of resource-efficient products [1], advancing both resource optimization and economic sustainability.

13.4.3 Strategies for waste minimization and resource recovery

Achieving a CE involves robust waste minimization strategies, which range from optimizing production processes to establishing closed-loop systems where waste from one industry serves as a resource for another. The concept of industrial symbiosis, demonstrated in eco-industrial parks, epitomizes this approach by creating networks where by-products are shared across industries, reducing waste and enhancing resource efficiency. This cross-sector collaboration exemplifies sustainable resource management, as seen in systems where heat, water, or materials generated as waste by one process are repurposed by another [20].

Advanced recovery technologies, including pyrolysis and hydrothermal carbonization, further enhance resource recovery by transforming waste into valuable secondary raw materials or energy. Such technologies support a zero-waste framework, converting end-of-life products into inputs for new production cycles. In addition, policy tools like deposit-refund systems incentivize consumer participation in recycling programs, significantly enhancing material recovery rates and supporting a closed-loop economy [65]. To fully realize these goals, collaboration between policymakers, industries, and consumers is essential, with public policies promoting tax incentives for recycled products and penalties for nonrecyclable waste. These measures underscore the importance of sustainable waste management and resource efficiency as pillars of the CE.

13.5 Challenges in current environmental risk assessment (ERA) approaches

13.5.1 Limitations of existing tools and data variability

The most critical issue, which ERA current methods and frameworks confront today, is the high variability and restricted application of tools and databases. These models, which have been established for evaluating typical pollutants, are not capable of providing precise solutions to multifaceted environmental processes and newly arising substances, such as microplastics and engineered nanomaterials. ERA methodologies are usually developed based on more or less simple assumptions [57], which do not always accurately reflect the interactions in the ecological systems. These assumptions pose problems when estimating interactions in different ecological environments whereby species' vulnerability and environmental conditions in different regions affect risk characterization. The kind of data variability often caused by differences in the exposure data from geographically and temporally different sites can add further layers of complexity in the assessments by reducing the extent of cross-ecosystem comparability of the results.

These weaknesses are even more evident when treating chemical mixtures and joint impacts of numerous stressors, as most of the current ERA tools are not aimed at complex matrices. Evaluating the effects of chemical combinations is particularly difficult because certain components of a mixture may enhance, inhibit, or simply summate the effects of another component. While there are strengths of the current system, this has four primary sources of weakness, including the following: Besides these factors, toxicological databases continually have issues with data deficiency especially when dealing with newly introduced materials in the market as well as new pollutants in the environment, which leads to overestimation of risks or underestimation of the same when conducting assessments. This deficiency in high-resolution information constrains the ERA, causing doubt about results concerning environmental security and effectiveness of the employed safeguards [2].

13.5.2 Methodological gaps and uncertainties

Current ERA methodologies face considerable uncertainties that arise from qualitative and quantitative limitations, impacting the robustness of risk predictions. For example, traditional toxicological models that focus on acute toxicity do not always adequately capture chronic exposure effects, nor do they account for the cumulative impact of multiple pollutants across varying environmental media. There is also a significant challenge in extrapolating laboratory results to real-world settings. Laboratory conditions often do not reflect the complexities of natural ecosystems, leading to uncertainty in how results translate to broader environmental contexts.

A key methodological gap is the slow adoption of advanced computational technologies, such as predictive toxicology and machine learning algorithms, which offer the potential for more accurate and faster risk assessments [18]. Although these tools are promising, their application is hindered by validation challenges, interpretability issues, and the need for comprehensive and high-quality input data. Regulatory frameworks often struggle to adapt to rapid advancements in scientific tools, creating a regulatory lag that leaves many ERA practices outdated or inconsistent across jurisdictions. This lag in regulatory adaptation contributes to inconsistencies in risk assessment protocols and results, making it difficult to establish clear and effective environmental protection standards.

13.5.3 Assessing impacts across circular economy phases

ERA faces unique challenges when applied within a CE framework, which promotes resource loops through recycling, reuse, and remanufacturing. Traditional risk assessment models are primarily designed for linear product lifecycles that extend from production to disposal. In contrast, CE models necessitate repeated product life cycles, requiring that ERA approaches adapt to consider cumulative environmental impacts across multiple reuse and recycling phases. For instance, recycled materials may contain residual contaminants or degrade over time, complicating risk assessments due to the dynamic changes in toxicity and environmental interaction as materials are repurposed [16].

Moreover, secondary materials, including those recycled or remanufactured often exhibit properties that diverge from virgin materials. These properties necessitate separate evaluations since standard toxicological data may not apply to recycled counterparts. Effective ERA in a circular context requires dynamic LCAs capable of tracking cumulative environmental impacts across multiple feedback loops. This transition toward integrated assessments calls for adopting holistic approaches, such as system dynamics modeling, which can capture the complex interplay of environmental, economic, and social factors within closed-loop systems. Such a shift is critical to ensure that CE practices do not inadvertently lead to new environmental hazards through the reintroduction of unassessed or inadequately managed risks.

13.6 Case studies

13.6.1 Green building materials: environmental and risk analysis

Green building materials are some of the most impactful tools in sustainable construction processes, selecting materials that decrease overall carbon emissions, use resources more efficiently, and providing higher indoor air quality than conventional building materials. The environmental assessment of these materials covers their life cycle from the extraction and processing stages to disposal. Some of them include the recycled steel, sustainable timber, and fly ash concrete. Most scrap iron and timber have the potential to reduce reliance on virgin materials and assist in decreasing the levels of emission of greenhouse gases, particularly in the construction industry, which is highly energy-consuming. Reducing landfill waste encapsulates fly ash, a combustion by-product of coal, which is creative in its use as a building material and resource-efficient [73].

However, there are some challenges (Table 13.1) that need to be overcome when practicing the developed strategies. Still, one of the risks is the incomprehensiveness of materials, which can create performance fluctuations and possible environmental deterioration [51, 52]. For example, if there are chemicals like residual processing chemicals or heavy metals, then they may, over time, leach and cause problems in the soil and water [10]. Hence, to assess such risks, overall ERAs should be made in order to avoid such risks (especially where the material in question is of a varying composition or contained additives). Through thorough LCA, green building projects can estimate the actual environmental impacts and integrate risk management methods in deciding on and using such materials in projects to impose only positive contributions to sustainable development without bringing about other unwanted ecological consequences [27].

Table 13.1: Key challenges and limitations in current environmental risk assessment (ERA) approaches.

Challenge area	Description	Common metrics	Statistics	Reference
Tool limitations	ERA models struggle to capture complex chemical mixtures and emerging contaminants.	Proportion of contaminants not covered by standard ERA tools; percentage of chemical mixtures requiring adjustments.	~50% of emerging contaminants lack specific risk profiles in standard ERA tools.	50
Data variability	Differences in geographical, temporal, and methodological data that affect assessment accuracy.	Range of exposure level variation by region; frequency of adjustments due to local ecological differences.	Exposure levels vary by 30–70% regionally for similar contaminants.	51
Uncertainties in exploration	Difficulty in applying lab-based results to real-world settings, leading to predictive uncertainties.	Rate of predictive error when moving from lab to field; percentage decrease in accuracy from lab to field assessments.	Predictive error ranges from 15–50% in toxicology extrapolations.	52
Computational gaps in emerging technology	Limited adoption of advanced tools like AI or ML for predictive toxicology due to validation issues.	Adoption rate of AI models in ERA; accuracy of AI models in limited studies vs. traditional models.	<20% of ERA assessments currently use AI; predictive accuracy gains estimated at 25% in limited AI-based studies.	53
Circular economy life cycle challenges	Complexities in assessing recycled materials with unknown contaminant profiles.	Percentage increase in assessment complexity for secondary vs virgin materials; cost increase for complex assessments.	Life cycle complexity in circular models increases costs by ~30%, with unknown contaminants requiring additional screening.	54
Regulatory lag	Slow adaptation of regulatory frameworks to new scientific advancements, limiting consistency.	Average time lag (years) for regulatory updates; proportion of ERA protocols awaiting regulatory review.	Regulatory frameworks average a 5-year lag for incorporating new ERA methodologies.	55
Toxicological data gaps	Lack of toxicological profiles for emerging pollutants, leading to risk underestimation.	Percentage of emerging pollutants with incomplete profiles; frequency of over- or underestimation in assessments.	40–60% of new materials lack comprehensive profiles, leading to underestimation of cumulative impacts in ~30% of cases.	56

13.6.2 Biodegradable polymers in practice

Biodegradable polymers are increasingly being used to address the persistent plastic waste problem, particularly in applications such as packaging, agriculture, and the medical field. Unlike conventional plastics, biodegradable polymers such as PLA and polyhydroxyalkanoates (PHAs) are designed to decompose into natural components under specific environmental conditions, minimizing long-term ecological impacts. These polymers are typically derived from renewable biomass sources, reducing reliance on fossil fuels and aligning with CE objectives [7].

However, the environmental benefits of biodegradable polymers depend heavily on the conditions of their disposal and degradation pathways. In natural environments lacking controlled composting conditions, the breakdown process may be incomplete, potentially leading to microplastics that can persist and accumulate in aquatic ecosystems. Research demonstrates that additives in biodegradable polymers may leach during decomposition, affecting microbial activity and potentially releasing toxic by-products into water systems. LCAs have proven instrumental in evaluating the true environmental impacts of biodegradable polymers, identifying the most suitable disposal environments, and ensuring that these materials fulfil their sustainability promise. By applying LCA, researchers can scrutinize each phase, from production to disposal, thereby confirming that the adoption of biodegradable polymers does not shift environmental burdens elsewhere.

13.6.3 Electronic waste recovery: green solutions

As consumer demand for electronics continues to surge, electronic waste (e-waste) management has become a crucial focus for sustainable material recovery. E-waste contains valuable resources [28], including gold, silver, and rare earth elements, yet improper disposal leads to the release of hazardous substances such as lead and cadmium, which pose serious health and environmental risks. Green recovery solutions seek to minimize these impacts by promoting strategies for resource conservation, such as recycling and reusing precious metals through environmentally responsible techniques [39]. Processes like pyrometallurgy and hydrometallurgy efficiently recover metals while minimizing environmental impact by reducing reliance on raw mineral extraction [4].

Despite advances, the safe management of e-waste still presents challenges. For instance, toxic residues can accumulate in recycled materials, leading to potential health risks if not carefully processed and contained. To address this, the implementation of safer materials, like lead-free solder and biodegradable circuit boards [42], is gaining traction among industry leaders who aim to reduce toxicity within electronic products. Effective e-waste recovery also requires robust recycling infrastructure and regulatory support, facilitating a closed-loop system where recovered materials are reincorporated into new products. By integrating these green solutions and adhering to international recycling standards, industries can better manage e-waste and reduce the overall environmental footprint of electronics manufacturing.

13.7 Strategies for risk mitigation

13.7.1 Innovative approaches for low-risk material design

To advance low-risk material development, industries are increasingly adopting innovative design approaches rooted in green chemistry and sustainable engineering [58]. Central to these innovations is the "safe-by-design" framework, which emphasizes reducing potential hazards at the molecular level before a material reaches production. This includes the development of bio-based polymers and biocompatible composites that break down into nontoxic by-products, effectively minimizing environmental persistence and toxicity. Advanced design also includes molecular simulation techniques that help scientists predict material behavior and optimize properties, such as biodegradability and resource efficiency [70, 74].

Nanotechnology enhances low risk design by the development of products that are efficient but have minimal impacts on the environment. For instance, nanostructured biocompatible coatings are under development for biodegradable polymers that will require increased stability in controlled conditions and the ability to degrade in natural environment where they are needed. Predictive toxicology, that is utilizing AI-based modeling, is one of them. It helps predict certain dangers inherent to new materials, including nanoparticles, thus minimizing such negative outcomes as too many adverse effects of mass production. All the above technological innovations show a move to predictive risk management contrary to the traditional risk management, which is defensive, in line with sustainable development.

13.7.2 Regulatory and policy guidelines

Regulations play a critical role in guiding industries towards safe and sustainable material use (Figure 13.2). One of the most comprehensive regulatory frameworks is the European Union's REACH (registration, evaluation, authorization, and restriction of chemicals) legislation, which mandates rigorous ERA for all chemicals produced or imported in large quantities [36]. This regulatory approach requires companies to gather extensive data on the properties and ecological impacts of substances, helping to ensure environmental safety across material lifecycles. Additionally, global standards like ISO 14001, which focus on environmental management systems, encourage companies to incorporate sustainability into their core operational frameworks, ensuring continuous monitoring and improvement of environmental impact [23].

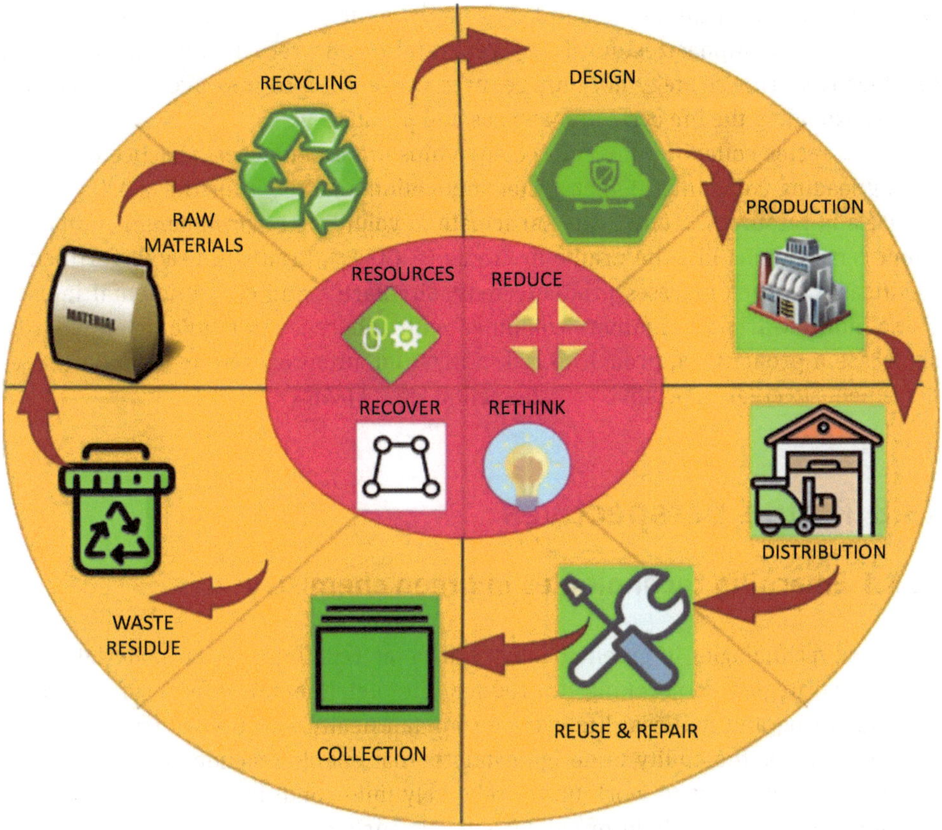

Figure 13.2: Green environment methods promoting sustainable development [17, 56].

Policies supporting EPR are also emerging as important, since they make manufacturers responsible for the final stage of the product's life cycle [11]. This makes the companies engage in designing for recyclability, reduction of wastes, and the impacts that they have on the environment thus completing the circularity within the CE. EPR policies promote sustainable material development and excellent waste management for which reasons they must be adopted to reduce the impact of industrial processes on the environment [34].

13.7.3 Best practices from industry leaders

Industry leaders like Patagonia, IKEA, and Philips have become benchmarks in environmental risk mitigation by integrating sustainable practices into their operations and value chains. For example, Patagonia utilizes recycled materials in product manufacturing, supporting closed-loop recycling, and reduced dependency on virgin resources. IKEA, on the other hand, has implemented a “circular product design” strategy, which emphasizes durability, reparability, and recyclability, aligning with CE objectives. This strategy not only reduces waste but also improves resource efficiency, extending the life cycle of materials and products [62].

Cross-sector collaborations and partnerships are another best practice observed among leading companies, which foster innovation and allow for the sharing of expertise and resources. Many firms participate in voluntary environmental audits and adopt certifications like the Cradle to Cradle (C2C) (e.g., Cradle to Cradle Certified™) certification, which assesses product safety, circularity, and environmental impact. These industry-driven initiatives, combined with adherence to regulatory standards, showcase a proactive approach to risk mitigation, often exceeding compliance requirements to create a positive environmental legacy [45].

13.8 Future perspectives

13.8.1 Emerging technologies in green chemistry

Advanced technologies in green chemistry are now redefining green material development strategies by seeking to reduce the detrimental effects of chemical manufacturing and material life cycle on human health and the environment [15]. Central to this is the ability to design catalysts that enable reactions that usually require harsh conditions to work under relatively mild conditions in terms of energy and solvent usage. The field of catalysis has branched into biocatalysis where enzymes are used to cause chemical reactions that normally require high heat or extreme chemicals and photocatalysis where light is used instead of heat to induce chemical reactions. One of the more important emerging trends is in biopolymers and biodegradable replacements for the more traditional plastics; these include PLA and PHAs, polymers which degrade naturally in the environment and that have been developed to lessen the use of plastics and dependence on petroleum. Additionally, biorefineries are integrating waste biomass to create renewable chemicals and biofuels, aligning with the CE by turning waste into valuable resources [50].

Electrochemical synthesis and green solvents such as supercritical fluids and ionic liquids are other pillars of green chemistry [14, 46]. These technologies reduce the emission of volatile organic compounds (VOCs) and can be reused, significantly lowering environmental toxicity. As these technologies continue to mature, their integration into mainstream production can help industry meet both economic and environmental objectives, particularly as regulatory frameworks around sustainable practices evolve [40]. Emerging tools in green chemistry are not just limited to improving processes but also extend to the assessment phase, with predictive toxicology tools that model environmental impact, facilitating the safer design of materials from the outset [31].

13.8.2 Nanotechnology and its impact on sustainability

Nanotechnology offers groundbreaking potential for enhancing sustainability, particularly through its applications in materials science and environmental remediation. Nanoparticles and nanoscale materials exhibit unique properties that make them highly effective in catalysis, pollutant removal, and energy storage, among other applications. For instance, nanostructured catalysts can improve reaction rates and selectivity in chemical processes, reducing energy consumption and waste production. Additionally, nanomaterials such as nano-silver, carbon nanotubes, and titanium dioxide are increasingly being used in water treatment and air purification, where they can capture and neutralize contaminants at a molecular level, offering solutions for persistent environmental pollutants.

Nevertheless, the application of nanomaterials in the production of consumer goods and various industries, bring up several questions concerning their effects on environment and health of people as the nanoparticles possess different characteristics in ecosystems in comparison with the large particles. Currently, new advanced and detailed ERAs that would take into consideration nanoscale materials are becoming more necessary. Researchers argue that *N*-related products should be subjected to LCAs so that any negative impact may be predicted and avoided. For example, some nanomaterials are potentially bio-accumulative or bio-persistent that necessitates that their behavior within ecosystems is assessed. Due to the current focus on the safe disposal and recycling of nanomaterials, nanotechnology has been found to have the potential to provide sustainable value proposition while addressing its risks [26].

13.8.3 Future directions for risk assessment in circular frameworks

As sustainable materials and processes become essential to addressing climate challenges, risk assessment frameworks must evolve to address the complexities of a CE. Traditional risk assessment models, which evaluate products linearly from production to disposal, need to expand to accommodate circular loops, where products are reused, recycled, or remanufactured [41]. Incorporating LCA within these frameworks allows for a broader perspective on environmental impacts across multiple life cycle stages, preventing "burden-shifting" where impacts are simply moved from one phase to another without reducing the overall environmental footprint [29].

Integrated systems risk assessment as well as digital twins are some promising solutions provided by the advanced modeling techniques to model interactions of the materials and closed-loop systems. Another emerging class of tools is the use of smart technologies, AI and ML, to analyze large sets of data in order to predict how materials will behave or their effects in different environmental conditions. Additionally, the use of biosensors for real-time monitoring that forms part of dynamic ERA is becoming increasingly important as materials recycle through industries in different forms including post-consumer recycled plastics or metals. As the limit of environmental laws tightens, integrating the ERA methods with circular approaches require that recycled or reused products undergo proper testing for safety, that is, sustainable development is achieved with no danger to the environment.

13.9 Conclusion

13.9.1 Key insights and takeaways

In examining the ERA of green materials within a CE context, several critical insights emerge that highlight both the challenges and opportunities associated with sustainable material innovation. Green materials, defined by their minimal environmental impact throughout production, use, and disposal, are instrumental in reducing the reliance on nonrenewable resources and mitigating pollution. Yet, as this chapter reveals, the development and adoption of these materials require a thorough, iterative approach to ERA to ensure that the transition to green alternatives does not inadvertently create new environmental risks. LCA is another important tool contributing to the common understanding by assessing all impacts pointing to the necessity of systematic approach to the assessment of green material so that it does not shift the burden from one life cycle phase to another phase without actually contributing to overall environmental improvement. Key advancements, such as catalytic processes in green chemistry and nanoscale innovations, demonstrate how science can drive material sustainability. However, these advancements also reinforce the need for advanced, dynamic ERA methods that can evolve with emerging technologies.

13.9.2 Bridging green chemistry with circular economy and ERA

The alignment of green chemistry principles with CE models and ERA frameworks marks a transformative shift in how sustainable materials are conceptualized and assessed. Green chemistry inherently supports the CE by focusing on the design of chemical processes and materials that eliminate waste, reduce toxicity, and maximize resource efficiency. This synergy allows for materials to be reused, recycled, or composted within a circular framework, thus reducing waste and conserving resources. ERA serves as the scientific foundation for this integration by providing systematic risk assessments that ensure materials remain environmentally safe throughout their extended lifecycles. Moreover, the incorporation of CE principles into ERA allows industries to consider the entire life cycle of a material, including post-use phases, thus fostering closed-loop systems that support sustainability at scale. As regulations increasingly favor sustainable materials, industries that integrate green chemistry with ERA and CE principles will be better positioned to meet both environmental standards and market demand for eco-friendly products.

13.9.3 Areas for future research

Future research in the ERA of green materials, especially within a CE context, should focus on several key areas to address emerging challenges and advance the field. First, there is a need to develop ERA models that can accommodate the complex life cycle loops of circular materials, accounting for factors like material degradation, toxicity shifts, and environmental accumulation in secondary applications. Further, as nanotechnology and other advanced materials become integral to green innovation, research should prioritize the creation of risk assessment tools specific to nanoscale materials and novel compounds. Additionally, the integration of real-time monitoring technologies, such as biosensors, and predictive modeling using AI could transform ERA by providing more accurate, up-to-date risk profiles. Exploring the socioeconomic dimensions of green materials, such as public acceptance, regulatory impacts, and cost-benefit analyses, will also be essential to guide effective policy and industry practices. By expanding ERA methodologies and embracing interdisciplinary research, future studies can enhance the sustainable deployment of green materials, paving the way for a more resilient and resource-efficient CE.

13.9.4 Perspective: integrating green materials within a circular economy for sustainable development

The relative adoption of green materials in the CE system constitutes a significant paradigm shift in regard to sustainable development for various reasons anchored on the requirement of green economy to strike a positive balance between the environmental, social, and economic. Bio-based composites are derived from natural resources or created to have reduced negative impact on the environment provide industries with the option to mitigate greenhouse emissions, save nonrenewable resources, and reduce waste. Their use is not limited to an environmental compliance – green material drives a paradigm shift across sectors by encouraging a more effective use of resources and stimulating innovations in design, production, and disposal. This method is reasonable since the principles of the CE, namely reuse, recycling, and resource management of products and materials, are in harmony with green materials. This synergy between green materials and CE practices holds promise not only for reducing ecological footprints but also for strengthening economic resilience against resource scarcity and fostering sustainable growth.

However, realizing these benefits hinges on the rigorous application of ERA methodologies to evaluate and manage potential ecological risks. Green materials may offer advantages, but their complex life cycles can vary significantly by application, raising the possibility of unintended environmental impacts, such as the formation of microplastics or toxic by-products during degradation. ERA frameworks, therefore, are essential in ensuring that green materials genuinely contribute to sustainability goals without introducing new risks. Collaboration among industries, policymakers, and researchers is crucial for refining these risk assessment models and establishing regulatory standards that uphold both environmental integrity and economic viability. Through a combined effort, the adoption of green materials within a CE framework can support a transition to a low-impact, resource-efficient global economy, positioning green materials as essential building blocks for a truly sustainable future.

References

[1] Akhimien, N. G., Latif, E., & Hou, S. S. (2021). Application of circular economy principles in buildings: A systematic review. *Journal of Building Engineering*, 38, 102041.

[2] Andrews, D. (2015). The circular economy, design thinking and education for sustainability. *Local Economy*, 30(3), 305–315.

[3] Ankley, G. T., Bennett, R. S., Erickson, R. J., Hoff, D. J., Hornung, M. W., Johnson, R. D., Mount, D. R., Nichols, J. W., Russom, C. L., Schmieder, P. K., Serrrano, J. A., Tietge, J. E., & Villeneuve, D. L. (2010). Adverse outcome pathways: A conceptual framework to support ecotoxicology research and risk assessment. *Environmental Toxicology and Chemistry*, 29(3), 730–741.

[4] Ascough, J. C., Maier, H. R., Ravalico, J. K., & Strudley, M. W. (2008). Future research challenges for incorporation of uncertainty in environmental and ecological decision-making. *Ecological Modelling* , 219(3–4), 383–399.

[5] Babita, Singh, V., & Mohan, C. (2023). Plant-derived compounds and their green synthesis in pharmaceuticals and nutraceuticals. In: *Book: Green Chemistry Approaches to Environmental Sustainability*. Elsevier, pp. 149–163. https://doi.org/10.1016/B978-0-443-18959-3.00001-X

[6] Badmus, S. O., Amusa, H. K., Oyehan, T. A., & Saleh, T. A. (2021). Environmental risks and toxicity of surfactants: overview of analysis, assessment, and remediation techniques. *Environmental Science and Pollution Research*, 28(44), 62085–62104.

[7] Bogen, K. T., & Spear, R. C. (1987). Integrating uncertainty and interindividual variability in environmental risk assessment. *Risk Analysis*, 7(4), 427–436.

[8] Buyle, M., Braet, J., & Audenaert, A. (2013). Life cycle assessment in the construction sector: A review. *Renewable and Sustainable Energy Reviews*, 26, 379–388.

[9] Chen, T. L., Kim, H., Pan, S. Y., Tseng, P. C., Lin, Y. P., & Chiang, P. C. (2020). Implementation of green chemistry principles in circular economy system towards sustainable development goals: Challenges and perspectives. *Science of the Total Environment*, 716, 136998. https://doi.org/10.1016/J.SCITOTENV.2020.136998

[10] Chiu, W. A., & Rusyn, I. (2018). Advancing chemical risk assessment decision-making with population variability data: Challenges and opportunities. *Mammalian Genome*, 29(1–2), 182.

[11] Chowdhury, S., Zhu, J., Rasoulkhani, K., Mostafavi, A., Jaselskis, E., Stoa, R., Li, Q., Banerjee, S., Alsharef, A., & Brannen, L. (2020). Guidelines for robust adaptation to environmental regulations in infrastructure projects. *Journal of Construction Engineering and Management*, 146(10), 04020121.

[12] Cooper, D. R., & Gutowski, T. G. (2017). The environmental impacts of reuse: A review. *Journal of Industrial Ecology*, 21(1), 38–56.

[13] Corona, B., Shen, L., Reike, D., Rosales Carreón, J., & Worrell, E. (2019). Towards sustainable development through the circular economy – A review and critical assessment on current circularity metrics. *Resources, Conservation and Recycling*, 151, 104498. https://doi.org/10.1016/J.RESCONREC.2019.104498

[14] Costa, S. P. F., Azevedo, A. M. O., Pinto, P. C. A. G., & Saraiva, M. L. M. F. S. (2017). Environmental impact of ionic liquids: Recent advances in (Eco)toxicology and (Bio)degradability. *ChemSusChem*, 10(11), 2321–2347. https://doi.org/10.1002/CSSC.201700261

[15] Crawford, S. E., Hartung, T., Hollert, H., Mathes, B., Van Ravenzwaay, B., Steger-Hartmann, T., Studer, C., & Krug, H. F. (2017). Green toxicology: A strategy for sustainable chemical and material development. *Environmental Sciences Europe*, 29(1), 1–16.

[16] Cullen, J. M. (2017). Circular economy: Theoretical benchmark or perpetual motion machine?. *Journal of Industrial Ecology*, 21(3), 483–486.

[17] D'Amato, D., & Korhonen, J. (2021). Integrating the green economy, circular economy and bioeconomy in a strategic sustainability framework. *Ecological Economics*, 188, 107143. https://doi.org/10.1016/J.ECOLECON.2021.107143

[18] Den Hollander, M. C., Bakker, C. A., & Hultink, E. J. (2017). Product design in a circular economy: Development of a typology of key concepts and terms. *Journal of Industrial Ecology*, 21(3), 517–525. https://doi.org/10.1111/JIEC.12610

[19] Dhawan, A., & Sharma, V. (2010). Toxicity assessment of nanomaterials: methods and challenges. *Analytical and Bioanalytical Chemistry*, 398(2), 589–605.

[20] Di Maio, F., Rem, P. C., Baldé, K., & Polder, M. (2017). Measuring resource efficiency and circular economy: A Market value approach. *Resources, Conservation and Recycling* , 122, 163–171.

[21] Dong, Y., Miraglia, S., Manzo, S., Georgiadis, S., Sørup, H. J. D., Boriani, E., Hald, T., Thöns, S., & Hauschild, M. Z. (2018). Environmental sustainable decision making– the need and obstacles for integration of LCA into decision analysis. *Environmental Science & Policy* , 87, 33–44.

[22] Gautam, R., Jindal, R., Mohan, C., & Mishra, A. K. (2024). Sustainable innovations as a solution to energy crisis and industrial pollution. In: *Book: Promoting Multi-Sector Sustainability with Policy and Innovation. IGI Global USA*, https://doi.org/10.4018/979-8-3693-2113-3.ch007

[23] Grafton, R. Q., Mclindin, M., Hussey, K., Wyrwoll, P., Wichelns, D., Ringler, C., Garrick, D., Pittock, J., Wheeler, S., Orr, S., Matthews, N., Ansink, E., Aureli, A., Connell, D., De Stefano, L., Dowsley, K., Farolfi, S., Hall, J., Katic, P., Lankford, B., Leckie, H., Mccartney, M., Pohlner, H., Ratna, N., Rubarenzya, M. H., Sai Raman, S. N., Wheeler, K., & Williams, J. (2016). Responding to global challenges in food, energy, environment and water: Risks and options assessment for decision-making. *Asia & the Pacific Policy Studies*, 3(2), 275–299.

[24] (2021). Green economy in sustainable development and improvement of resource efficiency. *Central European Business Review*, 10(1), 99–113.

[25] Gruiz, K., Fekete-Kertész, I., Kunglné-Nagy, Z., Hajdu, C., Feigl, V., Vaszita, E., & Molnár, M. (2016). Direct toxicity assessment – methods, evaluation, interpretation. *Science of the Total Environment*, 563–564, 803–812.

[26] Guan, L., Abbasi, A., & Ryan, M. J. (2020). Analyzing green building project risk interdependencies using interpretive structural modeling. *Journal of Cleaner Production*, 256, 120372.

[27] Hattisl, D. (1994). Assessment of variability practical risk analyses and uncertainty distributions for and David E. Burmaste. *Risk Analysis*, 14(5), 713–730.

[28] Hunka, A. D., Meli, M., Thit, A., Palmqvist, A., Thorbek, P., & Forbes, V. E. (2013). Stakeholders' perspective on ecological modeling in environmental risk assessment of pesticides: Challenges and opportunities. *Risk Analysis*, 33(1), 68–79.

[29] Isopescu, D. N. (2018). The impact of green building principles in the sustainable development of the built environment. *IOP Conference Series: Materials Science and Engineering* , 399(1), 012026.
[30] Jia, H., Ma, X. Y., Lin, Y., Yang, S., An, Y., Chen, Y., & Wang, X. C. (2023). Evolvement of carbamazepine under solar irradiation in the presence of chlorine: Efficiency, influences, degradation pathways and ecotoxicity. *Chemical Engineering Journal*, 461, 142106.
[31] Joshi, P. B., Durve, N. C., & Mohan, C. (2023). Full blown green metrics. In: *Book: Green Chemistry Approaches to Environmental Sustainability*. Elsevier, pp. 109–129. https://doi.org/10.1016/B978-0-443-18959-3.00013-6
[32] Kazancoglu, Y., Kazancoglu, I., & Sagnak, M. (2018). A new holistic conceptual framework for green supply chain management performance assessment based on circular economy. *Journal of Cleaner Production*, 195, 1282–1299. https://doi.org/10.1016/J.JCLEPRO.2018.06.015
[33] Kazancoglu, Y., Ozkan-Ozen, Y. D., Mangla, S. K., & Ram, M. (2022). Risk assessment for sustainability in e-waste recycling in circular economy. *Clean Technologies and Environmental Policy* , 24(4), 1145–1157. https://doi.org/10.1007/S10098-020-01901-3/TABLES/8
[34] Khoshnava, S. M., Rostami, R., Zin, R. M., Štreimikienė, D., Mardani, A., & Ismail, M. (2020). The role of green building materials in reducing environmental and human health impacts. *International Journal of Environmental Research and Public Health*, 17(7), 2589.
[35] Kumar, A., Thakur, A. K., Gaurav, G. K., Klemeš, J. J., Sandhwar, V. K., Pant, K. K., & Kumar, R. (2023). A critical review on sustainable hazardous waste management strategies: A step towards a circular economy. *Environmental Science and Pollution Research*, 30(48), 105030–105055. https://doi.org/10.1007/S11356-023-29511-8
[36] Kwan, M. (2016). Moving neighborhoods and health research forward: Using geographic methods to examine the role of spatial scale in neighborhood effects on health. 100–109.
[37] Länge, R., & Dietrich, D. (2002). Environmental risk assessment of pharmaceutical drug substances – conceptual considerations. *Toxicology Letters*, 131(1–2), 97–104.
[38] Li, J., Li, W., Liu, K., Guo, Y., Ding, C., Han, J., & Li, P. (2022). Global review of macrolide antibiotics in the aquatic environment: Sources, occurrence, fate, ecotoxicity, and risk assessment. *Journal of Hazardous Materials*, 439, 129628.
[39] Li, J., Song, G., Cai, M., Bian, J., & Sani Mohammed, B. (2022a). Green environment and circular economy: A state-of-the-art analysis. *Sustainable Energy Technologies and Assessments*, 52, 102106.
[40] Li, W., Liu, Q., Zhang, Y., Li, C., He, Z., Choy, W. C. H., Low, P. J., Sonar, P., & Kyaw, A. K. K. (2020). Biodegradable materials and green processing for green electronics. *Advanced Materials*, 32(33), 2001591.
[41] Luhar, S., & Luhar, I. (2019). Potential application of e-wastes in construction industry: A review. *Construction and Building Materials*, 203, 222–240.
[42] Maltby, L., Van den Brink, P. J., Faber, J. H., & Marshall, S. (2018). Advantages and challenges associated with implementing an ecosystem services approach to ecological risk assessment for chemicals. *Science of the Total Environment*, 621, 1342–1351.
[43] Melece, L. and others. (2016). Challenges and opportunities of circular economy and green economy. *Engineering for Rural Development*, 25, 1162–1169.
[44] Morseletto, P. (2020). Targets for a circular economy. *Resources, Conservation & Recycling*, 153, 104553.
[45] Nair, A. B., Sivasubramanian, P., Balakrishnan, P., Kumar, K. A. N. A., & Sreekala, M. S. (2013). Environmental effects, biodegradation, and life cycle analysis of fully biodegradable . *Green Composites. Polymer Composites, Biocomposites*, 3, 515–568.
[46] Nguyen, H. D., & Macchion, L. (2023). Risk management in green building: A review of the current state of research and future directions. *Environment, Development and Sustainability*, 25(3), 2136–2172.

[47] Notarnicola, B., Sala, S., Anton, A., McLaren, S. J., Saouter, E., & Sonesson, U. (2017). The role of life cycle assessment in supporting sustainable agri-food systems: A review of the challenges. *Journal of Cleaner Production*, 140, 399–409.

[48] Panigrahy, N., Priyadarshini, A., Sahoo, M. M., Verma, A. K., Daverey, A., & Sahoo, N. K. (2022). A comprehensive review on eco-toxicity and biodegradation of phenolics: recent progress and future outlook. *Environmental Technology & Innovation*, 27, 102423.

[49] Payraudeau, S., & Van Der Werf, H. M. G. (2005). Environmental impact assessment for a farming region: A review of methods. *Agriculture, Ecosystems & Environment*, 107(1), 1–19.

[50] Polat, G., Turkoglu, H., & Gurgun, A. P. (2017). Identification of material-related risks in green buildings. *Procedia Engineering*, 196, 956–963.

[51] Reddy, P. V. L., Kim, K. H., & Song, H. (2013). Emerging green chemical technologies for the conversion of CH4 to value added products. *Renewable and Sustainable Energy Reviews*, 24, 578–585.

[52] Resource Efficiency, the Circular Economy, Sustainable Materials Management and Trade in Metals and Minerals. (2021). 245.

[53] Rosenbaum, R. K., Bachmann, T. M., Gold, L. S., Huijbregts, M. A. J., Jolliet, O., Juraske, R., Koehler, A., Larsen, H. F., MacLeod, M., Margni, M., McKone, T. E., Payet, J., Schuhmacher, M., Van De Meent, D., & Hauschild, M. Z. (2008). USEtox – The UNEP-SETAC toxicity model: Recommended characterisation factors for human toxicity and freshwater ecotoxicity in life cycle impact assessment. *International Journal of Life Cycle Assessment*, 13(7), 532–546.

[54] Sala, S., Amadei, A. M., Beylot, A., & Ardente, F. (2021). The evolution of life cycle assessment in European Policies over three decades. *International Journal of Life Cycle Assessment*, 26(12), 2295–2314. https://doi.org/10.1007/S11367-021-01893-2/FIGURES/6

[55] Sassanelli, C., Urbinati, A., Rosa, P., Chiaroni, D., & Terzi, S. (2020). Addressing circular economy through design for X approaches: A systematic literature review. *Computers in Industry* , 120, 103245.

[56] Schwirn, K., Voelker, D., Galert, W., Quik, J., & Tietjen, L. (2020). Environmental risk assessment of nanomaterials in the light of new obligations under the REACH regulation: Which challenges remain and how to approach them?. *Integrated Environmental Assessment and Management*, 16(5), 706–717.

[57] Sheldon, R. A. (2016). Green chemistry and resource efficiency: Towards a green economy. *Green Chemistry*, 18(11), 3180–3183.

[58] Sherborne, N., Galic, N., & Ashauer, R. (2020). Sublethal effect modelling for environmental risk assessment of chemicals: Problem definition, model variants, application and challenges. *Science of the Total Environment*, 745, 141027.

[59] Simpson, D. (2021). *The Concept of Green, Blue and Circular Economy*. Journal Polska Akademia Nauk, University of Business and Administration in Gdynia.

[60] Smol, M., Marcinek, P., Duda, J., & Szołdrowska, D. (2020). Importance of sustainable mineral resource management in implementing the circular economy (CE) model and the European green deal strategy. *Resources*, 9(5), 55.

[61] Tambovceva, T., & Titko, J. (2020). *Introduction to Circular Economy*. Ekonomikas un kulturas augstskola.

[62] Thomassen, G., Van Dael, M., Van Passel, S., & You, F. (2019a). How to assess the potential of emerging green technologies? towards a prospective environmental and techno-economic assessment framework. *Green Chemistry*, 21(18), 4868–4886.

[63] Tiwari, A., Niveita, S. V., Mohan, C., & Bendi, A. (2024). Waste management from agricultural field: sustainable practices and framework. In: *Book: Exploring Waste Management in Sustainable Development Contexts. IGI Global, USA*, doi: 10.4018/979-8-3693-4264-0.ch011.

[64] Vafaeva, K. M., Chhetri, A., Sudan, P., Mishra, M., Pakkiraiah, B., & Mohan, C. (2024). Polymer matrix nanocomposites for sustainable packaging: A green approach. In: *E3S Web of Conferences*. vol. 511, pp. 01008. https://doi.org/10.1051/e3sconf/202451101008

[65] Van Ewijk, S. (2018). *Resource Efficiency and the Circular Economy: Concepts, Economic Benefits, Barriers, and Policies*. London, UK. : UCL Institute for Sustainable Resources.

[66] Vatin, N. I., Negi, G. S., Yellanki, S. V., Mohan, C., & Singla, N. (2024). Sustainability measures: An experimental analysis of AI and big data insights in industry 5.0. In: *BIO Web of Conferences*. vol. 86, pp. 01072. https://doi.org/10.1051/bioconf/20248601072

[67] Verma, V., Joshi, A., Kalra, S., Singh, A., Singh, A., & Mohan, C. (2024). Novel nanocatalysts for sustainable hydrogen production from renewable resources. In: *E3S Web of Conferences*. vol. 511, pp. 01006. https://doi.org/10.1051/e3sconf/202451101006

[68] Velenturf, A. P. M., & Purnell, P. (2021). Principles for a Sustainable Circular Economy. *Sustainable Production and Consumption*, 27, 1437–1457.

[69] Walker, W. C., Bosso, C. J., Eckelman, M., Isaacs, J. A., & Pourzahedi, L. (2015). Integrating life cycle assessment into managing potential EHS risks of engineered nanomaterials: Reviewing progress to date. *Journal of Nanoparticle Research*, 17(8), 1–16.

[70] Whaley, P., Halsall, C., Ågerstrand, M., Aiassa, E., Benford, D., Bilotta, G., Coggon, D., Collins, C., Dempsey, C., Duarte-Davidson, R., FitzGerald, R., Galay-Burgos, M., Gee, D., Hoffmann, S., Lam, J., Lasserson, T., Levy, L., Lipworth, S., Ross, S. M., Martin, O., Meads, C., Meyer-Baron, M., Miller, J., Pease, C., Rooney, A., Sapiets, A., Stewart, G., & Taylor, D. (2016). Implementing systematic review techniques in chemical risk assessment: Challenges, opportunities and recommendations. *Environment International*, 92–93,556–564.

[71] Yadav, S., Yadav, A., Mohan, C., Garg, V. K., & Kumari, N. (2023). Introduction to environmental and green chemistry. In: *Book: Green Chemistry Approaches to Environmental Sustainability*. Elsevier, pp. 1–22. https://doi.org/10.1016/B978-0-443-18959-3.00005-7

[72] Ying, J., & Li-jun, Z. (2012). Study on green supply chain management based on circular economy. *Physics Procedia*, 25, 1682–1688.

[73] Zeise, L., Bois, F. Y., Chiu, W. A., Hattis, D., Rusyn, I., & Guyton, K. Z. (2013). Addressing human variability in next-generation human health risk assessments of environmental chemicals. *Environmental Health Perspectives*, 121(1), 23–31.

[74] Zhang, X., & Mohandes, S. R. (2020). Occupational health and safety in green building construction projects: A Holistic Z-Numbers-Based risk management framework. *Journal of Cleaner Production*, 275, 122788.

Sandhya Mangla, Shalini Sharma, Yogendra Kumar Rajoria*, and Meenakshi Gusain

Chapter 14 A framework for analyzing growth, competition, and environmental impacts: A forest population dynamics through modelling

Abstract: Modeling of forest populations involves the use of mathematical techniques and frameworks to simulate, analyze, and predict the dynamics of forest ecosystems. This modeling approach helps in understanding the complex interactions between various biological, environmental, and anthropogenic factors that influence forest growth, species diversity, and population changes. Key aspects include modeling tree population dynamics, species competition, predator–prey relationships, and resource allocation within the forest. Models often incorporate differential equations, stochastic processes, and agent-based simulations to represent growth rates, mortality, reproduction, and migration patterns. These models serve as crucial tools for forest management, conservation efforts, and predicting the impacts of climate change, deforestation, and human activities on forest ecosystems. By providing quantitative insights, mathematical models help policymakers and environmentalists make informed decisions to preserve and restore forests, ensuring biodiversity and ecosystem sustainability.

Keywords: Forest population dynamics, species diversity, ecosystem sustainability, predator–prey relationships, tree growth models

14.1 Introduction

Forests are among the most complex and diverse ecosystems on the Earth, playing a vital role in global ecological stability by regulating carbon cycles, maintaining biodiversity, and supporting life [11]. Understanding forest dynamics, particularly popula-

***Corresponding author: Yogendra Kumar Rajoria**, Department of Mathematics, School of Basic and Applied Sciences, K. R. Mangalam University, Gurugram, Haryana, India,
e-mail: yogendrarajo@gmail.com
Sandhya Mangla, Shalini Sharma, Pt. Jawahar Lal Nehru Government College, Maharshi Dayanand University, Rohtak, India
Meenakshi Gusain, Center of Micro-Nano System, Fudan University, Shanghai, China

https://doi.org/10.1515/9783111618906-014

tion changes in tree species and other flora and fauna, is essential for effective forest management, conservation efforts, and predicting the impact of environmental changes such as climate change and deforestation [12]. One of the most effective approaches to gaining these insights is through mathematical modelling, which provides a framework to analyze, simulate, and predict forest population dynamics over time [13].

Mathematical models of forest populations allow researchers to describe the interactions among species, their responses to environmental conditions, and their overall contribution to ecosystem health [14]. These models are crucial for assessing the sustainability of forest ecosystems, optimizing management practices, and formulating policies to preserve biodiversity and mitigate the effects of climate change. These are important as forest ecosystems like forests cover approximately 31% of the Earth's surface and host more than half of the world's terrestrial species. They perform essential ecological functions such as sequestering carbon dioxide, purifying air and water, preventing soil erosion, and providing habitat for wildlife [2, 6, 7, 17–19]. Forests are also critical in stabilizing local and global climates, with tropical forests acting as significant carbon sinks. Despite their ecological importance, forests are under constant threat due to deforestation, illegal logging, habitat fragmentation, and climate change, all of which lead to loss of biodiversity and the disruption of ecosystem services [4, 8, 16].

Forest populations, particularly tree populations, are not static. They experience growth, competition for resources, reproduction, and mortality, influenced by both biotic factors (such as competition, predation, and mutualism) and abiotic factors (such as climate, soil composition, and water availability). Understanding how these factors interact over time requires sophisticated tools that can handle the complexity and variability of natural systems. This is where mathematical modelling becomes indispensable [3].

Mathematical modelling is the process of using mathematical equations and algorithms to represent and analyze real-world systems. In the context of forest population dynamics, mathematical models describe how different species in a forest interact with one another and with their environment [1]. These models can be deterministic or stochastic, depending on whether they predict a single outcome or a range of possible outcomes under varying conditions. There are different types of models used to represent forest populations, including:

a) **Deterministic models:** These use fixed values for parameters and initial conditions to produce predictable outcomes. For example, growth models based on differential equations can describe how a tree population increases over time under constant environmental conditions [15]
b) **Stochastic models:** These incorporate randomness and uncertainty, acknowledging that biological processes and environmental conditions are inherently variable. Stochastic models are often used to simulate unpredictable events like wildfires, disease outbreaks, or random seed dispersal patterns [20].

c) **Spatial models:** These models include a spatial dimension, allowing researchers to analyze how forest population dynamics vary across different locations. Spatial heterogeneity is critical in understanding phenomena such as species migration, forest fragmentation, and the spread of invasive species [21].

d) **Agent-based models:** These are computational models that simulate the actions and interactions of individual organisms (or agents) within a forest. They are particularly useful in studying how individual-level behaviors can lead to emergent, system-level patterns such as the formation of tree stands or the clustering of species in certain areas [5, 9, 10].

14.2 Model formulation

The formulation of a mathematical model for forest population dynamics typically involves defining variables, parameters, and equations that represent the growth, interaction, and behavior of tree species and other organisms in a forest ecosystem. Below is a step-by-step framework to develop a forest population model, focusing on tree population dynamics and incorporating key biological and environmental factors.

The graph of Figure 14.1 illustrates the complex interactions influencing forest population dynamics, including growth models, mortality factors, and climate effects. It demonstrates how variables like tree age, density, and environmental conditions collectively impact forest health and population rates.

14.2.1 Defining the variables

To create a model, we first need to define the main state variables, which represent the components of the forest population. These could include: Ni(t): The population density of species iii at time t, measured as the number of trees per unit area, Gi(t): The growth rate of species iii at time t, often modeled as a function of available resources and competition, Mi(t): The mortality rate of species i, which could depend on factors like age, competition, environmental conditions, or disturbances, Ri(t): The recruitment (or reproduction) rate of species i, which determines the number of new individuals entering the population over time, and Cij(t): The competition effect between species iii and species j, representing how species i influences the growth or survival of species.

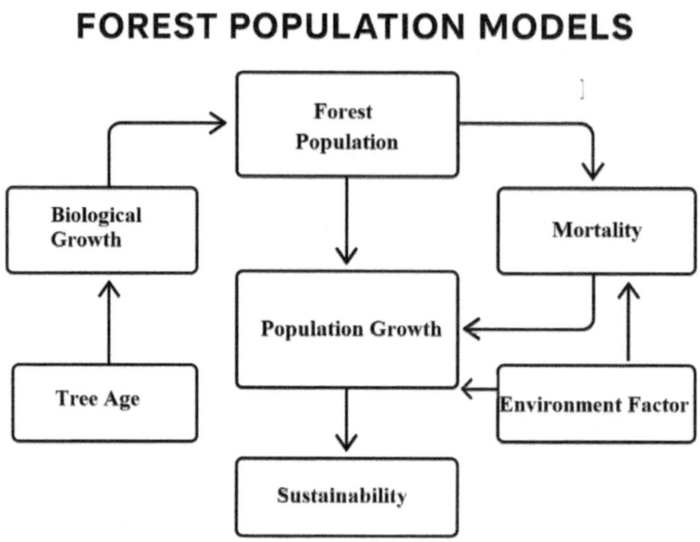

Figure 14.1: Dynamics of forest population growth and decline.

14.2.2 Basic population growth equation

The basic equation for population dynamics in forests is based on the balance of birth (recruitment), death (mortality), and growth. For a single species iii, the general form of the population equation can be written as

$$\frac{\mathrm{dNi}(x,t)}{\mathrm{d}t} = \mathrm{Ri}(t) - \mathrm{Mi}(t) + \mathrm{Gi}(t)$$

where $\frac{\mathrm{dNi}(x,t)}{\mathrm{d}t}$ is the rate of change in the population of species iii at time t, $\mathrm{Ri}(t)$ is the recruitment rate of new individuals into the population, $\mathrm{Mi}(t)$ is the mortality rate of individuals, and $\mathrm{Gi}(t)$ is the net growth rate (which can include aspects like resource acquisition, size, or biomass increase). This equation serves as the foundation for modelling the dynamics of individual species populations.

14.2.3 Incorporating competition and resource availability

Competition between species or individuals is a critical component of forest population dynamics. It affects resource availability (such as light, water, and nutrients) and ultimately the growth and survival rates. One common approach to model competition is to use the Lotka–Volterra competition equations or their modifications. The equation for two competing species, iii and jjj, is:

$$\frac{\mathrm{dNi}(x,t)}{\mathrm{d}t} = \mathrm{ri.Ni}(t)\left(1 - \mathrm{Ni}(t) + \frac{\alpha\mathrm{ij}\ \mathrm{Nj}(t)}{\mathrm{Ki}}\right)$$

where ri is the intrinsic growth rate of species I, Ni(*t*) are the population densities of species i and j, respectively, *α*ij represents the competitive impact of species j on species I, and Ki is the carrying capacity of the environment for species i, representing the maximum population size that the habitat can sustain. This equation shows that the population growth of species i is reduced not only by its own population (intraspecific competition) but also by the presence of competing species j (interspecific competition).

14.2.4 Recruitment and seed dispersal

Recruitment is often modelled as a function of the reproductive capacity of the population and the dispersal of seeds or offspring. A simple model for recruitment might be: $\mathrm{Ri}(t) = \mathrm{bi\ Ni}(t)\left(1 - \frac{\mathrm{Ni}(t)}{\mathrm{Ki}}\right)$

where bi is the per capita birth rate of species I, and Ki is the carrying capacity, which limits the growth of the population as it approaches saturation.

To model seed dispersal, spatial factors can be introduced. For instance, the probability Pd(*x*) that a seed disperses to a location x at a distance d from the parent tree can be modelled as an exponential decay function: $\mathrm{Pd}(x) = \mathrm{e} - \lambda\mathrm{d}$, where λ is a parameter controlling the spread of the seeds. This probability function can be integrated into spatially explicit models to track the distribution and establishment of new trees across a landscape.

14.2.5 Disturbances and mortality

Forests are subject to disturbances such as wildfires, storms, pests, and diseases, all of which contribute to tree mortality. To model disturbances, a stochastic element can be introduced. The mortality rate Mi(*t*) can be expressed as a function of both age and external factors like disturbance frequency: $\mathrm{Mi}(t) = m_0 + m_1\mathrm{Ai}(t) + D(t)$

where m_0 is the baseline mortality rate, Ai(*t*) is the age of species iii, with m1 representing the age-dependent increase in mortality, and *D*(*t*) is a disturbance function that can vary stochastically over time to represent unpredictable events like fire or pest outbreaks.

Disturbances can also be modelled as pulse events that reduce the population size suddenly, for example: $\mathrm{Ni}(t + \Delta t) = \mathrm{Ni}(t) - D(t)$, where $D(t)$ represents the number of individuals lost due to a disturbance at time t, and Δt is the time step.

14.2.6 Incorporating environmental factors and climate change

To account for environmental changes such as climate variability, temperature, and precipitation, external factors can be added to influence the parameters of the model. For example, the growth rate Gi(t) can be made dependent on environmental conditions E(t): $\mathrm{Gi}(t) = \mathrm{gi}.\, f(E(t))$

where gi is the baseline growth rate of species iii, and $f(E(t))$ is a function that modifies the growth rate based on environmental factors such as temperature or soil moisture. This function can be derived from empirical data or modelled using climate projections.

14.2.7 Spatially explicit models

Spatial models add an additional layer of complexity by incorporating the geographical distribution of species across a landscape. These models divide the forest into grid cells or patches, with each cell representing a local population. The population dynamics in each cell are governed by local interactions (competition, reproduction, and mortality) as well as movement between cells (dispersal). The basic equation for a spatial model might look like:

$$\frac{\mathrm{dNi}(x,\mathrm{t})}{\mathrm{d}t} = \mathrm{Ri}(x,t) - \mathrm{Mi}(x,t) + \mathrm{Gi}(x,t) + \mathrm{Di}(x,t)$$

where Ni(x, t) is the population density of species iii at location x and time t, and Di(x, t) represents dispersal between different locations.

14.3 Derivation of model

Spatially explicit models are mathematical representations used to describe population dynamics over space and time by dividing a landscape into discrete cells or patches. Each cell represents a local population, and the dynamics are determined by both local interactions (e.g., birth, death, and competition) and dispersal between cells. These models are especially useful for studying ecological processes where the spatial arrangement of individuals or populations plays a significant role, such as in fragmented habitats or species with limited dispersal abilities. The general form of the spatially explicit population model is expressed as:

$$\frac{\mathrm{dNi}(x,t)}{\mathrm{d}t} = \mathrm{Ri}(x,t) - \mathrm{Mi}(x,t) + \mathrm{Gi}(x,t) + \mathrm{Di}(x,t)$$

where $\mathrm{Ni}(x,t)$ is the population density of species iii at location x and time t, $\mathrm{Ri}(x,t)$ represents the local population growth (reproduction), $\mathrm{Mi}(x,t)$ represents local mortality or losses, $\mathrm{Gi}(x,t)$ accounts for other local population changes (e.g., gains due to migration into the population), and $\mathrm{Di}(x,t)$ represents dispersal between different locations.

14.3.1 Local population growth Ri(x, t)

The local population growth can be modelled using standard logistic growth, which accounts for reproduction and competition: $\mathrm{Ri}(x,t) = \mathrm{ri.Ni}(x,t)\left(1 - \frac{\mathrm{Ni}(x,t)}{\mathrm{Ki}(x)}\right)$

where ri is the intrinsic growth rate of species i, and $\mathrm{Ki}(x)$ is the carrying capacity of location x for species i, accounting for spatial heterogeneity in resources.

This equation implies that population growth is density-dependent, with growth slowing as the population approaches the carrying capacity.

14.3.2 Local mortality Mi(x, t)

Mortality can be expressed as a function of population density and environmental factors:

$$\mathrm{Mi}(x,t) = \mathrm{mi.Ni}(x,t)$$

where mi is the per capita mortality rate, which may depend on environmental factors such as predation or resource scarcity.

14.3.3 Migration or gain from immigration Gi(x, t)

Population gains can occur due to immigration from other locations or through recruitment. This term may include the effects of external immigration: $\mathrm{Gi}(x,t) = \mathrm{gi}(x,t)\mathrm{Ni}(x,t)$

where gi(x,t) represents an immigration function that could depend on environmental factors, distance from other populations, or other spatial variables.

14.3.4 Dispersal Di(x, t)

Dispersal reflects the movement of individuals between different locations. In many spatial models, dispersal can be represented by a diffusion-like term:

$$\mathrm{Di}(\mathrm{x},\mathrm{t}) = \mathrm{Di}.\nabla^2\mathrm{Ni}(\mathrm{x},\mathrm{t})$$

where Di is the diffusion coefficient, which determines the rate of dispersal,$\nabla^2\mathrm{Ni}(x,t)$ is the Laplacian operator, representing the spatial diffusion of the population. This term describes how individuals move from areas of high population density to areas of low population density. It assumes a continuous spatial domain, but in practice, dispersal is often modelled discretely in grid-based models, where individuals can move between adjacent cells.

14.3.5 Combining the terms

The full spatially explicit model can now be written as:

$$\frac{\mathrm{dNi}(x,t)}{\mathrm{dt}} = \mathrm{ri.Ni}(x,t)\{1 - \mathrm{Ni}(x,t)\mathrm{Ki}(x)\} - \mathrm{miNi}(x,t) + \mathrm{gi}(x,t)\mathrm{Ni}(x,t) + \mathrm{Di}.\nabla^2\mathrm{Ni}(x,t)$$

This equation describes how the population density of species iii changes over time and space, considering both local processes (reproduction, competition, mortality) and spatial processes (dispersal and immigration). The Steps involved in forest population modelling are illustrated in a systematic flowchart (figure 14.2).

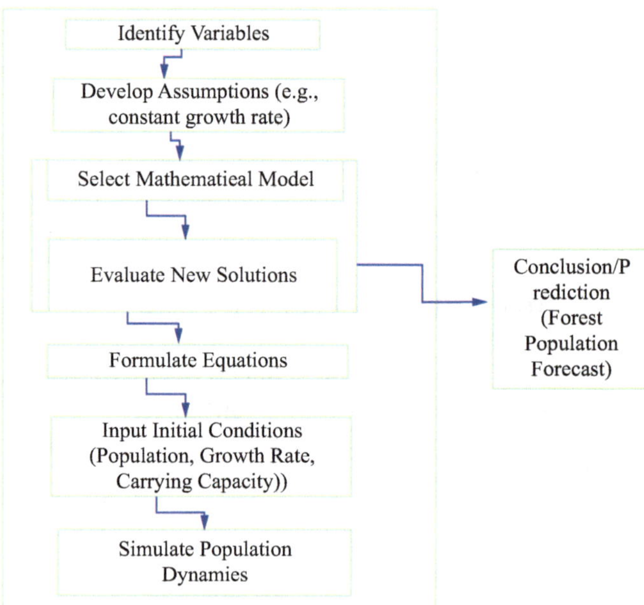

Figure 14.2: Flowchart on forest population of models.

14.3.6 Discretization

In practice, the landscape is often divided into discrete grid cells, and the population dynamics in each cell are modelled using difference equations or cellular automata. In a discretized version, each cell's population at time t is updated based on its current population, interactions with neighboring cells, and dispersal.

14.3.7 Stochastic elements

Since forest population dynamics are subject to random environmental fluctuations and unpredictable disturbances, stochasticity is often incorporated into models. Stochastic models include random variables or noise terms that affect growth rates, recruitment, or disturbances. For example, the growth rate might vary as: $Gi(t) = gi + (t)$

where (t) is a random noise term with a given probability distribution.

14.4 Model analysis

The first step in analyzing the model is to identify its equilibrium points where the population of each species remains constant over time. Mathematically, equilibrium points occur when the rate of change of the population is zero: $\frac{\mathrm{dNi}(t)}{\mathrm{d}t} = 0$

For the general population growth equation: $\frac{\mathrm{dNi}(t)}{\mathrm{d}t} = \mathrm{Ri}(t) - \mathrm{Mi}(t) + \mathrm{Gi}(t)$

The equilibrium occurs when the recruitment, growth, and mortality are balanced. Setting the right-hand side of the equation to zero gives:

$$\mathrm{Ri}(t) = \mathrm{Mi}(t) - \mathrm{Gi}(t)$$

At equilibrium, the species population Ni (the steady-state population of species i) can be found by solving this equation for the population variable. Depending on the model, this might involve solving algebraic equations or, in more complex models, systems of coupled nonlinear equations.

For example, in a competition model between two species, the equilibrium can be found by solving the following system:

$$\frac{\mathrm{dN1(t)}}{\mathrm{dt}} = r_1 N_1 \left(1 - \frac{N_1 + \alpha_{12}\, N_2}{K_1}\right), \frac{\mathrm{dN2}(t)}{\mathrm{d}t} = r_2 N_2 \left(1 - \frac{N_1 + \alpha_{12}\, N_2}{K_1}\right)$$

These equations provide insight into whether the populations will coexist, whether one species will outcompete the other, or whether they will reach a stable equilibrium. Solving these equations gives the population sizes of species 1 and 2 at equilibrium.

14.5 Conclusion

Mathematical modeling of forest population dynamics provides valuable insights into the complex interactions that shape forest ecosystems. Through these models, we can better understand the growth, competition, reproduction, and mortality of different species, as well as the effects of disturbances and environmental factors such as climate change. By developing equations that represent these biological processes, we can predict the future behavior of forests, allowing us to make informed decisions about management, conservation, and restoration. The model formulation helps capture the fundamental components of forest ecosystems, including competition between species, resource limitations, and spatial dispersal. It also accounts for stochastic events like disturbances and environmental variability, which play crucial roles in forest population dynamics. The analysis of these models – particularly through equilibrium and stability studies, sensitivity analysis, and bifurcation analysis – reveals critical thresholds and tipping points that could lead to significant changes in forest structure and composition. Incorporating stochastic elements and spatial considerations into forest population models enhances their realism, allowing for the simulation of real-world forest conditions and providing more accurate predictions. These models are not only useful for theoretical exploration but also for practical applications such as assessing the impact of human activities, climate change, or natural disturbances on forest ecosystems.

References

[1] Agarwal, M., & Pathak, R. (2015). Conservation of forestry biomass with the use of alternative resource. *Open Journal of Ecology*, 5(4), 87–109. doi: 10.4236/oje.2015.54009

[2] Andhika, Z. R. N., Nursanti, A., & Asep, S. (2020). A mathematical model to study the effects of population pressure on two-patch forest resources. *AIP Conference Proceedings*, *2264*, article 050003.

[3] Berec, L. (2011). "Habilitation thesis Mathematical modeling in *ecology and epidemiology*".

[4] Fengsheng, C., & Stanford, S. (2021). Volterra–Lyapunov stability analysis of the aolutions of babesiosis disease model. *Symmetry*, 13(7), 1272,1–14. https://doi.org/10.3390/sym13071272

[5] Halim, N. B. T. A. (2015). Tuberculosis Model: A mathematical analysis.

[6] Hatsey, A. (2015). Tend of forest cover change in Ethiopia; *South Africa Georg-August University Gottingen*, *September*, *2015.Math 4330*, Numerical Solution of ODEs Using MATLAB.

[7] Kubsa, A., & Tadesse, T. (2003). Granting exclusive user rights to the forest dwellers in the state owned forest: the WAJIB approach in Ethiopia. *Environmental Science, Law*, 130453679.

[8] Kumari, S. S., Jayantika, P., & Kumari, J. (2022). A dynamical study on the adverse effects of industrial activities in the forest and wildlife region through modelling. *Modeling Earth Systems and Environment*, 8(4), 1–13.

[9] Mangla, S., Sharma, S., Rajoria, Y. K., & Chikara, D. (2024). Modelling the Combined Impact of Population Density and its Precursor on Forest Resources. *International Journal of Global Warming*, 32(4), 399–417. doi: 10.1504/IJGW.2024.10060144

[10] Mangla, S., Sharma, S., Boadh, R., Rajoria, Y. K., & Chikara, D. (2022). A Mathematical Model of the Impact of Deforestation on the Growth of Forest Resources. *NeuroQuantology, 20*(17), 223.

[11] Misra, A. K., & Lata, K. (2015a). A mathematical model to achieve sustainable forest management. *International Journal of Modeling, Simulation, and Scientific Computing*, 6(4), 1550040.

[12] Misra, A. K., & Lata, K. (2015b). Depletion and Conservation of Forestry Resources: A Mathematical Model. *Differential Equations and Dynamical Systems*, 23(1), 25–41. https://doi.org/10.1007/s12591-013-0177-3

[13] Misra, A. K., Lata, K., & Shukla, J. B. (2014a). Effects of population and population pressureon forest resources and their conservation: A modeling study. *Environment, Development and Sustainability*, 16, 361–374. https://doi.org/10.1007/s10668-013-9481-x

[14] Misra, A. K., Lata, K., & Shukla, J. B. (2014b). A mathematical model for the depletion offorestry resources due to population and population pressure augmented industrialization. *International Journal of Modeling, Simulation, and Scientific Computing*, 5(1), 1350022. https://doi.org/10.1142/S1793962313500220

[15] Prabir, P. (2020). Is the forest biomass a key regulator of global warm-ing?: A mathematical modelling study. *Geology, Ecology, and Landscapes*, 6(1), 66–74. https://doi.org/10.1080/24749508.2020.1752021

[16] Rachana, P., Vinay, V., & Manju, A. (2021). Impact of human activities on forest resources and wildlife population, Computational Ecology and Software. *Computational Ecology and Software*, 11(2), 83–99.

[17] Regassa, A., Diriba, D., & Negera, G. (2019). Bale Mountains Eco-region Reduction of Emission from Deforestation and Forest Degradation (REDD+) Project Ethiopia, *Developed by: Oromia Forest and Wildlife Enterprise (OFWE), Farm Africa and SOS Sahel Ethiopia.*

[18] Shukla, J. B., Lata, K., & Misra, A. K. (2011). Modelling the depletion of a renewable resource by population and industrialization: Effect of technology on its conservation. *Natural Resource Modeling*, 24(2), 242–267. https://doi.org/10.1111/j.1939-7445.2011.00090.x

[19] Vanessa, L. D., Vinicius, R. A., Zimermann, O. L., Luis, D. G. A., & Christian, V. A. (2020). Modelling changes in forest attributes driven by human activities at different spatial scales in the sub-tropical Atlantic Forest. *Biodiversity and Conservation*, 29(4), 1283–1299.

[20] Wilkie, M. L., & Caracalla, V. T. (2010). Global forest resources assessment 2010, country report, Ethiopa, Italy: FAO Forestry Department Rome 00153.

[21] Zhang, L., Ameca, E. I., & Jimenez, B. O. (2022). Human-induced deforestation increases extinction risk faster than cli-mate pressures: Evidence from long-term monitoring of the globally endangered Milne-Edward's sifaka. *Biological Conservation*, 274, 109716.

Index

https://doi.org/10.1515/9783111618906-015

De Gruyter Series in Green Chemical Processing

Volume 1
Mark Anthony Benventuo (Ed.)
Sustainable Green Chemistry, 2017
ISBN 978-3-11-044189-5, e-ISBN 978-3-11-043585-6

Volume 2
Mark Anthony Benventuo (Ed.)
Green Chemical Processes: Developments in Research and Education, 2017
ISBN 978-3-11-044487-2, e-ISBN 978-3-11-044592-3

Volume 3
Mark Anthony Benventuo, Heinz Plaumann (Eds.)
Green Chemistry in Industry, 2018
ISBN 978-3-11-056113-5, e-ISBN 978-3-11-056278-1

Volume 4
Mark Anthony Benventuo, George Ruger (Eds.)
Green Chemistry Education: Recent Developments, 2018
ISBN 978-3-11-056578-2, e-ISBN 978-3-11-056649-9

Volume 5
Mark Anthony Benventuo, Heinz Plaumann (Eds.)
Green Chemistry in Government and Industry, 2020
ISBN 978-3-11-059728-8, e-ISBN 978-3-11-059778-3

Volume 6
Mark Anthony Benventuo, George Ruger (Eds.)
Green Chemistry and Technology, 2021
ISBN 978-3-11-066991-6, e-ISBN 978-3-11-066998-5

Volume 7
Mark Anthony Benventuo, Heinz Plaumann (Eds.)
Green Chemistry: Water and its Treatment, 2021
ISBN 978-3-11-059730-1, e-ISBN 978-3-11-059782-0

Volume 8
Mark Anthony Benventuo, Heinz Plaumann (Eds.)
Green Chemistry: Advances in Alternative Energy, 2022
ISBN 978-3-11-070638-3, e-ISBN 978-3-11-070663-5

Volume 9
Mark Anthony Benventuo, Steven Kosmas (Eds.)
Green Chemistry and UN Sustainability Development Goals, 2022
ISBN 978-3-11-072386-1, e-ISBN 978-3-11-072396-0

www.degruyterbrill.com

Volume 10
Mark Anthony Benventuo, Lindsey Welch (Eds.)
Green Chemistry: Research and Connections to Climate Change, 2023
ISBN 978-3-11-074560-3, e-ISBN 978-3-11-074565-8

Volume 11
Mark Anthony Benventuo, Lawrence Kolopajlo (Eds.)
Green Chemical Processes: Developments in Science, Math, Engineering and Technology, 2024
ISBN 978-3-11-079907-1, e-ISBN 978-3-11-079908-8

Volume 12
Chandra Mohan, Neeraj Kumari, Mark Anthony Benvenuto and Asha Anish Madhavan (Eds.)
Green Materials for Eco-Friendly Environment: Sustainable Materials, Biopolymers, Renewables, Fuel Cells, Biofuels, Biorefineries, 2025
ISBN 978-3-11-161883-8, e-ISBN 978-3-11-161890-6